Mastering™ AutoCAD® 2000

for Mechanical Engineers

George Omura

SYBEX®

San Francisco • Paris • Düsseldorf • Soest • London

Associate Publisher: Amy Romanoff
Contracts and Licensing Manager: Kristine O'Callaghan
Acquisitions & Developmental Editor: Melanie Spiller
Editor: Valerie Haynes Perry
Project Editors: Elizabeth Hurley-Clevenger, Dann McDorman
Technical Editor: Priscilla Mills
Book Designer: Kris Warrenburg
Graphic Illustrators: Tony Jonick, Jerry Williams
Electronic Publishing Specialists: Franz Baumhackl, Kate Kaminski,
Cyndy Johnsen, Robin Kibby, and Nila Nichols
Project Team Leaders: Shannon Murphy, Teresa Trego
Proofreaders: Davina Baum, Susan Berge, Bonnie Hart,
  Catherine Morris
Indexer: Ted Laux
Companion CD: Ginger Warner
Cover Designer: Design Site
Cover Illustrator/Photographer: Sergie Loobkoff, Design Site

SYBEX is a registered trademark of SYBEX Inc.
Mastering is a trademark of SYBEX Inc.

Screen reproductions produced with Collage Complete.
Collage Complete is a trademark of Inner Media Inc.

The CD Interface music is from GIRA Sound AURIA Music Library
GIRA Sound 1996.

TRADEMARKS: SYBEX has attempted throughout this book to
distinguish proprietary trademarks from descriptive terms by following the capitalization style used by the manufacturer.

The author and publisher have made their best efforts to prepare
this book, and the content is based upon final release software
whenever possible. Portions of the manuscript may be based upon
pre-release versions supplied by software manufacturer(s). The
author and the publisher make no representation or warranties of
any kind with regard to the completeness or accuracy of the contents herein and accept no liability of any kind including but not
limited to performance, merchantability, fitness for any particular
purpose, or any losses or damages of any kind caused or alleged to
be caused directly or indirectly from this book.

Library of Congress Card Number: 99-61818
ISBN: 0-7821-2500-X

Manufactured in the United States of America
10 9 8 7 6 5 4 3 2 1

To my family and my teachers
—George Omura

ACKNOWLEDGMENTS

Book production is a complicated process, so I'm always amazed at how quickly *Mastering AutoCAD for Mechanical Engineers* evolves from manuscript to finished product. In many ways, it's quite magical; but behind the magic, there are many hard-working people giving their best effort. I'd like to thank those people who helped bring this book to you.

Heartfelt thanks go to the editorial and production teams at Sybex for their efforts in getting this book to press on an incredible schedule. Developmental Editor Melanie Spiller got things going and offered many great suggestions. Project Editor Elizabeth Hurley-Clevenger and Editor Valerie Haynes Perry made the frantic schedule bearable with humor and encouragement. Priscilla Mills, the Technical Editor, provided helpful suggestions as she carefully reviewed the book. Electronic Publishing Specialist, Franz Baumhackl, created the pages you see before you with speed and expertise. Ginger Warner compiled our CD and made it easy and fun to use. Project Team Leader Shannon Murphy and proofreaders Davina Baum, Susan Berge, and Catherine Morris deserve a big round of applause for checking every word (twice!) and each piece of art. Dan Schiff archived and retrieved all of the art at a moment's notice with unbelievable patience. Finally, a big thanks to Mike Gunderloy for his great work on ActiveX Automation.

And finally, a great big thanks to my wife and sons, who are always behind my work 100 percent.

George Omura

CONTENTS AT A GLANCE

TABLE OF CONTENTS

INTRODUCTION

Welcome to *Mastering AutoCAD 2000 for Mechanical Engineers*. As many readers have already discovered, *Mastering AutoCAD* offers a unique blend of tutorial and source book that offers everything you need to get started and stay ahead with AutoCAD.

How to Use This Book

Rather than just showing you how each command works, *Mastering AutoCAD 2000 for Mechanical Engineers* shows you AutoCAD in the context of a meaningful activity. You will learn how to use commands while working on an actual project and progressing toward a goal. It also provides a foundation on which you can build your own methods for using AutoCAD and become an AutoCAD expert yourself. For this reason, we haven't covered every single command or every permutation of a command response. The *AutoCAD 2000 Instant Reference*, which we've included on the companion CD-ROM, will fill that purpose nicely. This online resource will help you quickly locate the commands you need. You should think of *Mastering AutoCAD 2000 for Mechanical Engineers* as a way to get a detailed look at AutoCAD as it is used on a real project. As you follow the exercises, we encourage you to explore AutoCAD on your own, applying the techniques you learn to your own work.

If you are not an experienced user, you may want to read *Mastering AutoCAD 2000 for Mechanical Engineers* as a tutorial. You'll find that each chapter builds on the skills and information you learned in the previous one. To help you navigate, the exercises are shown in numbered steps. This book can also be used as a ready reference for your day-to-day problems and questions about commands. Optional exercises at the end of each chapter will help you review what you have learned and look at different ways to apply these skills.

Getting Information Fast

We've also included plenty of Notes, Tips, and Warnings. *Notes* supplement the main text; *Tips* are designed to make practice easier; and *Warnings* steer you away from pitfalls. Also, in each chapter you will find more extensive tips and discussions in the form of specially screened *sidebars*. Together the Notes, Tips, Warnings, and sidebars provide a wealth of information gathered over years of using AutoCAD on a variety of projects in different environments. You may want to browse through the book, looking at the projects and sidebars, to get an idea of how the information they contain might be useful to you.

Another quick reference you'll find yourself turning to often is Appendix D. This appendix contains tables of all the system settings and comments on their use.

What to Expect

Mastering AutoCAD 2000 for Mechanical Engineers is divided into six parts, each representing a milestone in your progress toward becoming an expert AutoCAD user. Here is a description of those parts and what each will show you.

Part I: The Basics

As with any major endeavor, you must begin by tackling small, manageable tasks. In this first part, you will become familiar with the way AutoCAD looks and feels. Chapter 1, "This Is AutoCAD," shows you how to get around in AutoCAD. In Chapter 2, "Creating Your First Drawing," you will learn how to start and exit the program and how to respond to AutoCAD command prompts. Chapter 3, "Learning the Tools of the Trade," tells you how to set up a work area, edit objects, and lay out a drawing. In Chapter 4, "Organizing Your Work," you will explore some tools unique to CAD: symbols, blocks, and layers. As you are introduced to AutoCAD, you will also get a chance to make some drawings that you can use later in the book and perhaps even in future projects of your own.

Part II: Building on the Basics

Once you have the basics down, you will begin to explore some of AutoCAD's more subtle qualities. Chapter 5, "Editing for Productivity," tells you how to reuse drawing setup information and parts of an existing drawing. In Chapter 6, "Enhancing Your Drawing Skills," you will learn how to assemble and edit a large drawing file. In Chapter 7, "Adding Text to Drawings," you will learn how to annotate your drawing and edit your notes. Chapter 8, "Using Dimensions," gives you practice in using automatic dimensioning, another unique AutoCAD capability. Chapter 9, "Advanced Productivity Tools," adds to and refines your editing toolbox. In Chapter 10, "Drawing Curves and Solid Fills," you will get an in-depth look at some special drawing objects, such as spline and fitted curves. Along the way, we will be giving you tips on editing and problems you may encounter as you begin to use AutoCAD for more complex tasks.

Part III: Modeling and Imaging in 3D

While 2D drafting is AutoCAD's workhorse application, AutoCAD's 3D capabilities give you a chance to expand your ideas and look at them in a new light. Chapter 11, "Introducing 3D," explores solid modeling and the User Coordinate System. Chapter 12, "Mastering 3D Solids," extends your knowledge of solids and shows you how to create a 2D drawing almost automatically from your 3D model. You'll also learn how to use AutoCAD to calculate the mass properties of a solid object. Chapter 13, "Using 3D Surfaces," sorts out some of the mysteries of creating and managing surfaces.

Part IV: Printing and Plotting as an Expert

This part demonstrates the wide variety of hardcopy (paper) output available from AutoCAD. Chapter 14, "Printing and Plotting," shows you how to get your drawing onto paper. In Chapter 15, "3D Rendering in AutoCAD," you'll see how you can use the AutoCAD rendering tool to produce lifelike views of your 3D drawings. Chapter 16, "Working with Existing Drawings and Raster Images," explains techniques for transferring paper drawings to AutoCAD.

Part V: Customization: Taking AutoCAD to the Limit

In this part of the book, you will learn how you can take full control of AutoCAD. In Chapter 17, "Storing and Linking Data with Graphics," you'll learn how to attach information to drawing objects and how to link your drawing to database files. Chapter 18, "Getting and Exchanging Data from Drawings," will give you practice getting information about a drawing, and you'll learn how AutoCAD can interact with other applications such as spreadsheets and desktop publishing programs. Chapter 19, "Introduction to Customization," gives you a gentle introduction to the world of AutoCAD customization. You'll learn how to load and use existing utilities that come with AutoCAD. Chapter 20, "Using ActiveX Automation with AutoCAD," shows you how you can tap the power of automation to add new functions to AutoCAD and link AutoCAD to other applications. Chapter 21, "Integrating AutoCAD into Your Projects and Organization," shows you how you can adapt AutoCAD to your own work style. Customizing menus, linetypes, and screens are only three of the many topics. You will also find out how you can publish high-resolution drawings on the World Wide Web.

Part VI: Appendices

Finally, this book has four appendices. Appendix A, "Hardware and Software Tips," offers information on hardware related to AutoCAD. It also provides tips on improving AutoCAD's performance and on troubleshooting. Appendix B, "Installing and Setting Up AutoCAD," contains an installation and configuration tutorial. If AutoCAD is not already installed on your system, you should follow this tutorial before starting Chapter 1. Appendix C, "What's on the Companion CD-ROM," describes the utilities available on the companion CD-ROM. Appendix D, "System Variables," will illuminate the references to the system variables scattered throughout the book.

The Minimum System Requirements

This book assumes you have an IBM-compatible Pentium computer that will run AutoCAD and support a mouse. Your computer should have at least one CD-ROM drive, and a hard disk with 140MB or more free space after AutoCAD is installed (about 70MB for AutoCAD to work with and another 30MB available for drawing files). In addition to these requirements, you should also have enough free disk space to allow for a Windows virtual memory page file of at least 128MB. Consult your Windows manual or Appendix A of this book for more on virtual memory.

AutoCAD 2000 runs best on systems with at least 64MB of RAM for Windows 95/98 and with 128MB for Windows NT. Your computer should also have a high-resolution monitor and a color display card. The current standard is the Super Video Graphics Array (SVGA) display. This is quite adequate for most AutoCAD work. The computer should also have at least one serial port. If you have only one, you may want to consider having another one installed. We also assume you are using a mouse and have the use of a printer or a plotter. A relatively fast CD is needed to load the AutoCAD 2000 software. Most computers come equipped with a sound card, though you won't necessarily need one to use this book.

If you want a more detailed explanation of hardware options with AutoCAD, see Appendix A. You will find a general description of the available hardware options and their significance to AutoCAD.

Doing Things in Style

Much care has been taken to see that the stylistic conventions in this book—the use of uppercase or lowercase letters, italic or boldface type, and so on—will be the ones most likely to help you learn AutoCAD. On the whole, their effect should be subliminal. However, you may find it useful to be conscious of the rules that we have followed:

- Pull-down selections are shown by a series of menu options separated by the ➤ symbol (for example, choose File ➤ New).

- Keyboard entries are shown in boldface (for example, enter **rotate↵**). AutoCAD is not case-sensitive, so keyboard entries are shown in lowercase to make typing easier.

- Command-line prompts are shown in a monospace font (for example, `Select object:`).

For most functions, we describe how to select options from toolbars and the menu bar. In addition, where applicable, we include related keyboard shortcuts and command names. By providing command names, we have provided continuity for those readers already familiar with earlier releases of AutoCAD.

All This, and Software Too

Finally, we have included a CD-ROM containing a wealth of utilities, drawings, and sample programs that can greatly enhance your use of AutoCAD. We have also included two online books: the *AutoCAD 2000 Instant Reference* and the *ABCs of AutoLISP*. These easy-to-use online references complement *Mastering AutoCAD 2000 for Mechanical Engineers* and will prove invaluable for quick command searches and customization tips. Appendix C gives you detailed information about the CD-ROM, but here's a brief rundown of what's available. Check it out!

Software You Can Use Right Away

An example of great software available to you on the CD-ROM is Eye2eye, a utility that makes perspective viewing of your 3D work a simple matter of moving camera and target objects. This utility lets you easily fine-tune your perspective views so you can use them to create rendered images with AutoCAD's enhanced rendering tools. Also included is Whip 4.0, a Netscape plug-in that allows you to view AutoCAD drawings over the Internet.

Additional Resources

If you just need to quickly find information about a command, the online version of the *AutoCAD 2000 Instant Reference* is here to help you. It is a comprehensive guidebook that walks you through every feature and command of AutoCAD 2000. *Mastering AutoCAD 2000 for Mechanical Engineers* and the *AutoCAD 2000 Instant Reference* have always been a great combination. We have included an electronic version of this best-selling reference so that you can have the best AutoCAD resources in one place.

And if you want in-depth coverage of AutoLISP, AutoCAD's macro programming language, you can delve into the *ABCs of AutoLISP*. This book is an online AutoLISP reference and tutorial. AutoCAD users and developers alike have found the original *ABCs of AutoLISP* book an indispensable resource in their customization efforts. Now in its new HTML format, it's even easier to use.

Finally, a bonus chapter, "Using VBA to Create AutoCAD Applications," has been included on the CD to show you how the Visual Basic for Applications programming language can expand your work experience with AutoCAD.

Drawing Files for the Exercises

We have also included drawing files from the exercises in this book. These are provided so that you can pick up an exercise anywhere in the book, without having to work through the book from front to back. You can also use these sample files to repeat exercises or just to explore how files are organized and put together.

ActiveX Samples

Samples of ActiveX code from the VBA and ActiveX chapters have also been included to help you explore the newest customization feature of AutoCAD.

New Features of AutoCAD 2000

AutoCAD 2000 offers a higher level of speed, accuracy, and ease of use. It has always provided drawing accuracy to 16 decimal places. With this kind of accuracy, you can create a computer model of the earth and include details down to sub-micron levels. It also means that no matter how often you edit an AutoCAD drawing, its dimensions will remain true. And AutoCAD 2000 has greatly improved its overall speed. The interface is more consistent than in earlier releases, so learning and using AutoCAD is easier than ever.

3D Solid Modeling

- 3D Orbit

- ShadeEdge—improved shading and rendering

- Separate User Coordinate System per viewport

Heads-Up Design Environment

- AutoCAD DesignCenter—Microsoft Windows Explorer with an AutoCAD "attitude"

- Multiple Document Interface (MDI) allows you to have several drawings open at the same time.

- AutoSnap and AutoTrack enhancements

Streamlined Output

- Nonrectangular viewports

- Layouts—multiple Paper Space configurations

- Lineweight control without using polylines

Improved Access and Usability

- Improved right-click context menus

- Improved modeless Properties command allows for real-time object properties changes

- Find and replace text

- Partial open and partial load options speed drawing performance

- QDIM (quick dimensioning) automatically dimensions a series of features

- In-place editing of blocks and external references (Xrefs), without exiting the drawing

Up and Running in No Time

- Register on the Web, and AutoCAD will install the authorization code

Finally, perhaps the most important feature is what AutoCAD doesn't offer. You will not see an AutoCAD 2000 version for DOS, Macintosh, SGI, or UNIX. By eliminating these other platforms since Release 14, and concentrating on Windows 95/98 and NT, Autodesk is able to produce a leaner, meaner AutoCAD. It uses less memory than its previous Windows version and is faster than the previous DOS version. In many ways, this is the AutoCAD you've been waiting for.

The AutoCAD Package

This book assumes you are using AutoCAD Release 2000. If you are using an earlier version of AutoCAD, you will want to refer to *Mastering AutoCAD Release 14 for Windows 95 and NT* or its appropriate predecessor. You can find out about other books on AutoCAD published by Sybex by visiting the Sybex Web site at `http://www.sybex.com`.

With AutoCAD 2000, you receive the two following manuals in both hardcopy and electronic formats:

- The *AutoCAD User's Guide*

- The *Installation Guide*

You may also send away for and purchase hardcopies of the *AutoCAD Command Reference* and *Customization Guide* (which are both on the CD-ROM).

In addition, the AutoCAD package contains the AutoCAD Learning Assistant. This is a CD-ROM-based multimedia training and reference tool designed for those users who are upgrading from earlier versions of AutoCAD. It offers animated video clips, tips, and tutorials on a variety of topics. You'll need a CD player and sound card to take full advantage of the Learning Assistant.

You'll probably want to read the *Installation Guide* for Windows first, and then browse through the *Command Reference* and *User's Guide* to get a feel for the kind of information available there. You may want to save the *Customization Guide* for when you've become more familiar with AutoCAD.

AutoCAD comes on a CD-ROM and offers several levels of installation. This book assumes that you will use the full installation, which includes the Internet and Bonus utilities. You'll also want to install the ActiveX Automation software, included on the AutoCAD CD-ROM, if you plan to explore this new feature.

The Digitizer Template

If you intend to use a digitizer tablet in place of a mouse, Autodesk also provides you with a digitizer template. Commands can be selected directly from the template by pointing at the command on the template and pressing the pick button. Each command is shown clearly by name and a simple icon. Commands are grouped on the template by the type of operation the command performs. Before you can use the digitizer template, you must configure the digitizer. See Appendix A for a more detailed description of digitizing tablets and Appendix B for instructions on configuring the digitizer.

NOTE   We won't specifically discuss using the digitizer for selecting commands because the process is straightforward. If you are using a digitizer, you can use its puck like a mouse for all of the exercises in this book.

A Note From the Author

I hope that *Mastering AutoCAD 2000 for Mechanical Engineers* will be of benefit to you and that, once you have completed the tutorials, you will continue to use the book as a reference. If you have comments, criticisms, or ideas about how the book can be improved, you can write to mes, or send e-mail to the address below. And thanks for choosing *Mastering AutoCAD 2000 for Mechanical Engineers*.

George Omura
P.O. Box 6357
Albany, CA 94706-0357
Gomura@sirius.com

PART I

The Basics

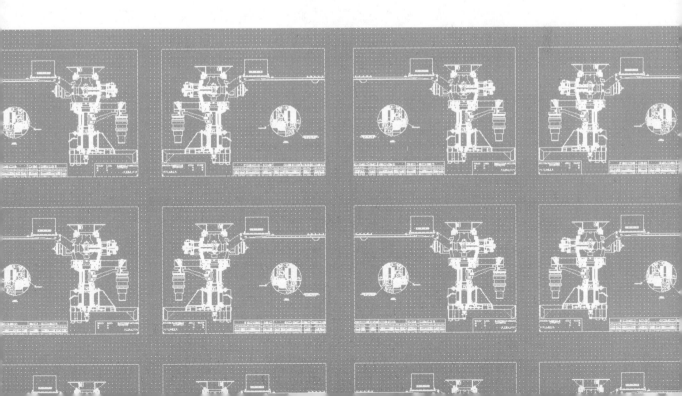

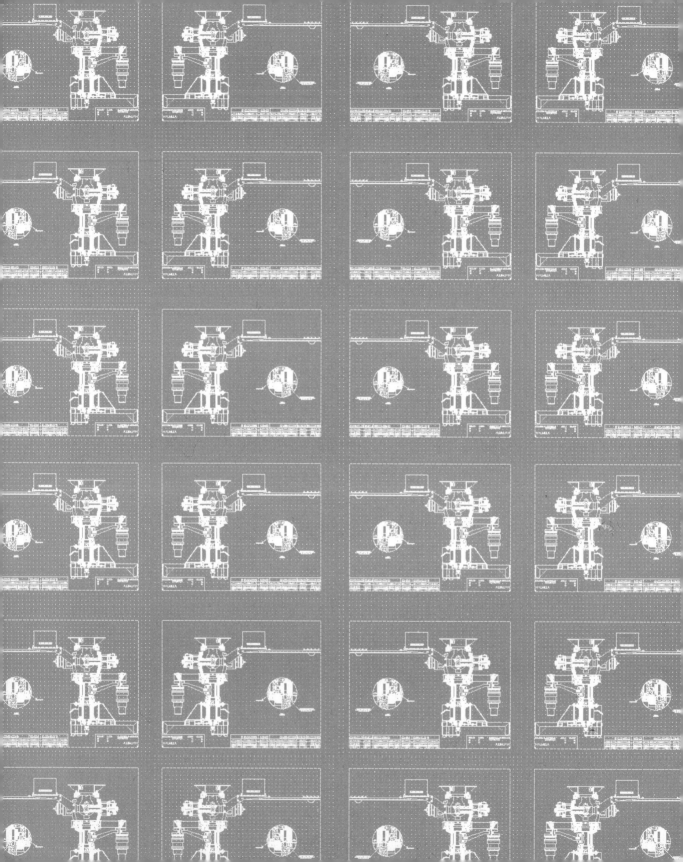

CHAPTER
ONE

This Is AutoCAD

- Navigating the AutoCAD Window

- Opening an Existing File

- Saving a File

- Making Changes

- Opening Multiple Files

- Exiting AutoCAD

Over the years, AutoCAD has evolved from a DOS-based, command-line driven program to a full-fledged Windows application. AutoCAD 2000 continues this trend with a new look and a wealth of new features that allow you to work more efficiently and with less effort. AutoCAD 2000 is strictly a Windows 95/98/NT program, so there are no UNIX or DOS versions.

By concentrating on a single operating system, Autodesk is able to create a more efficient, faster AutoCAD. AutoCAD 2000 offers the speed you demand with the convenience of a Windows multitasking environment. You'll also find that AutoCAD makes great use of the Windows environment. For example, you can use Windows' OLE features to paste documents directly into AutoCAD from Excel, Windows Paint, or any other programs that support OLE as a server application. Additionally, as in Releases 13 and 14, you can also export AutoCAD drawings directly to other OLE clients. This means no more messy conversions and reworking to get spreadsheet, database, text, or other data into AutoCAD. It also means that if you want to include a photograph in your AutoCAD drawing, all you have to do is cut and paste. Text-based data can also be cut and pasted, saving you time in transferring data, such as layer or block names.

NOTE   OLE stands for *Object Linking and Embedding*—a Windows feature that lets different applications share documents. See Chapter 18, "Getting and Exchanging Data from Drawings," for a more detailed discussion of OLE.

Windows gives you the freedom to arrange AutoCAD's screen by clicking and dragging its components. And AutoCAD 2000 now sports a look that is more in line with the Microsoft Office suite of applications, including Excel-like sheet tabs and borderless toolbar buttons. The changes in AutoCAD 2000 are not only skin deep, either. Among the many new features of AutoCAD 2000, you now have the ability to open multiple documents within a single session of AutoCAD. This means an easier exchange of data between different files, and the ability to compare files more easily. AutoCAD 2000 also introduces a wealth of new tools that will help you manage your drawing projects. If you're new to AutoCAD, this is the version you may have been waiting for. Even with its many new features, the programmers at Autodesk have managed to make AutoCAD easier to use than previous releases. AutoCAD's interface has been trimmed down and is more consistent than prior versions. They have even improved the messages you receive from AutoCAD to make them more understandable.

But in one sense, AutoCAD 2000 represents a return to an old feature of Auto-CAD: Autodesk has begun to listen to the user community again and has clearly incorporated many of the "wish-list" features that users have been asking for over the years. These include enhancements such as the ability to open multiple drawing sessions, a powerful improvement similar to the Windows Explorer in the form of the new Design Center, and more. So whether you're an old hand at AutoCAD or whether you're just starting out, AutoCAD 2000 offers a powerful drawing and design tool that is easier to use than ever.

So let's get started. In this first chapter, we'll look at many of AutoCAD's basic operations, such as opening and closing files, getting a close-up look at part of a drawing, and making changes to a drawing.

Taking a Guided Tour

First, you'll get a chance to familiarize yourself with the AutoCAD screen and how you communicate with AutoCAD. Along the way, you'll also get a feel for how to work with this book. Don't worry about understanding or remembering everything that you see in this chapter. You'll get plenty of opportunities to probe the finer details of the program as you work through the later chapters. If you're already familiar with earlier versions of AutoCAD, you may want to read through this chapter anyway, to get acquainted with new features and the graphical interface. To help you remember the material, you'll find a brief exercise at the end of each chapter. For now, just enjoy your first excursion into AutoCAD.

TIP   You might also consider purchasing *Mastering Windows 98* by Robert Cowart or *ABCs of Windows 98* by Sharon Crawford and Neil J. Salkind, both published by Sybex.

If you already installed AutoCAD, and you're ready to jump in and take a look at the program, then proceed with the following steps to launch it.

1. Click the Start button in the lower-left corner of the Windows 95/98 or NT 4 screen. Then choose Program ➢ AutoCAD 2000 ➢ AutoCAD 2000. You can also double-click the AutoCAD 2000 icon on your Windows Desktop.

2. You'll see an opening greeting, called a *splash screen,* telling you which version of AutoCAD you're using, to whom the program is registered, and the AutoCAD dealer's name and phone number should you need help.

3. Next, you'll see the Startup dialog box. This dialog is a convenient tool for setting up new drawings. You'll learn more about this tool in later chapters. For now, click Cancel.

Message to Veteran AutoCAD Users

Autodesk is committed to the Windows operating environment. The result is a graphical user interface (GUI) that is easier on the AutoCAD neophyte, but perhaps a bit foreign to a veteran AutoCAD user.

If you've been using AutoCAD for a while, and you prefer the older interface, you can still enter AutoCAD commands through the keyboard, and you can still mold AutoCAD's interface into one that is more familiar to you.

You can, for example, restore the side menu that appears in the DOS version of AutoCAD. Here's how it's done:

1. Choose Tools ➤ Options.

2. At the Options dialog box, click the Display tab.

3. Click the check box labeled Display AutoCAD Screen Menu in Drawing Window.

4. Finally, click OK. The side menu will appear.

Continued on next page

A word of caution: If you're accustomed to pressing Ctrl+C to cancel an operation, you must now retrain yourself to press the Escape (Esc) key. Ctrl+C now conforms to the Windows standard, making this key combination a shortcut for saving marked items to the Clipboard. Similarly, instead of using F1 to view the full text window, you must use F2. F1 is most commonly reserved for the Help function in Windows applications.

If you prefer entering commands through the keyboard, you'll also want to know about some changes to specific commands in AutoCAD 2000. Several commands that usually invoke dialog boxes can be used through the Command window prompt. Here is a list of those commands:

Bhatch	Boundary	Group	Hatchedit	Image
Layer	Linetype	Mtext	Pan	Block
XBind	Style	Osnap	Xref	Wblock

When you enter these commands through the keyboard, you'll normally see a dialog box. In the case of Pan, you'll see the Realtime Pan hand graphic. To utilize these commands from the command prompt, add a minus sign (-) to the beginning of the command name. For example, to use the Layer command in the older command-line method, enter **–layer** at the command prompt. To use the old Pan command, enter **–pan** at the command prompt.

Even if you don't care to enter commands through the keyboard, knowing about the use of the minus sign can help you create custom macros. See Chapter 19, "Introduction to Customization," for more on AutoCAD customization.

The AutoCAD Window

The AutoCAD program window is divided into five parts:

- Pull-down menu bar
- Docked and floating toolbars
- Drawing area
- Command window
- Status bar

NOTE
A sixth hidden component, the Aerial View window, displays your entire drawing and lets you select close-up views of parts of your drawing. After you've gotten more familiar with AutoCAD, consult Chapter 6, "Enhancing Your Drawing Skills," for more on this feature.

Figure 1.1 shows a typical layout of the AutoCAD program window. Along the top is the *menu bar,* and at the bottom are the *Command window* and the *status bar.* Just below the menu bar and to the left of the window are the *toolbars.* The rest of the screen is occupied by the *drawing area.*

FIGURE 1.1:

A typical arrangement of the elements in the Auto-CAD window

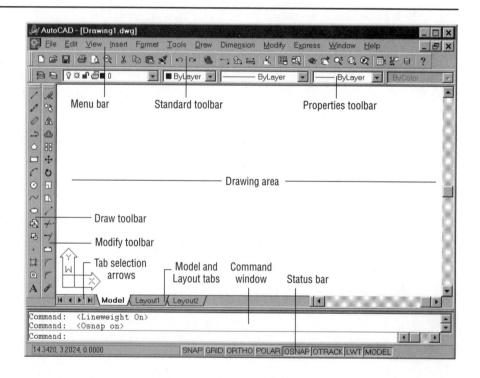

Many of the elements within the AutoCAD window can be easily moved and reshaped. Figure 1.2 demonstrates how different AutoCAD can look after some simple rearranging of window components. Toolbars can be moved from their default locations to any location on the screen. When they are in their default locations, they are in their *docked* positions. When they are moved to locations where they are free-floating, they are *floating.*

FIGURE 1.2:

An alternative arrangement of the elements in the AutoCAD window

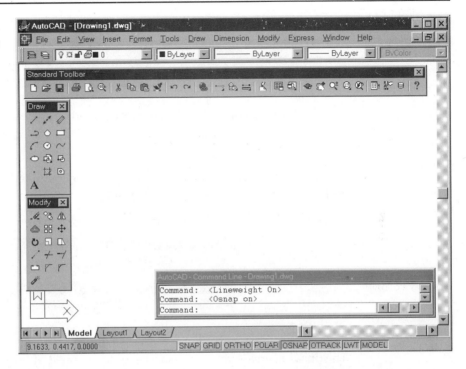

The menu bar at the top of the drawing area (as shown in Figure 1.3) offers pull-down menus from which you select commands in a typical Windows fashion. The toolbars offer a variety of commands through tool buttons and drop-down lists. For example, the *current layer* name or number you're presently working on where new objects are placed is displayed in a drop-down list in the Object Properties toolbar. The layer name is preceded by tools that inform you of the status of the layer. You'll learn more about the Layer tool and all of its options in Chapter 4, "Organizing Your Work." The tools and lists in the toolbar are plentiful, and you'll learn more about all of them later in this chapter in "The Toolbars" section and as you work through this book.

NOTE A *layer* is like an overlay that allows you to separate different types of information. AutoCAD allows an unlimited number of layers. On new drawings, the default layer is 0. You'll get a detailed look at layers and the meaning of the Layer tools in Chapter 4, "Organizing Your Work."

FIGURE 1.3:

The components of the menu bar and toolbar

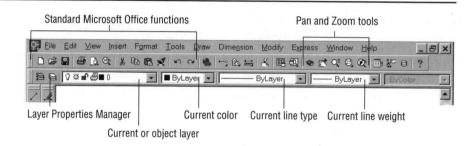

Standard Microsoft Office functions

Pan and Zoom tools

Layer Properties Manager

Current color

Current line type

Current line weight

Current or object layer

The Draw and Modify toolbars (Figure 1.4) offer commands that create new objects and edit existing ones. These are just two of many toolbars available to you.

FIGURE 1.4:

The Modify and Draw toolbars as they appear when they are floating

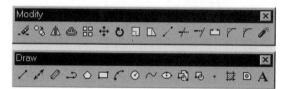

The drawing area—your workspace—occupies most of the screen. Everything you draw appears in this area. As you move your mouse around, you'll see crosshairs that appear to move within the drawing area. This is your drawing cursor that lets you point to locations in the drawing area. At the bottom of the drawing area, the status bar (see Figure 1.5) gives you information about the drawing at a glance. For example, the coordinate readout toward the far left of the status bar tells you the location of your cursor. The Command window can be moved and resized in a manner similar to toolbars. By default, the Command window is in its docked position, shown here. Let's practice using the coordinate readout and drawing cursor by picking some points on the screen.

FIGURE 1.5:

The status bar and Command window

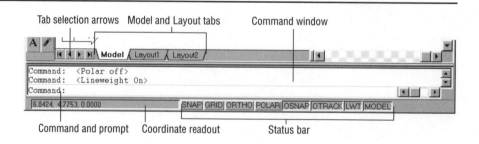

Tab selection arrows   Model and Layout tabs   Command window

Command and prompt   Coordinate readout   Status bar

NOTE   Some AutoCAD users prefer to turn off the scrollbars to the right and at the bottom of the drawing area. This allows you to maximize the size of the drawing area. Settings that control the scroll bars and other screen-related functions can be found in the Options dialog box. See Appendix B, "Installing and Setting Up AutoCAD," for more on the Options dialog box.

Picking Points

Follow these steps to pick points.

1. Move the cursor around in the drawing area. As you move, note how the coordinate readout changes to tell you the cursor's location. It shows the coordinates in an X,Y format.

2. Now place the cursor in the middle of the drawing area and press and immediately release the left mouse button. You've just picked a point. Move the cursor, and a rectangle follows. This is a *Selection window*; you'll learn more about this window in Chapter 2, "Creating Your First Drawing."

3. Move the cursor a bit in any direction; then press and let go of the left mouse button again. Notice that the rectangle disappears.

4. Try picking several more points in the drawing area.

If you accidentally press the right mouse button, a pop-up menu appears. This will be a surprise to both new and experienced AutoCAD users. In AutoCAD 2000, a right mouse click will frequently bring up a pop-up menu with options that are context sensitive. This means that the contents of the pop-up menu will depend on where you right-click as well as the command that is active at the time of your right-click. If options are not appropriate at the time of the right-click, AutoCAD will treat the right-click as a ↵. You'll learn more about these options as you progress through the book. For now, if you happen to open this menu by accident, press the Escape (Esc) key to dismiss it.

NOTE   The carriage-return symbol (↵) is used in this book to represent pressing the Enter key.

The User Coordinate System (UCS) Icon

In the lower-left corner of the drawing area, you'll see a thick, L-shaped arrow outline. This is the *User Coordinate System* (UCS) icon, which tells you the orientation of the drawing. This icon becomes helpful as you start to work with complex 2D drawings and 3D models. The X and Y inside the icon indicate the X and Y axes of your drawing. The W tells you that you're in what is called the *World Coordinate System* (WCS). For now, you can use it as a reference to tell you the direction of the X and Y axes.

NOTE **If you can't find the UCS icon...** The UCS icon can be turned on and off, so if you're on someone else's system and you don't see the icon, don't panic. It also changes shape depending on whether you're in Paper Space or Model Space mode. If you don't see the icon or it doesn't look like it does in this chapter, see "Understanding Model Space and Paper Space" in Chapter 9. Also see Chapter 11, "Introducing 3D" for more on the UCS icon.

The Command Window

At the bottom of the screen, just above the status bar, is a small horizontal window, which is the Command window. Here, AutoCAD displays responses to your input. It shows three lines of text. The bottom shows the current messages; the top two lines show messages that have scrolled by, or in some cases, components of the current message that do not fit in a single line. Right now, the bottom line displays the message Command (see Figure 1.5). This tells you that AutoCAD is waiting for your instructions. As you click a point in the drawing area, you'll see the message Other corner. At the same time, the cursor starts to draw a Selection window that disappears when you click another point.

As a new user, it's important to pay special attention to messages displayed in the Command window because this is how AutoCAD communicates with you. Besides giving you messages, the Command window records your activity in AutoCAD. You can use the scroll bar to the right of the Command window to review previous messages. You can also enlarge the window for a better view. (We'll discuss this in more detail in Chapter 2, "Creating Your First Drawing.")

NOTE As you become more familiar with AutoCAD, you may find you don't need to rely on the Command window as much. However, for new users it can be quite helpful in understanding what steps to take as you work.

Now let's look at AutoCAD's window components in detail.

The Pull-Down Menus

Like most Windows programs, the pull-down menus available on the menu bar offer an easy-to-understand way to access the general controls and settings for AutoCAD. Within these menus you'll find the commands and functions that are the heart of AutoCAD. By clicking menu items, you can cut and paste them to and from AutoCAD, change the settings that make AutoCAD work the way you want it to, set up the measurement system you want to use, access the help system, and much more.

TIP To close a pull-down menu without selecting anything, press the Escape (Esc) key. You can also click any other part of the AutoCAD window or another pull-down menu.

The pull-down menu options perform four basic functions:

- Displaying additional menu choices
- Displaying a dialog box that contains settings you can change
- Issuing a command that requires keyboard or drawing input
- Offering an expanded set of the same tools found in the Draw and Modify toolbars

As you point to commands and options in the menus or toolbars, AutoCAD provides additional help for you in the form of brief descriptions of each menu option, which appear in the status bar.

Here's an exercise to let you practice with the pull-down menus and get acquainted with AutoCAD's interface.

1. Click View in the menu bar. The items that appear are the commands and settings that let you control the way AutoCAD displays your drawings. Don't worry if you don't understand them; you'll get to know them in later chapters.

2. Move the highlight cursor slowly down the list of menu items. As you highlight each item, notice that a description of it appears in the status bar at the bottom of the AutoCAD window. These descriptions help you choose the menu option you need.

3. Some of the menu items have triangular pointers to their right. This means the command has additional choices. For instance, highlight the Zoom item, and you'll see another set of options appear to the right of the menu.

NOTE   If you look carefully at the command descriptions in the status bar, you'll see an odd word at the end, such as *LINE*, *CIRCLE*, or whatever command you may be using. This is the keyboard command equivalent to the highlighted option in the menu or toolbar. You can actually type in these keyboard commands to start the tool or menu item that you're pointing to. You don't have to memorize these command names, but knowing them will be helpful to you later if you want to customize AutoCAD.

This second set of options is called a *cascading menu*. Whenever you see a pull-down menu item with the triangular pointer, you know that this item opens a cascading menu offering a more detailed set of options.

You might have noticed that other pull-down menu options are followed by an ellipsis (…). This indicates that the option brings up a dialog box, as the following exercise demonstrates:

1. Move the highlight cursor to the Tools option in the menu bar.

NOTE   If you prefer, you can click and drag the highlight cursor over the pull-down menu to select an option.

---

2. Click the Options item. The Options dialog box appears.

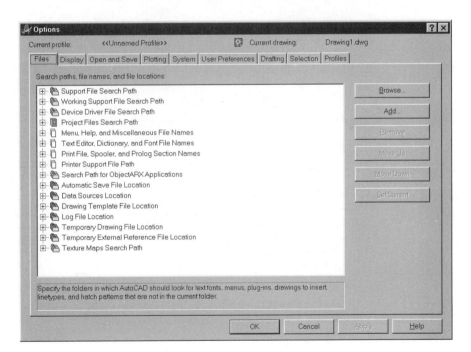

NOTE   If you're familiar with Windows 95/98 Explorer, you should feel at home with the Files tab of the Options dialog box. The plus sign (+) to the left of the items in the list expands the option to display more detail.

This dialog box contains several *pages*, indicated by the tabs across the top, that contain settings for controlling what AutoCAD shows you on its screens, where you want it to look for special files, and other "housekeeping" settings. You needn't worry about what these options mean at this point. Appendix B, "Installing and Setting Up AutoCAD," describes this dialog box in more detail.

3.   In the Options dialog box, click the tab labeled Open and Save. The options change to reveal new options.

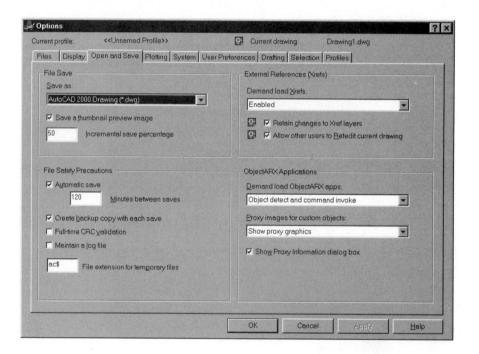

In the middle-left side of the dialog box, you'll see a check box labeled Automatic Save, with the Minutes between Saves setting input box set to 120 minutes. This setting controls how frequently AutoCAD performs an automatic save.

4.   Change the 120 to 20, and then click OK. You've just changed AutoCAD's automatic save feature to automatically save files every 20 minutes instead of every two hours. (Let this be a reminder to give your eyes a rest.)

5.   Finally, click the OK button at the bottom of the dialog to save your changes and dismiss the dialog box.

Another type of item you'll find in pull-down menus is a command that directly executes an AutoCAD operation. Let's try an exercise to explore a drawing command.

1. Click the Draw option in the menu bar, and then click the Rectangle command. Notice that the Command window now shows the comment

    ```
    Specify first corner point or
    [Chamfer/Elevation/Fillet/Thickness/Width]:
    ```

    AutoCAD is asking you to select the first corner for the rectangle and, in brackets, it is offering a few options that you can take advantage of at this point in the command. Don't worry about those options right now. You'll have an opportunity to learn about command options in the next chapter.

2. Click a point roughly in the lower-left corner of the drawing area, as shown in Figure 1.6. Now as you move your mouse, you'll see a rectangle follow the cursor with one corner fixed at the position you just selected. You'll also see the following message in the Command window:

    ```
    Specify other corner point:
    ```

3. Click another point anywhere in the upper-right region of the drawing area. A rectangle appears (see Figure 1.7). You'll learn more about the different cursor shapes and what they mean in Chapter 2, "Creating Your First Drawing."

FIGURE 1.6:

Selecting points to define a rectangle

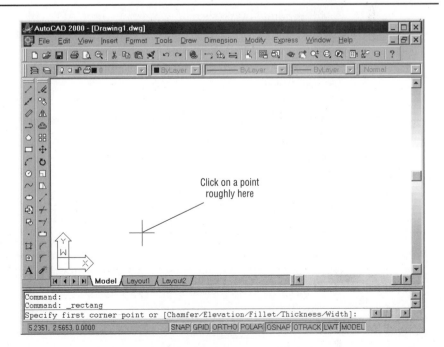

FIGURE 1.7:

Once you've selected your first point of the shape, the cursor disappears and you see a rectangle follow the motion of your mouse.

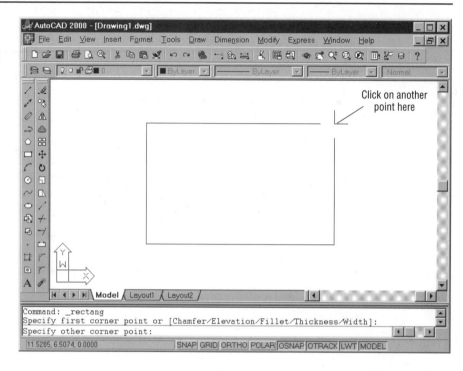

At this point you've seen how most of AutoCAD's commands work. You'll find that dialog boxes are offered when you want to change settings, while many drawing and editing functions present messages in the Command window. Also, be aware that many of the pull-down menu items are duplicated in the toolbars that you'll explore next.

Communicating with AutoCAD

AutoCAD is the perfect servant: It does everything you tell it to, and no more. You communicate with AutoCAD using the pull-down menus and the toolbars. These devices invoke AutoCAD commands. A *command* is a single-word instruction you give to AutoCAD telling it to do something, such as draw a line (using the Line tool in the Draw toolbar) or erase an object (with the Erase tool in the Modify toolbar). Whenever you invoke a command, by either typing it in or selecting a menu or toolbar item, AutoCAD responds by presenting messages to you in the Command window, or by displaying a dialog box.

Continued on next page

The messages in the Command window often tell you what to do next, or they offer a list of options. A single command will often present several messages, which you answer to complete the command. These messages serve as an aid to new users who need a little help. If you ever get lost while using a command, or forget what you're supposed to do, look at the Command window for clues. As you become more comfortable with Auto-CAD, you'll find that you won't need to refer to these messages as frequently.

As an additional aid, you can right-click to display a context-sensitive menu. If you're in the middle of a command, and are not selecting points, this menu will offer a list of options specifically related to that command. For example, if you had right-clicked your mouse before picking the first point for the Rectangle command in the previous exercise, a pop-up menu would have appeared, offering the same options that were listed in the Command window, plus some additional options.

A dialog box is like a form you fill out on the computer screen. It lets you adjust settings or make selections from a set of options pertaining to a command.

The Toolbars

While the pull-down menus offer a full range of easy-to-understand options, they require some effort to navigate. The toolbars, on the other hand, offer quick, single-click access to the most commonly used AutoCAD features. In the default AutoCAD window arrangement, you see the most commonly used toolbars. Other toolbars are available but hidden from view until you open them.

The tools in the toolbars perform three types of actions, just like the pull-down menu commands:

- Displaying further options
- Opening dialog boxes
- Issuing commands that require keyboard or cursor input

Read on for details.

The Toolbar Tool Tips

AutoCAD's toolbars contain tools that represent commands. To help you understand each tool, a *tool tip* appears just below the arrow cursor when you rest the cursor on a tool. Each tool tip helps you identify the tool with its function. A tool tip appears when you follow these steps:

1. Move the arrow cursor onto one of the toolbar tools and leave it there for a second or two. Notice that the command's name appears nearby—this is the tool tip. In the status bar, a brief description of the button's purpose appears (see Figure 1.8).

2. Move the cursor across the toolbar. As you do, notice that the tool tips and status bar descriptions change to describe each tool.

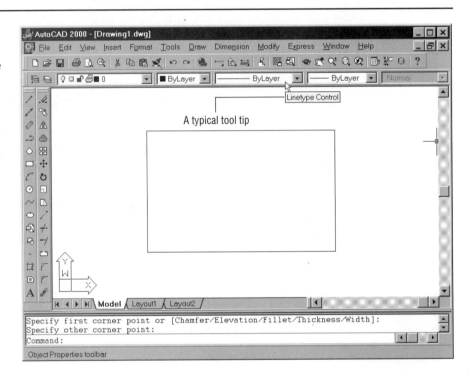

Toolbar Flyout Menus

Most toolbar tools start a command as soon as you click them, but other tools will display a set of additional tools (similar to the menus in the menu bar) that are related to the tool you've selected. This set of additional tools is called a toolbar *flyout*. If you've used other Windows graphics programs, chances are you've seen flyouts. Look closely at the tools just below the Help pull-down menu option on your screen or in Figure 1.8. You'll be able to identify which toolbar tools have flyouts; they'll have a small right-pointing arrow in the lower-right corner of the tool.

When an instruction says *click*, you should lightly press the left mouse button until you feel a click; then immediately let it go. Don't hold down the button, unless you don't wish to start the command.

The following steps show you how a flyout works.

1. Move the cursor to the Zoom Window tool in the Standard toolbar. Click and hold the left mouse button to display the flyout. Don't release the mouse button.

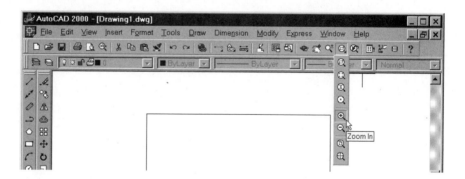

2. Still holding down the left mouse button, move the cursor over the flyout; notice that the tool tips appear here as well. Also, notice the description in the status bar.

3. Move the cursor to the Zoom Window tool at the top of the flyout and release the mouse button.

4. You won't need to use this tool yet, so press the Escape (Esc) key to cancel this tool.

As you can see from this exercise, you get a lot of feedback from AutoCAD.

Moving the Toolbars

One unique characteristic of AutoCAD's toolbars is their mobility. They can either float anywhere in the AutoCAD window or be in a docked position. *Docked* means it is placed against the top and side borders of the AutoCAD window, so that the toolbar occupies a minimal amount of space. If you want to, you can move the toolbar to any location on your Desktop, thus turning it into a *floating toolbar*.

Try the following exercise to move the Object Properties toolbar away from its current position in the AutoCAD window:

1. Move the arrow cursor so that it points to the vertical bars, called *grab bars*, to the far left of the Object Properties toolbar, as shown here.

2. Press and hold down the left mouse button. Notice that a gray rectangle appears by the cursor.

3. Still holding down the mouse button, move the mouse downward. The gray box follows the cursor.

4. When the gray box is over the drawing area, release the mouse button, and the Object Properties toolbar—now a floating toolbar—moves to its new location.

You can now move the Object Properties toolbar to any location on the screen that suits you. You can also change the shape of the toolbar. To do so, continue with the following steps.

1. Place the cursor on the bottom-edge border of the Object Properties toolbar. The cursor becomes a double-headed arrow, as shown here.

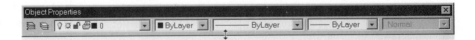

2. Click and drag the border downward. The gray rectangle jumps to a new, taller rectangle as you move the cursor.

3. When the gray rectangle changes to the shape you want, release the mouse button to reshape the toolbar.

4. To move the toolbar back into its docked position, place the arrow cursor on (point to) the toolbar's title bar and slowly click and drag the toolbar so the cursor is in position in the upper-left corner of the AutoCAD window. Notice how the gray outline of the toolbar changes as it approaches its docked position.

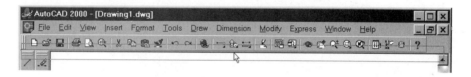

5. When the outline of the Object Properties toolbar is near its docked position, release the mouse button. The toolbar moves back into its previous position in the AutoCAD window.

You can move and reshape any of AutoCAD's toolbars to place them out of the way, yet still have them at the ready to give you quick access to commands. You can also put them away altogether when you don't need them and bring them back at will, as shown in these next steps.

1. Click and drag the Draw toolbar from its position at the left of the AutoCAD window to a point near the center of the drawing area. Remember to click and drag the grab bars at the top of the toolbar.

2. Click the Close button in the upper-left corner of the Draw floating toolbar. This is the small square button with the X in it. The toolbar disappears.

3. To recover the Draw toolbar, right-click any toolbar—the Object Properties toolbar, for example. A pop-up list of toolbars appears.

4. Locate and select Draw in the pop-up list of toolbars. The Draw toolbar reappears.

5. Click and drag the Draw toolbar back to its docked position in the far-left side of the AutoCAD window.

AutoCAD will remember your toolbar arrangement between sessions. When you exit and then reopen AutoCAD later, the AutoCAD window will appear just as you left it.

Identifying the Toolbars

You may have noticed that many of the toolbars listed in the pop-up list of tool-bars don't appear automatically in the AutoCAD window. You can display any toolbar by selecting it from the list. Here are brief descriptions of all the toolbars.

3D Orbit   Tools to control 3D views.

Dimension   Commands that help you dimension your drawings. Many of these commands are duplicated in the Dimension pull-down menu.

Draw   Commands for creating common objects, including lines, arcs, circles, curves, ellipses, and text. This toolbar appears in the AutoCAD window by default. Many of these commands are duplicated in the Draw pull-down menu.

Inquiry   Commands for finding distances, point coordinates, object properties, mass properties, and areas.

Insert   Commands for importing other drawings, raster images, and OLE objects. Many of these commands are duplicated in the Insert pull-down menu.

Layouts   Tools that let you set up drawing layouts for viewing, printing, and plotting.

Modify   Commands for editing existing objects. You can move, copy, rotate, erase, trim, extend, and so on. Many of these commands are duplicated in the Modify pull-down menu.

Modify II   Commands for editing complex objects such as polylines, multilines, 3D solids, and hatches.

Object Properties   Commands for manipulating the properties of objects. This toolbar is normally docked below the Standard toolbar.

Object Snap   Tools to help you select specific points on objects, such as endpoints and midpoints.

Refedit   Tools that allow you to make changes to symbols or background drawings that are imported as external reference drawings.

Reference   Commands that control cross-referencing of drawings.

Render   Commands to operate AutoCAD's rendering feature.

Shade   Tools to control the way 3D models are displayed.

Solids   Commands for creating 3D solids.

Solids Editing   Commands for editing 3D solids.

Standard   The most frequently used commands for view control, file management, and editing. This toolbar is normally docked below the pull-down menu bar.

Surfaces   Commands for creating 3D surfaces.

UCS   Tools for setting up a plane on which to work. UCS stands for *User Coordinate System.* This toolbar is most useful for 3D modeling, but it can be helpful in 2D drafting, as well.

UCS II   Tools for selecting from a set of predefined user coordinate systems.

View   Tools to control the way you view 3D models.

Viewports   Tools that let you create and edit multiple views of your drawing.

Web   Tools for accessing the World Wide Web.

Zoom   Commands that allow you to enlarge and reduce your view of portions of your drawing.

You'll get a chance to use the various toolbars as you work through this book. Or, if you plan to use the book as a reference rather than working through it as a chapter-by-chapter tutorial, any exercise you try will tell you which toolbar to use for performing a specific operation.

Menus versus the Keyboard

Throughout this book, you'll be told to select commands and command options from the pull-down menus and toolbars. For new and experienced users alike, menus and toolbars offer an easy-to-remember method for accessing commands. If you're an experienced AutoCAD user who is accustomed to the earlier versions of AutoCAD, you still have the option of entering commands directly through the keyboard. Most of the commands you know and love still work as they did from the keyboard.

Another method for accessing commands is to use *accelerator keys,* which are special keystrokes that open and activate pull-down menu options. You might have noticed that the commands in the menu bar and the items in the pull-down menus all have an underlined character. By pressing the Alt key followed by the key corresponding to the underlined character, you activate that command or option, without having to engage the mouse. For example, to issue File ➣ Open, press Alt, then F, then finally O (Alt+F+O).

Many tools and commands have keyboard shortcuts: one-, two-, or three-letter abbreviations of a command name. As you become more proficient with AutoCAD, you may find these shortcuts helpful. As you work through this book, we'll point out the shortcuts for your reference.

Finally, if you're feeling adventurous, you can create your own accelerator keys and keyboard shortcuts for executing commands by adding them to the AutoCAD support files. We'll discuss customization of the menus, toolbars, and keyboard shortcuts in Chapter 19, "Introduction to Customization."

Working with AutoCAD

Now that you've been introduced to the AutoCAD window, try using a few of AutoCAD's commands. First, you'll open a sample file and make a few simple modifications to it. In the process, you'll get familiar with some common methods of operation in AutoCAD.

Opening an Existing File

In this exercise, you'll get a chance to use a typical Select File dialog box. To start with, you'll open an existing file.

1. From the menu bar, choose File ➢ Open. A message appears asking you if you want to save the changes you've made to the current drawing. Click No.

2. Next, a Select File dialog box appears. This is a typical Windows file dialog box, with an added twist. The large Preview box on the right allows you to preview a drawing before you open it, thereby saving time while searching for files.

3. In the Select File dialog box, go to the Directories list and locate the directory named Figures (you may need to scroll down the list to find it). Point to it and then double-click. (If you're having trouble opening files with a double-click, here's another way to do it until you're more proficient with the mouse: Click the file once to highlight it, and then click the OK button.) The file list on the left changes to show the contents of the Sample directory.

4. Move the arrow to the file named `Nozzle3d`, and click it. Notice that the name now appears in the File Name input box below the file list. Also, the Preview box now shows a thumbnail image of the file.

NOTE

The `Nozzle3d` drawing is included on the companion CD-ROM. If you cannot find this file, be sure you've installed the sample drawings from the CD. See Appendix C, "What's on the Companion CD-ROM," for installation instructions.

---

5. Click the OK button at the bottom of the dialog box. AutoCAD opens the `Nozzle3d` file, as shown in Figure 1.9.

The `Nozzle3d` file opens to display the entire drawing. Also, the AutoCAD window's title bar displays the name of the drawing. This offers easy identification of the file. This particular file contains both a 2D and 3D model of a fire-hose nozzle. The opening view is actually a 3D view.

FIGURE 1.9:

In the early days, this nozzle drawing became the unofficial symbol of Auto-CAD, frequently appearing in ads for AutoCAD third-party products.

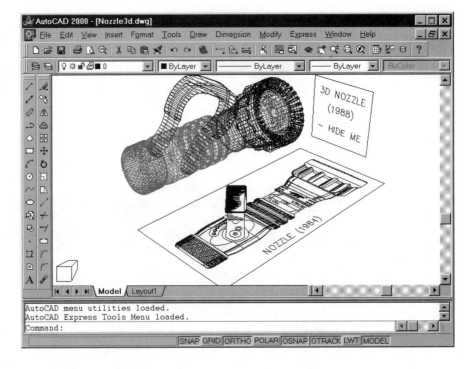

Getting a Closer Look

Zoom is one of the most frequently used commands. It lets you get a closer look at a part of your drawing and also offers a variety of ways to control your view. Now you'll enlarge a portion of the Nozzle3d drawing to get a more detailed look. To tell AutoCAD what area you wish to enlarge, you'll use a *window*.

1. Choose View ➣ 3D Views ➣ Plan View ➣ World UCS. Your view changes to display a two-dimensional view looking down on the drawing.

2. Click the Zoom Window button on the Standard toolbar.

You can also choose View ➣ Zoom ➣ Window from the pull-down menu or type z ↵ w↵.

3. The Command window displays the prompt First corner:. Look at the top image of Figure 1.10. Move the crosshair cursor to a location similar to the one shown in the figure; then click the left mouse button. Move the cursor and you'll see the rectangle appear, with one corner fixed on the point you just picked, while the other corner follows the cursor.

FIGURE 1.10:

Placing the Zoom window around the nozzle handle. After clicking the Zoom Window button in the Standard toolbar, select the two points shown in this figure.

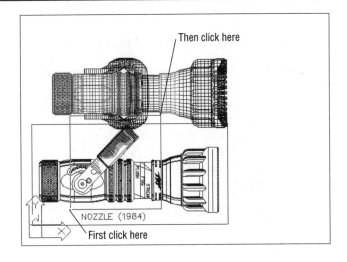

4. The Command window now displays the prompt `Specify first corner:` `Specify opposite corner:`. Position the other corner of the window so it encloses the handle of the nozzle, as shown in the figure, and press the mouse button. The handle enlarges to fill the screen (see the bottom image of Figure 1.10).

NOTE   You'll notice that tiny crosses appear where you picked points. These are called *blips*—markers that show where you've selected points. They do not become a permanent part of your drawing, nor do they print onto hard copy output.

In this exercise, you used the Window option of the Zoom command to define an area to enlarge for your close-up view. You saw how AutoCAD prompts you to indicate first one corner of the window, and then the other. These messages are helpful for first-time users of AutoCAD. You'll be using the Window option frequently—not just to define views, but also to select objects for editing.

Getting a close-up view of your drawing is crucial to working accurately with a drawing, but you'll often want to return to a previous view to get the overall picture. To do so, click the Zoom Previous button on the Standard toolbar.

Do this now, and the previous view—one showing the entire nozzle—returns to the screen. You can also get there by choosing View ➢ Zoom ➢ Previous.

NOTE   You can also zoom in and out using the Zoom In and Zoom Out buttons in the Zoom Window flyout of the Standard toolbar. The Zoom In button shows a magnifying glass with a plus sign (+); Zoom Out shows a minus sign (−). If you have a mouse equipped with a scroll wheel, you can zoom in and out just by turning the wheel. The location of the cursor at the time you move the wheel will determine the center of the zoom. A click-and-drag of the scroll wheel will let you pan your view.

You can quickly enlarge or reduce your view using the Zoom Realtime button on the Standard toolbar.

1. Click the Zoom Realtime button in the Standard toolbar.

The cursor changes into a magnifying glass.

2. Place the Zoom Realtime cursor slightly above the center of the drawing area, and then click and drag downward. Your view zooms out to show more of the drawing.

3. While still holding the left mouse button, move the cursor upward. Your view zooms in to enlarge itself. When you have a view similar to the one shown in Figure 1.11, release the mouse button. (Don't worry if you don't get the exact same view as the figure. This is just for practice.)

4. You're still in Zoom Realtime mode. Click and drag the mouse again to see how you can further adjust your view. To exit, you can select another command besides a Zoom or Pan command, press the Escape (Esc) key, or right-click your mouse.

5. Go ahead and right-click now. A pop-up menu appears.

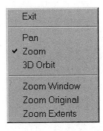

This menu lets you select other display-related options.

6. Click Exit from the pop-up menu to exit the Zoom Realtime command.

As you can see from this exercise, you have a wide range of options for viewing your drawing, just by using a few buttons. In fact, these three buttons, along with the scroll bars at the right side and bottom of the AutoCAD window, are all you need to control the display of your 2D drawings.

FIGURE 1.11:

The final view you want to achieve in step 3 of the exercise

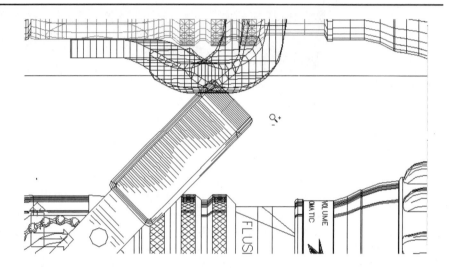

The Aerial View Window

The Aerial View window is an optional AutoCAD display tool. It gives you an overall view of your drawing, no matter how much magnification you may be using for the drawing editor. Aerial View also makes it easier to get around in a large-scale drawing. You'll find that this feature is best suited to more complex drawings that cover great areas, such as site plans, topographical maps, or city-planning documents.

We won't discuss this view much in the first chapter, as it can be a bit confusing for the first-time AutoCAD user. However, as you become more comfortable with AutoCAD, you may want to try it out. You'll find a detailed description of the Aerial View window in Chapter 6, "Enhancing Your Drawing Skills."

Saving a File As You Work

It's a good idea to periodically save your file as you work on it. You can save it under its original name (with File ➢ Save) or under a different name (with File ➢ Save As), thereby creating a new file.

By default, AutoCAD automatically saves your work at 120-minute intervals under the name AUTO.SV$; this is known as the *Autosave* feature. Using settings in the Options dialog box or system variables, you can change the name of the autosaved file and control the time between autosaves. See the sidebar, "Using AutoCAD's Automatic Save Feature," in Chapter 3, "Learning the Tools of the Trade," for details.

Let's first try the Save command. This quickly saves the drawing in its current state without exiting the program.

Choose File ➢ Save. You'll notice some disk activity while AutoCAD saves the file to the hard disk. As an alternative to picking File ➢ Save from the menus, you can type **Alt+f s**. This is the accelerator key, also called *hotkey*, for the File ➢ Save command.

Now try the Save As command. This command brings up a dialog box that allows you to save the current file under a new name.

1. Choose File ➢ Save As, or type **saveas** ↵ at the command prompt. The Select File dialog box appears. Note that the current filename, Nozzle3d.dwg, is highlighted in the File Name input box at the bottom of the dialog box.

2. Type **myfirst**. As you type, the name Nozzle3d disappears from the input box and is replaced by Myfirst. You don't need to enter the .dwg filename extension. AutoCAD adds it to the filename automatically when it saves the file.

3. Click the Save button. The dialog box disappears, and you will notice some disk activity.

You now have a copy of the Nozzle3d file under the name Myfirst.dwg, and the name of the file displayed in the AutoCAD window's title bar has changed to Myfirst. From now on, when you use the File ➢ Save option, your drawing will be saved under its new name. Saving files under a different name can be useful when you're creating alternatives or when you just want to save one of several ideas you're trying out.

TIP   If you're working with a monitor that is on the small side, you may want to consider closing the Draw and Modify toolbars. The Draw and Modify pull-down menus offer the same commands, so you won't lose any functionality by closing these toolbars. If you really want to maximize your drawing area, you can also turn off the scroll bars and reduce the Command window to a single line. See Appendix B for details on how to do this.

Making Changes

You'll be making frequent changes to your drawings. In fact, one of AutoCAD's chief advantages is the ease with which you can make changes. The following exercise shows you a typical sequence of operations involved in making a change to a drawing.

1. From the Modify toolbar, click the Erase tool (the one with a pencil eraser touching paper). This activates the Erase command. You can also choose Modify ➤ Erase from the pull-down menu.

Notice that the cursor has turned into a small square; this square is called the *pickbox*. You also see `Select object:` in the command-prompt area. This message helps remind new users what to do.

2. Place the pickbox on the diagonal pattern of the nozzle handle (see Figure 1.12) and click it. The 2D image of the nozzle becomes highlighted. The pickbox and the `Select object` prompt remain, telling you that you can continue to select objects.

3. Now press ↵. The nozzle and the rectangle disappear. You've just erased a part of the drawing.

FIGURE 1.12:

Erasing a portion of the nozzle handle

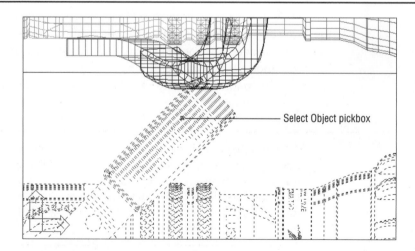

Select Object pickbox

In this exercise, you first issued the Erase command and then selected an object by clicking it using a pickbox. The pickbox tells you that you must select items on the screen. Once you've done that, you press ↵ to move on to the next step. This sequence of steps is common to many of the commands you'll work with in AutoCAD.

Opening Multiple Files

AutoCAD 2000 allows you to have multiple documents open at the same time. This can be especially helpful if you want to exchange parts of drawings between files or if you just want to have another file open for reference. Try the following exercise to see how multiple documents work in AutoCAD:

1. Choose File ➤ New.

2. In the Create New Drawing dialog box, click the Start from Scratch button at the top of the dialog box.

3. Click the English radio button, then click OK. You'll get a blank drawing file.

4. Choose Window ➤ Tile Vertically to get a view of both drawing files. The options in the Window pull-down menu act just like their counterparts in other Windows programs that allow multiple document editing.

5. Click the Window with the `Nozzle3d` drawing to make it active.

6. Choose View ➤ Zoom ➤ All to get an overall view of the drawing.

7. Click the words *Nozzle 1984* near the bottom of the drawing.

8. Click and drag on the words *Nozzle 1984*. You'll see a small rectangle appear next to the cursor.

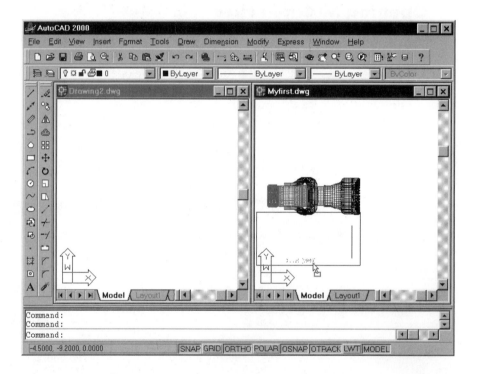

9. While still holding the left mouse button, drag the cursor to the new file window. When you see the words *Nozzle 1984* appear in the new drawing window, release the mouse button. You've just copied a text object from one file to another.

Now you've opened two files at once. You can have as many files open as you want, as long as you have adequate memory to accommodate them. You can control the individual document windows as you would any window using the

Window pull-down menu or the window control buttons in the upper-right corner of the document window.

Closing AutoCAD

When you're done with your work on one drawing, you can open another drawing, temporarily leave AutoCAD, or close AutoCAD entirely. To close all of the open files at once and exit AutoCAD, use the Exit option on the File menu.

1. Choose File ➤ Exit, which is the last item in the menu. A dialog box appears, asking you if you want to "Save Changes to `Myfirst.dwg`?" and offering three buttons labeled Yes, No, and Cancel.

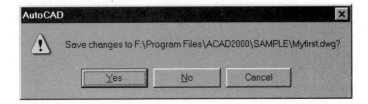

2. Click the No button.

3. AutoCAD will display another message asking you if you want to save `Drawing1.dwg`, which is the new drawing you opened in the last exercise. Click the No button again. AutoCAD exits both the nozzle drawing and the new drawing and closes without saving your changes.

Whenever you attempt to exit a drawing that has been changed, you will get this same dialog box. This request for confirmation is a safety feature that lets you change your mind and save your changes before you exit AutoCAD. In the previous exercise, you discarded the changes you made, so the nozzle drawing reverted back to its state before you erased the handle. The new drawing is completely discarded with no file saved.

If you only want to exit AutoCAD temporarily, you can minimize it so it appears as a button on the Windows 95/98 or NT 4 toolbar. You can do this by clicking the Minimize button in the upper-right corner of the AutoCAD window; the Minimize button is the title-bar button that looks like an underscore (_). Alternatively, you can use the Alt+Tab key combination to switch to another program.

If You Want to Experiment...

Try opening and closing some of the sample drawing files.

1. Start AutoCAD by choosing Start ➤ Programs ➤ AutoCAD 2000 ➤ AutoCAD 2000.

2. Choose File ➤ Open.

3. Use the dialog box to open the Myfirst file again. Notice that the drawing appears on the screen with the handle enlarged. This is the view you had on-screen when you used the Save command in the earlier exercise.

4. Erase the handle, as you did in the earlier exercise.

5. Choose File ➤ Open again. This time, open the Dhouse file from the companion CD-ROM. Notice that you get the Save Changes dialog box you saw when you used the Exit option earlier. File ➤ Open acts just like Exit, but instead of exiting AutoCAD altogether, it closes the current file and then opens a different one.

6. Click the No button. The 3D Dhouse drawing opens.

7. Choose File ➤ Exit. Notice that you exit AutoCAD without getting the Save Changes dialog box. This is because you didn't make any changes to the Dhouse file.

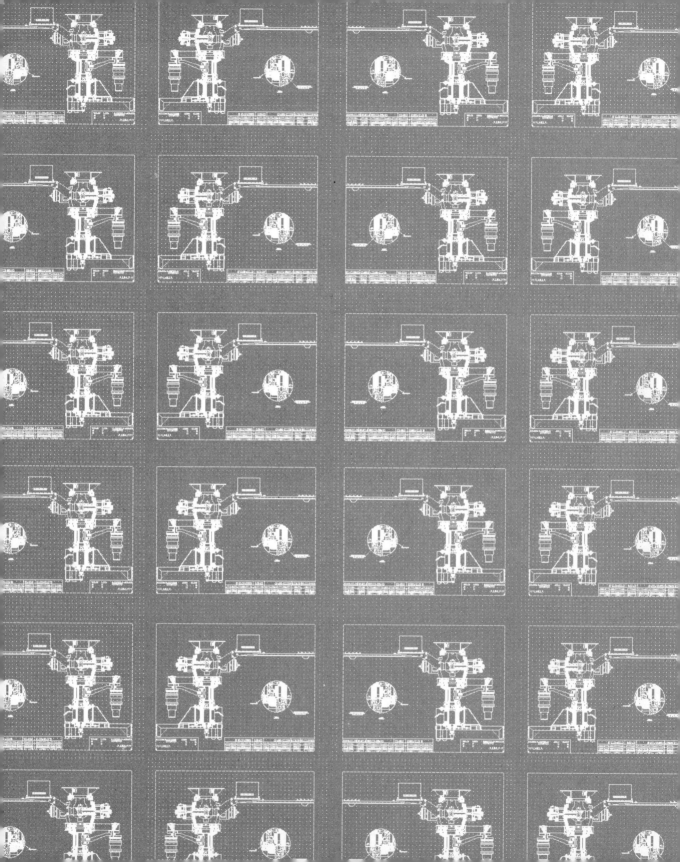

CHAPTER

TWO

Creating Your First Drawing

- Understanding the AutoCAD Interface

- Drawing Lines

- Drawing Arcs

- Selecting Objects for Editing

- Specifying Point Locations for Moves and Copies

- Using Grips

- Selecting Specific Geometric Elements

- Getting Help

2

In this chapter we'll examine some of AutoCAD's basic functions and practice with the drawing editor by building a simple drawing to use in later exercises. We'll discuss giving input to AutoCAD, interpreting prompts, and getting help when you need it. We'll also cover the use of coordinate systems to give AutoCAD exact measurements for objects. You'll see how to select objects you've drawn, and how to specify base points for moving and copying.

If you're not a beginning AutoCAD user, you may want to move on to the more complex material in Chapter 3, "Learning the Tools of the Trade." You can use the files supplied on the companion CD-ROM to this book to continue the tutorials at that point.

Getting to Know the Draw Toolbar

Your first task in learning how to draw in AutoCAD is to try drawing a line. But before you begin drawing, take a moment to familiarize yourself with the toolbar you'll be using more than any other to create objects with AutoCAD: the Draw toolbar.

1. Start AutoCAD just as you did in the first chapter, by clicking Start ➤ Programs ➤ AutoCAD 2000 ➤ AutoCAD 2000. (You can also start AutoCAD by double-clicking the AutoCAD 2000 icon on your Windows Desktop.)

2. When the Startup Drawing Wizard appears, click the Cancel button. You'll learn about setting up a drawing in "Starting Your First Drawing" later in this chapter.

3. In the AutoCAD window, move the arrow cursor to the top icon in the Draw toolbar, and rest it there so that the tool tip appears.

4. Slowly move the arrow cursor downward over the other tools in the Draw toolbar, and read each tool tip.

In most cases, you'll be able to guess what each tool does by looking at its icon. The icon with an arc, for instance, indicates that the tool draws arcs; the one with the ellipse shows that the tool draws ellipses, and so on. The tool tip gives you the name of the tool for further clarification. For further help, you can look at the status bar at the bottom of the AutoCAD window. For example, if you point to the Arc icon just below the Rectangle icon, the status bar reads Creates an Arc. It also shows you the actual AutoCAD command name: Arc. This name is what you would type in the Command window to invoke the Arc tool. You also use this word if you are writing a macro or creating your own custom tools.

Figure 2.1 and Table 2.1 will aid you in navigating the two main toolbars—Draw and Modify—and you'll get experience with many of AutoCAD's tools as you work through this book.

FIGURE 2.1:

The Draw and Modify toolbars. The options available from each toolbar tool are listed by number in Table 2.1.

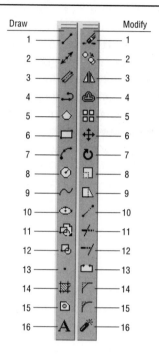

TABLE 2.1: The Options That Appear in the Draw and Modify Toolbars and Flyouts

Draw Toolbar		Modify Toolbar	
1	Line	1	Erase
2	Construction Line (Xline)	2	Copy Object
3	Multiline (Mline)	3	Mirror
4	Polyline (Pline)	4	Offset
5	Polygon	5	Array
6	Rectangle	6	Move
7	Arc	7	Rotate
8	Circle	8	Scale
9	Spline	9	Stretch
10	Ellipse	10	Lengthen
11	Insert Block	11	Trim
12	Make Block	12	Extend
13	Point	13	Break
14	Hatch	14	Chamfer
15	Region	15	Fillet
16	Multiline Text	16	Explode

As you saw in Chapter 1, clicking a tool issues a command. Some tools allow *clicking and dragging*, which opens a *flyout*. A flyout offers further options for that tool. You can identify flyout tools by a small triangle located in the lower-right corner of the tool.

1. Click and drag the Distance tool in the Standard toolbar. A flyout appears with a set of tools. As you can see, there are a number of additional tools for gathering information about your drawing.

2. Move the cursor down the flyout to the second-to-last tool, until the tool tip reads *List*, and then let go of the mouse button. Notice that the icon representing the Distance tool now changes and becomes the icon from the flyout that represents List. By releasing the mouse you've also issued the List command. This command lists the properties of an object.

3. Press Esc twice to exit the List command.

TIP   If you find you're working a lot with one particular flyout, you can easily turn that flyout into a custom toolbar, so that all the flyout options are readily available with a single-click. See Chapter 21, "Integrating AutoCAD into Your Projects and Organization," for details on how to do this.

By making the most recently selected option on a flyout the default option for the toolbar tool, AutoCAD gives you quick access to frequently used commands. A word of caution, however: This feature can confuse the first-time AutoCAD user. Also, the grouping of options on the flyout menus is not always self-explanatory— even to a veteran AutoCAD user.

Working with Toolbars

As you work through the exercises, this book will show you graphics of the tools to choose, along with the toolbar or flyout that contains the tool. Don't be alarmed if the toolbars you see in the examples don't look exactly like those on your screen. To save page space, we've horizontally oriented the toolbars and flyouts for the illustrations; the ones on your screen may be oriented vertically, like the Draw and Modify toolbars to the left of the AutoCAD window. Although the shape of your toolbars and flyouts may differ from the ones you see in this book, the contents are the same. So, when you see a graphic showing a tool, focus on the tool icon itself with its tool-tip name, along with the name of the toolbar in which it is shown.

Starting Your First Drawing

In Chapter 1, you looked at a preexisting sample drawing. This time, you'll begin to draw on your own by creating a part that will be used in later exercises. First, though, you must learn how to tell AutoCAD what you want and let the program help you. Even more important, you must learn what AutoCad wants from you.

1. Choose File ➤ Close to close the current file. Notice that the toolbars disappear and the AutoCAD drawing window appears blank when no drawings are open.

2. Choose File ➢ New. Click the Use a Wizard button at the top of the dialog box.

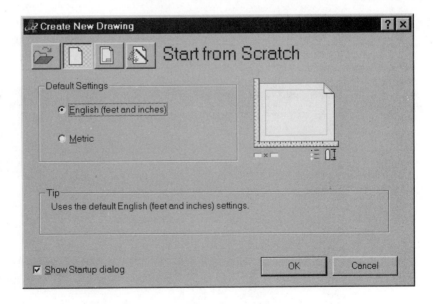

3. In the Create New Drawing dialog box, choose Quick Setup from the list box. This will bring you to a QuickSetup window where you can choose the unit of measurement.

4. Click Next. The Units dialog box appears.

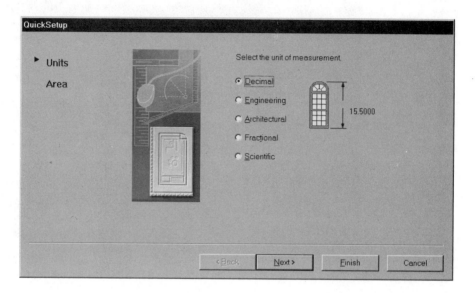

5. For now, you'll use the default decimal units as indicated by the radio buttons. You'll learn about these options in the next chapter. Go ahead and click Next again in the Units dialog box. The Area dialog box appears.

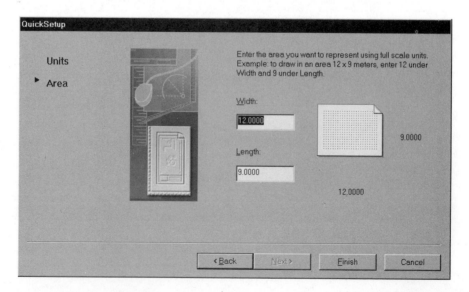

6. Double-click the Width input box and enter **16**.

7. Press the Tab key to move to the next input box and enter **9** for the length.

8. Click Finish. A new drawing file will appear in the AutoCAD window.

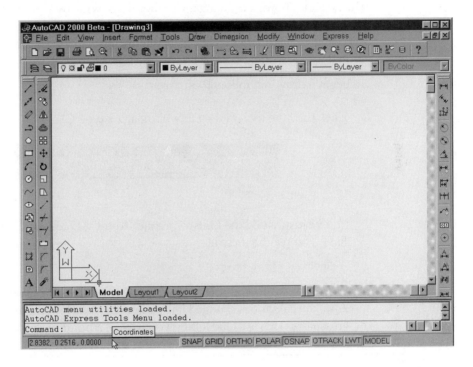

9. To give your new file a unique name, choose File ≻ Save As.

10. At the Save Drawing As dialog box, type **Base**. As you type, the name appears in the File Name input box.

TIP

You can use up to 255 characters and spaces to name files in Windows 95, 98, and NT. If you're going to save AutoCAD 2000 drawings in any previous version, you may want to use the 8.3-character file-naming convention for compatibility with other file systems. Make your filename eight or fewer characters. In all cases, a dot (.) and the final three characters are reserved for the .dwg file type. (The so-called *8.3 short filenames* have another advantage. Seven characters or less are easier to remember, and cause fewer typing errors. The eighth character can be used for a revision number.)

---

11. Double-click the Sample folder shown in the main file list of the dialog box. By doing this, you open the Sample subdirectory.

12. Click Save. You now have a file called Base.dwg, located in the Sample sub-directory of your AutoCAD2000 directory. Of course, your drawing doesn't contain anything yet. You'll take care of that next.

The new file shows a drawing area roughly 16" wide by 9" high. To check this for yourself, move the crosshair cursor to the upper-right corner of the screen, and observe the value shown in the coordinate readout. This is the standard AutoCAD default drawing area for new drawings.

To begin a drawing, follow these steps:

1. Click the Line tool on the Draw toolbar, or type L↵.

You've just issued the Line command. AutoCAD responds in two ways. First, you see the message

    Specify first point:

in the command prompt, asking you to select a point to begin your line. Also, the cursor has changed its appearance; it no longer has a square in the crosshairs. This is a clue telling you to pick a point to start a line.

NOTE   You can also type **line.**↵ in the Command window to start the Line command.

---

2. Using the left mouse button, select a point on the screen near the center. As you select the point, AutoCAD changes the prompt to

    Specify next point or [Undo]:

Now as you move the mouse around, you'll notice a line with one end fixed on the point you just selected, and the other end following the cursor (see the top image of Figure 2.2). This action is called *rubber-banding*.

Now continue with the Line command:

3. Move the cursor to a point below and to the right of the first point you selected, and press the left mouse button again. The first rubber-banding line is now fixed between the two points you selected, and a second rubber-banding line appears (see the bottom image of Figure 2.2).

Two rubber-banding lines

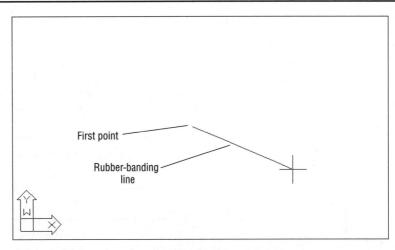

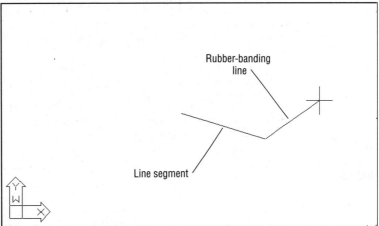

4. If the line you drew isn't the exact length you want, you can back up during the Line command and change it. To do this, click Undo in the Standard toolbar, or type **u**↵ from the keyboard. Now the line you drew previously will rubber-band as if you hadn't selected the second point to fix its length.

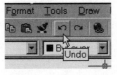

You've just drawn a line at an arbitrary length, and then while still in the command, used Undo to remove the line. The Line command is still active. Two things tell you that you're in the middle of a command. If the area in the bottom of the Command window where the **Command:** prompt is located is not empty, you know a command is still active. In fact, one of its prompts may be visible. For example, the line command will ask you to either `Specify first point:` or `Specify next point:`. The cursor will be the plain crosshair without the little box at its intersection.

NOTE   From now on, we'll refer to the crosshair cursor without the small box as the *Point Selection mode* of the cursor.

Getting Out of Trouble

Beginners and experts alike are bound to make a few mistakes. Before you get too far into the tutorial, here are some powerful, yet easy-to-use tools to help you recover from accidents.

Backspace ←   If you make a typing error, you can use the Backspace key to back up to your error, and then retype your command or response. Backspace is located in the upper-right corner of the main keyboard area.

Escape [Esc]   This is perhaps the single most important key on your keyboard for getting out of trouble. When you need to quickly exit a command or dialog box without making changes, just press the Escape key in the upper-left corner of your keyboard. Press it twice if you want to cancel a selection set of objects or to make absolutely sure you've canceled a command.

Tip: Use the Escape key before editing with grips or issuing commands through the keyboard. You can also press Esc twice to clear grip selections.

U↵   If you accidentally change something in the drawing and want to reverse that change, click the Undo tool in the Standard toolbar (the left-pointing curved arrow). Or type **u↵** at the Command prompt. Each time you do this, AutoCAD will undo one operation at a time, in reverse order so the last command performed will be undone first, then the next to last, and so on. The prompt will display the name of the command being undone, and the drawing will revert to its state prior to that command. If you need to, you can undo everything back to the beginning of an editing session.

Redo   If you accidentally undo one too many commands, you can redo the last undone command by clicking the Redo tool (the right-pointing curved arrow) in the Standard toolbar by typing **redo↵**. Unfortunately, Redo only restores one command, and it can only be invoked immediately after an Undo.

Specifying Distances with Coordinates

Next, you'll continue with the Line command to draw a *plan view* (an overhead view) of a base. The base will be nine units long and 6.5 units wide. To specify these exact distances in AutoCAD, you can use either *absolute coordinates*, *relative polar coordinates*, or *polar coordinates*.

AutoCAD 2000 Coordinate Systems

AutoCAD 2000 uses vector notation and Cartesian coordinates to locate its objects. Vectors require a starting point, endpoint, and direction. Coordinates can be absolute, relative, or polar.

Absolute coordinates (X,Y,Z)   These coordinates are measured from a starting point called the *drawing origin* (0,0,0,), usually located in the lower-left corner. Measurements are horizontal along the X axis and vertical along the Y axis. The Z coordinate is assumed to be 0.00 (unless you specify another coordinate). Positive values are to the right of and above the origin. For example, a point four units to the right of the origin and six units above it is indicated as 4.00,6.00 using absolute coordinates.

Relative coordinates (@X,Y,Z)   These coordinates are like absolute coordinates, but the starting point is the last point entered. You indicate relative coordinates by preceding them with the at sign (@). For example, to draw a line six units long, starting from the previous line's final endpoint, enter @6,0. The line's endpoint will be six units to the right along the X axis. It will be a horizontal line, because the Y-axis coordinate was specified as zero (for no change in the Y coordinate).

Polar coordinates (@*distance*,<*angle*)   Like relative coordinates, polar coordinates are located from the last point entered (indicated by the @ sign). However, the second coordinate, preceded by a left angle bracket (<), is an angular value measured from the X axis (AutoCAD uses the default angular units from the Units command). For example, to draw a line six units long at an angle of 45~o from the previous line's final endpoint, enter @6<45.

Specifying Polar Coordinates

To enter the exact distance of nine units to the right of the last point you selected, do the following:

1. Type **@9<0**. As you type, the characters appear in the command prompt.

2. Press ↵. A line appears, starting from the first point you picked and ending nine units to the right of it (see Figure 2.3). You have just entered a relative polar coordinate.

FIGURE 2.3:

A line nine units long. Notice that the rubber-banding line now starts from the last point selected. This means that you can continue to add more line segments.

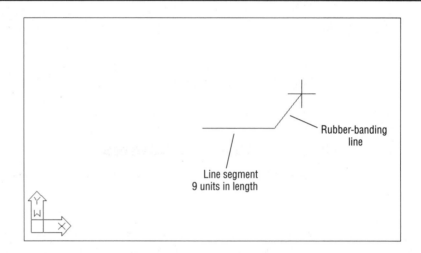

The "at" sign (@) you entered tells AutoCAD that the distance you're specifying is from the last point you selected. The 9 is the distance, and the less-than symbol (<) tells AutoCAD that you're designating the angle at which the line is to be drawn. The last part is the value for the angle, which in this case is 0. This is how to use polar coordinates to communicate distances and direction to AutoCAD.

NOTE   If you're accustomed to a different method for describing directions, you can set AutoCAD to use a vertical direction or downward direction as 0°. See Chapter 3, "Learning the Tools of the Trade," for details.

Angles are given based on the system shown in Figure 2.4, where 0° is a horizontal direction from left to right, 90° is straight up, 180° is horizontal from right to left, and so on. You can specify degrees, minutes, and seconds of arc if you want to be that exact. We'll discuss angle formats in more detail in Chapter 3, "Learning the Tools of the Trade."

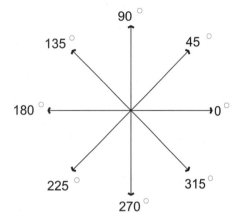

Specifying Relative Coordinates

For the next line segment, let's try another method of specifying exact distances.

1. Enter **@0,6.5**↵. A line appears above the endpoint of the last line.

   Once again, the @ tells AutoCAD that the distance you specify is from the last point picked. But in this example, you give the distance in X and Y values. The X distance, 0, is given first, followed by a comma (,), and then the Y distance, 6.5. This is how to specify distances in relative coordinates.

2. Enter **@-9,0**↵. The result is a drawing that looks like Figure 2.5.

   The distance you entered in step 2 was also in X,Y values, but here you used a negative value to specify the X distance. Positive values in any Cartesian coordinate system are to the right of and above the origin (see Figure 2.6). (You may remember this from your high-school geometry class.) If you want to draw a line from right to left, you must designate a negative value.

NOTE   If you have trouble remembering which of the two angle symbols is used to specify a polar coordinate, the correct one is on the same key as the comma (,).Use a comma for Cartesian coordinates and a shifted comma (<) for polar coordinates.

FIGURE 2.5:

Three sides of the base drawn using the Line tool. Points are specified using either relative or polar coordinates.

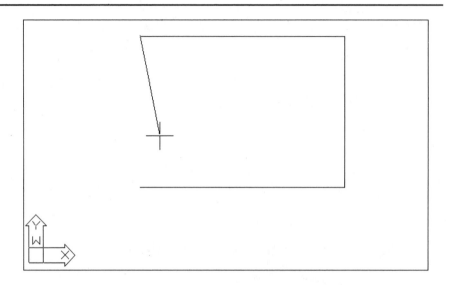

FIGURE 2.6:

Positive and negative Cartesian coordinate directions

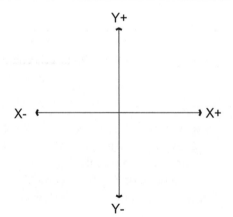

3.   Now type **c↵**. This *c* stands for Close. It closes a sequence of line segments. A line connecting the first and last point in a sequence of lines is drawn (see

Figure 2.7), and the Line command terminates. The rubber-banding line also disappears, telling you that AutoCAD has finished drawing line segments. You can also use the rubber-banding line to indicate direction while simultaneously entering the distance through the keyboard. See the sidebar below, "A Fast Way to Enter Distances."

FIGURE 2.7:

Distance and direction input for the base

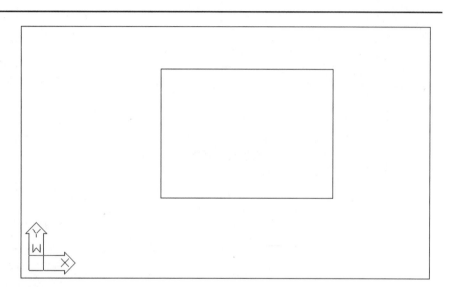

A Fast Way to Enter Distances

A third method for entering distances is to simply point in a direction with a rubber-banding line, and then enter the distance through the keyboard. For example, to draw a line nine units long from left to right, click the Line tool in the Draw toolbar, click a start point, and then move the cursor so the rubber-banding line points to the right at some arbitrary distance. While holding the cursor in the direction you want, type **9↵**. The rubber-banding line becomes a fixed line nine units long.

Using this method, called the *direct-distance method,* along with the Ortho mode described in Chapter 3, can be a fast way to draw objects of specific lengths. Use the standard Cartesian or polar coordinate methods when you need to enter exact distances at angles other than those that are exactly horizontal or vertical.

Cleaning Up the Screen

On some systems, the AutoCAD Blipmode setting may be turned on. This will cause tiny cross-shaped markers, called *blips*, to appear where you've selected points. These blips can be helpful to keep track of the points you've selected on the screen.

Blips aren't actually part of your drawing and they won't print. Still, they can be annoying. To clear the screen of blips, click the Redraw tool in the Standard toolbar (it's the one that looks like a pencil point drawing an arc), or type **r↵**. The screen quickly redraws the objects, clearing the screen of the blips. You can also choose View ➤ Redraw View to accomplish the same thing. As you'll see in Chapter 6, "Enhancing Your Drawing Skills," Redraw can also clear up other display problems.

Another command, Regen, does the same thing as Redraw, but also updates the drawing display database—which means it takes a bit longer to restore the drawing. In general, you'll want to avoid Regen, though at times using it is unavoidable. You'll examine Regen in Chapter 6.

Still another method involves the Grid tool, which we'll look at in Chapter 3, "Learning the Tools of the Trade." The grid responds to the function key F7. This key will toggle the grid dots on or off. When the grid is toggled off, AutoCAD performs a redraw, and thereby cleans the screen of grid dots and blips.

To turn Blipmode on and off, type **blipmode↵** at the command prompt, and then enter **on↵** or **off↵**.

Interpreting the Cursor Modes and Understanding Prompts

The key to working with AutoCAD successfully is in understanding the way it interacts with you. In this section you'll become familiar with some of the ways AutoCAD prompts you for input. Understanding the format of the messages in the Command window and recognizing other events on the screen will help you learn the program more easily.

As the Command window aids you with messages, the cursor also gives you clues about what to do. Figure 2.8 illustrates the various modes of the cursor and gives a brief description of the role of each mode. Take a moment to study this figure.

FIGURE 2.8:

The drawing cursor's modes

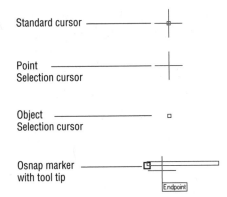

Standard cursor

Point
Selection cursor

Object
Selection cursor

Osnap marker
with tool tip

Endpoint

The Standard cursor tells you that AutoCAD is waiting for instructions. You can also edit objects using grips when you see this cursor. The Point Selection cursor appears whenever AutoCAD expects point input. It can also appear in conjunction with a rubber-banding line. You can either click a point or enter a coordinate through the keyboard. The Object Selection cursor tells you that you must select objects—either by clicking them or by using any of the object-selection options available. The object snap (Osnap) marker appears along with the Point Selection cursor when you invoke an Osnap. Osnaps let you accurately select specific points on an object, such as endpoints or midpoints.

TIP   If you're an experienced AutoCAD user and would prefer to work with the older style crosshair cursor that crosses the entire screen, you can use the Crosshair Size slider in the lower-left area of the Display tab of the Options dialog box (Tools ➢ Options) to set the cursor size. Set the percent of screen size to 100. The cursor will then appear as it did in prior versions of AutoCAD. As the option implies, you can set the cursor size to any percentage of the screen you want. The default is 5 percent.

Choosing Command Options

Many commands in AutoCAD offer several options, which are often presented to you in the Command window in the form of a prompt. Here, we'll use the Arc command to illustrate the format of AutoCAD's prompts.

The part we're drawing has a square hole with rounded corners. You'll draw the hole as an arc followed by a line, then another arc, then another line, and so on until you close the figure. Figure 2.9 shows how this will look when you're finished.

FIGURE 2.9:

Part drawn with an arc.

Let's draw the arc for the hole.

1. Click the Arc tool in the Draw toolbar. The Specify start point of arc or [CEnter]: prompt appears, and the cursor changes to Point Selection mode.

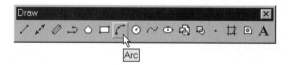

Let's examine this prompt. It contains two options. The default option is the one stated in the main part of the prompt. In this case, the default option is to "Specify start point of arc". If other options are available, they will appear within brackets. In the Arc command, you see the word *CEnter* within brackets telling you that if you prefer, you can also start your arc by selecting a center point instead of a start point. If multiple options are available, they appear within the brackets and are separated by slashes (/).

2. Type **ce↵** to select the Center option. The Specify center point of arc: prompt appears. Notice that you only had to type in the ce and not the whole word *Center*.

NOTE When you see a set of options in the Command-window prompt, note their capitalization. If you choose to respond to prompts using the keyboard, these capitalized letters are all you need to enter to select that option. In some cases, the first two letters are capitalized to differentiate two options that begin with the same letter, such as LAyer and LType.

3. Now pick a point representing the center of the arc near the middle of the base (see the top image of Figure 2.10). The Specify start point of arc: prompt appears.

FIGURE 2.10:

Using the Arc command

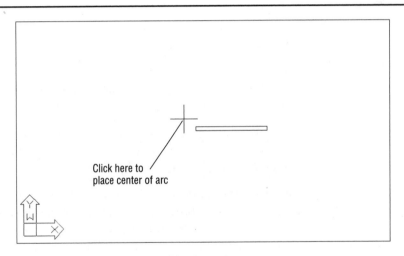

Click here to
place center of arc

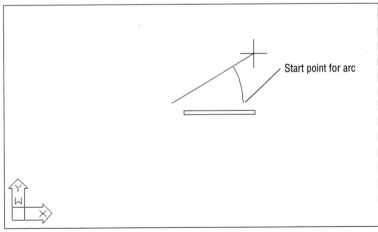

Start point for arc

FIGURE 2.10:
CONTINUED

Using the Arc command

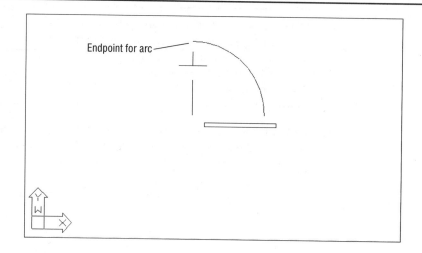

Endpoint for arc

4. Type **@.125<90**↲. The Specify end point of arc or [Angle/chord Length]: prompt appears.

5. Move the mouse and you'll see a temporary arc originating from a point 0.125 units above the center point you selected and rotating about that center (see the second image of Figure 2.10).

   As the prompt indicates, you now have three options. You can enter an angle, a length of chord, or the endpoint of the arc. The default, indicated by specify end point in the prompt, is to pick the arc's endpoint. Again, the cursor is in Point Selection mode, telling you it is waiting for point input. To select this default option, you only need to pick a point on the screen indicating where you want the endpoint.

6. Type **a**↲ to select the Angle option. At the Specify included Angle: prompt, type **90**↲ to draw a quarter-round arc. The arc is now fixed in place, as in the last image of Figure 2.10.

7. To continue with the hole, click the Line tool in the Draw toolbar. At the Specify first point: prompt, press ↲ to use the last point that was at the end of the arc.

• The line obtained its starting point in the same way that it did when you drew the large rectangle for the base. AutoCAD was expecting a value for the first point. When you did not give AutoCAD one, it used the last one in memory.

8. The width of the recess is 0.75 units between the centers. At the `Specify next point or [Undo]:` prompt, type **.75**↵ and then press ↵ again to end the Line command.

NOTE   You didn't have to tell AutoCAD what direction to go when you gave it the length. The line obtained its vector, or direction, from the direction of the arc at the end where the line began.

---

9. Another arc must be drawn at the end of this line. Click the Arc tool again. At the `Specify start point of arc or [CEnter]:` prompt, press ↵ to use the endpoint of the last line. At the `Specify end point of arc:` prompt, type **@.375,-.375**↵ to create an arc 0.375 units across and 90° around.

10. Draw a line from the end of the last arc. Type **L**↵ to begin the Line command and then press ↵ again to begin the line at the end of the last arc.

11. The length of the recess is 1.75 units between the centers. At the `Length of line:` prompt, type **1.75**↵ and then press ↵ again to end the Line command.

12. Another arc must be drawn at the end of this line. Click the Arc tool again. At the `Specify start point of arc or [CEnter]:` prompt, press ↵ to use the endpoint of the last line. At the `Specify end point of arc:` prompt, type **@.375,.375**↵ to create an arc 0.375 units across and 90° around.

13. To continue the figure, click the Line tool and press ↵ to begin the line at the end of the last arc. The width of the recess is 0.75 units between the centers. At the `Length of line:` prompt, type **.75**↵ and then press ↵ again to end the Line command.

14. Another arc must be drawn at the end of this line. Click the Arc tool again. At the `Specify start point of arc or [CEnter]:` prompt, press ↵ to use the endpoint of the last line. At the `Specify end point of arc:` prompt, type **@-.125,.125**↵ to create an arc with a radius of 0.125 and 90° included angle.

15. Finally, you'll draw the line that closes the slot. Click the Line tool again. At the From point: prompt, press ↵ to start the line at the end of the arc. At the Length of line: prompt, type **2.25**↵↵ to draw a line 2.25 units long back to the beginning of the first arc and end the command.

This exercise has given you some practice working with AutoCAD's Command window prompts and entering keyboard commands—a skill you'll need when you start to use some of the more advanced AutoCAD functions.

As you can see, AutoCAD has a distinct structure in its prompt messages. You first issue a command, which in turn offers options in the form of a prompt. Depending on the option you select, you'll get another set of options, or you'll be prompted to take some action, such as picking a point, selecting objects, or entering a value.

As shown in Figure 2.11, the sequence is something like a tree. As you work through the exercises, you'll become intimately familiar with this routine. Once you understand the workings of the toolbars, Command window prompts, and dialog boxes, you can almost teach yourself the rest of the program.

FIGURE 2.11:

A typical command structure, using the Arc command as an example. You'll see different messages, depending on the options you choose as you progress through the command. This figure shows the various pathways to creating an arc.

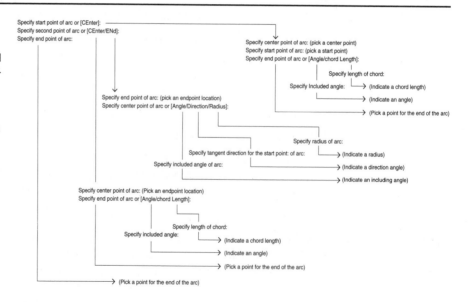

Selecting Options from a Pop-up Menu

Any time during a command, you can right-click to open a pop-up menu containing that command's options. For example, in step 2 of the previous exercise, you typed **ce**↵ to tell AutoCAD that you wanted to select the center of the arc. Instead of typing, you could have right-clicked the mouse to open a pop-up menu of options applicable to the Arc command at that time.

You'll notice that in addition to the options shown in the command prompt, the pop-up menu also shows you a few more options, namely Enter, Cancel, Pan, and Zoom. The Enter option is the same as pressing ↵. Cancel will cancel the current command. Pan and Zoom allow you to make adjustments to your view as you're working through the current command.

As you work with AutoCAD, you'll find that you can right-click at any time to get a list of options. This list is context sensitive, so you'll only see options that pertain to the command or activity that is currently in progress. Also, when AutoCAD is expecting a point, object selection, or numeric value, you won't get a pop-up menu with a right-click. Instead, AutoCAD will treat a right-click as a ↵.

Be aware that the location of your cursor when you right-click will determine the contents of the pop-up list. You've already seen that you can right-click a toolbar to get a list of other toolbars. A right-click on the Command window will display a list of operations you can apply to the command line, such as repeating one of the last five commands you've used, or copying the most recent history of command activity to the Clipboard.

A right-click in the drawing area when no command is active will give you a set of basic options for editing your file, such as Cut, Paste, Undo, Repeat (the Last Command), Pan, and Zoom, to name a few.

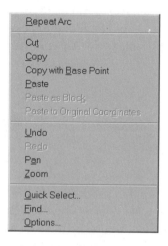

If you're ever in doubt about what to do in AutoCAD, you can right-click any time to see a list of options. You'll learn more about these options as we progress through the book. You'll find many commands that have been enhanced this way. For now, let's move on to the topic of selecting objects.

TIP   If you're a veteran AutoCAD user, and you prefer to have the right-click issue a ↵ at all times (as in previous versions) instead of opening the pop-up menu, you can set up AutoCAD to do just that. See "Configuring AutoCAD" in Appendix B for details on how to set up the mouse's right-click action. But be aware, however, that the tutorials in this book assume that AutoCAD is set up for the new right-click pop-up menu.

Selecting Objects

AutoCAD provides many options for selecting objects. The first part of this section deals with object-selection methods unique to AutoCAD. The second part deals with the more common selection method used in most popular graphic programs, the *Noun/Verb* method. Because these two methods play a major role in working with AutoCAD, it's a good idea to familiarize yourself with them early on.

TIP   If you need to select objects by their characteristics rather than by their locations, Chapter 9, "Advanced Productivity Tools," describes the Object Selection Filter tool. This feature lets you easily select a set of objects based on their properties, including object type, color, layer assignment, and so on.

Selecting Objects in AutoCAD

Many AutoCAD commands involve the to `Select objects:` prompt. Along with this prompt, the cursor will change from crosshairs to a small square (look back at Figure 2.8). Whenever you see this object-selection prompt and the square cursor, you have several options while making your selection. Often, as you select objects on the screen, you'll change your mind about a selection or accidentally pick an object you don't want. Let's take a look at most of the selection options available in AutoCAD and learn what to do when you make the wrong selection.

Before you continue, you'll turn off two features that, while extremely useful, can be confusing to new users. These features are called Running Osnap, and Object Snap Tracking. You'll get a chance to explore these features in Chapter 6, "Enhancing Your Drawing Skills."

1. First, check to see if either Running Osnaps or Object Snap Tracking is turned on. To do this, look at the buttons labeled Osnap and Otrack in the status bar at the bottom of the AutoCAD window. If they are turned on, they will look like they are depressed.

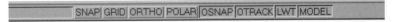

If they look like this graphic, proceed with the next step.

2. To turn off Running Osnap or Object Snap Tracking, click the button labeled Osnap or Otrack in the status bar at the bottom of the AutoCAD window. When turned off, they will look like they are not depressed.

Now let's go ahead and see how to select object in AutoCAD.

1. Choose Move from the Modify toolbar.

2.   At the Select objects: prompt, click the two horizontal lines in the hole drawing you created in the previous exercise. As you saw in Chapter 1, whenever AutoCAD wants you to select objects, the cursor turns into the small, square pickbox. This tells you that you are in object-selection mode. When you pick an object, it is highlighted, as shown in Figure 2.12.

NOTE   *Highlighting* means that an object changes from a solid image to one composed of dots. When you see an object highlighted on the screen, you know that you've chosen that object to be acted upon by whatever command you're currently using.

FIGURE 2.12:

Selecting the lines of the hole and seeing them highlighted

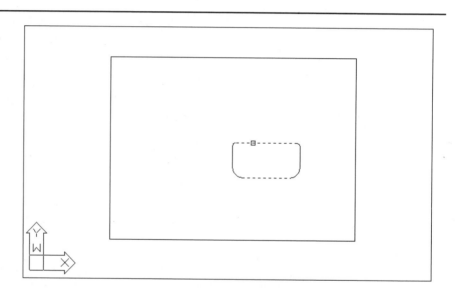

3.   After making your selections, you may decide to deselect some items. Click Undo in the Standard toolbar, or enter **u**↵ from the keyboard.

Notice that one line is no longer highlighted. The Undo option deselects objects, one at a time, in reverse order of selection.

4. There is another way to deselect objects: Hold down the Shift key and click the remaining highlighted line. It reverts to a solid line, showing you that it is no longer selected for editing.

Now you've deselected both lines. Let's try using another method for selecting groups of objects.

1. Another option for selecting objects is to *window* them. Type **w.**┘. The cursor changes to a Point Selection cursor, and the prompt changes to

       Specify first corner:

2. Click a point below and to the left of the hole. As you move your cursor across the screen, the window appears and stretches across the drawing area.

3. Once the window completely encloses the arc and both adjacent lines, but not the other arcs, click this location. One arc and two lines will be highlighted. This window selects only objects that are completely enclosed by the window, as shown in Figure 2.13.

FIGURE 2.13:

Selecting parts of the hole within a window

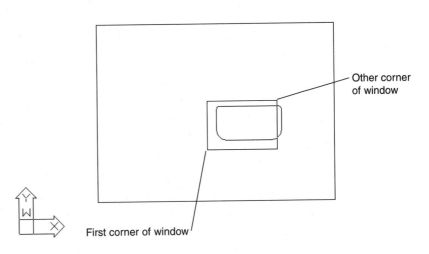

TIP

You may remember that you used a window with the Zoom command in Chapter 1. The Window option under the Zoom command does not select objects. Rather, it defines an area of the drawing you want to enlarge. Remember that the Window option works differently under the Zoom command than it does for other editing commands.

NOTE

If you're using a mouse you're not familiar with, it's quite easy to accidentally click the right mouse button when you really wanted to click the left mouse button, and vice versa. If you click the wrong button, you'll get the wrong results. On a two-button mouse, the right button will either act like the ↵ key or open a context-sensitive pop-up menu, depending on your current operation. A ↵ will be issued if you're selecting objects, but otherwise, the pop-up menu appears.

4. You can add the other arcs by simply clicking them. Now that you've selected the entire hole, press ↵. It is important to remember to press ↵ as soon as you've finished selecting the objects you want to edit. Pressing ↵ tells AutoCAD when you've finished selecting objects. A new prompt, `Specify base point or displacement:`, appears. The cursor changes to its Point Selection mode.

Now you've seen how the selection process works in AutoCAD—but we've left you in the middle of the Move command. In the next section, we'll discuss the prompt that's now on your screen and see how to input base points and displacement distances.

Providing Base Points

When you move or copy objects, AutoCAD prompts you for a *base point*, which is a difficult concept to grasp. AutoCAD must be told specifically *from* where and *to* where the move occurs. The base point is the exact location from which you determine the distance and direction of the move. Once the base point is determined, you can tell AutoCAD where to move the object in relation to that point.

1. To select a base point, hold down the Shift key and press the right mouse button. A menu opens up on the screen. This is the Object Snap (Osnap) menu.

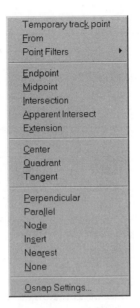

WARNING   When right-clicking the mouse, make sure that the cursor is within the AutoCAD drawing area. Otherwise, you won't get the results described in this book.

2. Choose Center. The Osnap menu disappears.

3. Move the cursor to the lower-left arc of the recess. Notice that as you approach the arc, a small X-shaped graphic appears on the corner. This is called an *Osnap marker.*

4. After the Osnap marker appears, hold the mouse motionless for a second or two. A tool tip appears, telling you the current Osnap point AutoCAD has selected.

5. Now press the left mouse button to select the center indicated by the Osnap marker. Whenever you see the Osnap marker at the point you wish to select, you don't have to point exactly to the location with your cursor. Just left-click and the exact Osnap point is selected (see Figure 2.14). In this case, you selected the exact center of the arc.

FIGURE 2.14:

Using the Osnap marker

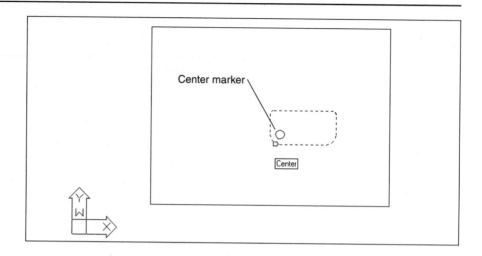

AutoCAD can use Osnap to locate the center of an arc, ellipse, or circle. A common error made by those new to AutoCAD is to place the cursor at the center of the object. AutoCAD does not identify this space as part of the object. You must place the cursor on the circumference of the circle or ellipse or along the length of the arc. An Osnap tool tip will come on indicating by its position and name where it is and what it is.

6. At the `Specify second point of displacement or <use first point as displacement>:` prompt, hold down the Shift key and press the right mouse button again. You'll use the From Osnap option this time, but instead of clicking the option with the mouse, type **f.↵**.

Step 6 demonstrates another method for selecting a Running or temporary Osnap. With the Osnap menu open, enter the first letter of the Osnap's name.

7. You're going to locate the arc center of the recess, three units to the right of and 2.75 units down from the upper-left corner of the part. At the `_from Base point:` prompt, once again hold down the Shift key and press the right mouse button. You'll use the Endpoint Osnap option this time; type **e.↵**.

8. Now move the cursor anywhere along the top line or the left-side line. A box-shaped marker appears. The end of the line that AutoCAD chooses depends on which side of the line's midpoint you are over. Move the cursor until the

box marker appears at the upper-left corner and press the left mouse button. You did not have to place the cursor exactly at the corner, because you had already told AutoCAD to go to the endpoint.

9. You'll now tell AutoCAD how far from the corner of the base you want to place the center of the arc circle. At the <Offset>: prompt, type @3,-2.75↲ (see Figure 2.15).

FIGURE 2.15:

The hole in its new position using the From, Center, and Endpoint Osnaps

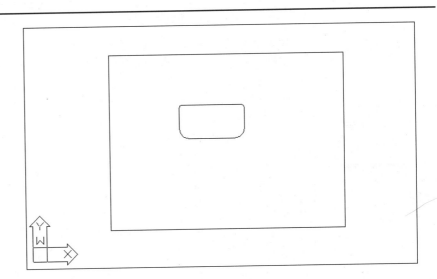

TIP   You may have noticed the statement use first point as displacement in the prompt in step 6. This means that if you press ↲ instead of clicking a point, the object will move a distance based on the coordinates of the point you selected as a base point. If, for example, the point you click for the base point is at coordinate 2,4, the object will move two units in the X axis and four in the Y axis.

As you can see, the Osnap options allow you to select specific points on an object. You used From, Center, and Endpoint in this exercise, but other options are available. You may also have noticed that the Osnap marker is different for each of the options you used. You'll learn more about Osnaps in Chapter 3. Now let's continue with our look at point selection.

If you want to specify an exact distance and direction by typing in a value, you can select any point on the screen as a base point. Or you can just type @ followed by ↲ at the Specify base point: prompt; then enter the second point's location in relative coordinates. Remember that @ means the last point selected. In the next exercise, you'll move the entire hole an exact distance of one unit in a 45° angle.

1. Click the Move tool in the Modify toolbar.

2. Type **p↵**. The set of objects you selected in the previous command is high-lighted. **P** selects the previously selected set of objects.

3. Now press ↵ to tell AutoCAD you've finished your selection. The cursor changes to Point Selection mode.

4. At the Specify base point or displacement: prompt, pick a point on the screen between the hole and the left side of the screen (see Figure 2.16).

FIGURE 2.16:

The highlighted hole and the base point just left of the hole. Note that the base point does not need to be on the object you're moving.

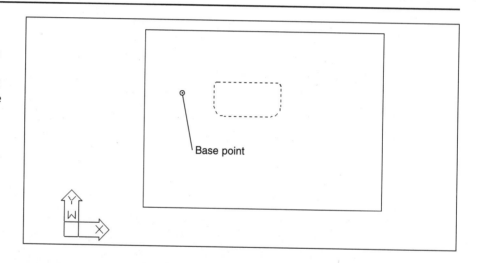

Base point

5. Move the cursor around slowly and notice that the hole moves as if the base point you selected were attached to the hole. The hole moves with the cursor, at a fixed distance from it. This demonstrates how the base point relates to the objects you select.

6. Now type **@1<45↵**. The hole will move to a new location on the screen at a distance of 1.25 units from its previous location and at an angle of 270°.

TIP   If AutoCAD is waiting for a command, you can repeat the last command used by pressing the spacebar or the ↵ key. You can also right-click in the drawing area and select the option at the top of the list. If you right-click on the Command window, a pop-up list offers the most recent commands.

This exercise illustrates that the base point does not have to be on the object you're manipulating. The base point can be virtually anywhere on your drawing. You also saw how you can reselect objects that were selected previously, without having to duplicate the selection process.

NOTE   In performing a Move command, you'll often pick both the `From:` and `To:` points with your mouse. A common error made by those new to AutoCAD is to attempt to pick the `To:` point with the right mouse button. If you do this, AutoCAD treats the `From:` point as a displacement, and typically places the object far beyond your currently displayed view. Your moved object seems to disappear. Remember that the left mouse button is used for picking points, and the right mouse button acts like the ↵ key.

Other Selection Options

There are several other selection options you haven't tried yet. This sidebar describes them. You'll see how these options work in exercises throughout this book. Or if you're adventurous, try them out now on your own. To use these options, type their keyboard abbreviations (shown in brackets in the following list) at any `Select objects:` prompt.

All [all↵]   Selects all of the objects in a drawing except those in frozen or locked layers (see Chapter 4 for more on layers).

Crossing [c↵]   Similar to the Select Window option but will select anything that crosses through the window you define.

Crossing Polygon [cp↵]   Acts exactly like WPolygon (Window Polygon) but, like the Select Crossing option, will select anything that crosses through a polygon boundary.

Fence [f↵]   Selects objects that are crossed over by a temporary line called a *fence*. The operation is like crossing out the objects you want to select with a line. When you invoke this option, you can then pick points, as when you're drawing a series of line segments. When you're finished drawing the fence, press ↵, then go on to select other objects, or press ↵ again to finish your selection.

Multiple [m↵]   Lets you select several objects first, before AutoCAD highlights them. In a very large file, picking objects individually can cause AutoCAD to pause after each pick, while it locates and highlights each object. The Multiple option can speed up things by letting you first pick all of the objects quickly, and then highlight them all by pressing ↵. This has no menu equivalent.

Continued on next page

Previous [p↵]   Selects the last object or set of objects that was edited or changed.

Window [w↵]   Forces a standard selection window. This option is useful when your drawing area is too crowded to use the Autoselect feature to place a window around a set of objects (see the Auto entry in this sidebar). It prevents you from accidentally selecting an object with a single pick when you're placing your window.

WPolygon (Window Polygon) [wp↵]   Lets you select objects by enclosing them in an irregularly shaped polygon boundary. When you use this option, you'll see the `First polygon point:` prompt. You then pick points to define the polygon boundary. As you pick points, the `Specify endpoint of line or [Undo]:` prompt appears. Select as many points as you need to define the boundary. You can undo boundary line segments as you go by clicking the Undo tool in the Standard toolbar or by pressing **u↵**. With the boundary defined, press ↵. The bounded objects are highlighted and the `Select objects:` prompt returns, allowing you to use more selection options.

If you need more control over the selection of objects, you'll find the Add/Remove selection mode setting useful. This setting lets you deselect a set of objects within a set of objects you've already selected. Both settings must be used in Object Selection mode.

Add [a↵]   Continues to add more objects to the selection set.

Remove [r↵]   Removes objects from the selection set. Or, if you need to deselect only a single object, hold Shift and click the object.

The following two selection options are also available, but are seldom used. They're intended for use in creating custom menu options or custom toolbar tools.

Auto [au↵]   Forces the standard automatic window or crossing window when a point is picked and no object is found (see "Using Autoselect" later in this chapter). A standard window is produced when the two window corners are picked from left to right. A crossing window is produced when the two corners are picked from right to left. Once this option is selected, it remains active for the duration of the current command. Auto is intended for use on systems where the Autoselect feature has been turned off.

Single [si↵]   Forces the current command to select only a single object. If you use this option, you can pick a single object; then the current command will act on that object as though you had pressed ↵ immediately after selecting the object. This has no menu equivalent.

Selecting Objects before the Command with Noun/Verb

Nearly all graphics programs today have tacitly acknowledged the *Noun/Verb* method for selecting objects. This method requires you to select objects *before* you issue a command to edit them. The next set of exercises shows you how to use the Noun/Verb method in AutoCAD.

TIP   This chapter presents the standard AutoCAD method of object selection. AutoCAD also offers selection methods with which you may be more familiar. Refer to Appendix B, "Installing and Setting Up AutoCAD," to learn how you can control object-selection methods.

You've seen that when AutoCAD is waiting for a command, it displays the crosshair cursor with the small square. This square is actually a pickbox superimposed on the cursor. It tells you that you can select objects, even while the command prompt appears at the bottom of the screen and no command is currently active. The square momentarily disappears when you're in a command that asks you to select points. From now on, we'll refer to this crosshair cursor with the small box as the *Standard cursor*.

Now try moving objects by first selecting them and then using the Move command.

1. First, press Esc twice to make sure AutoCAD isn't in the middle of a command you may have accidentally issued. Then click the arc at the lower-right end of the hole. The arc is highlighted, and you may also see squares appear at its endpoints and midpoint. These squares are called *grips*. You may know them as *workpoints* from other graphics programs.

2. Choose Move from the Modify toolbar. The cursor changes to Point Selection mode. Notice that the grips on the arc disappear, but the arc is still selected.

3. At the Specify base point or displacement: prompt, pick any point on the screen. The prompt Specify second point of displacement: prompt appears.

4. Type @.5<0.⌐. The arc moves to a new location 0.5 units to the right.

NOTE If you find that this exercise doesn't work as described here, chances are the Noun/Verb setting has been turned off on your copy of AutoCAD. Refer to Appendix B, "Installing and Setting Up AutoCAD," to find out how to activate this setting.

In this exercise, you picked the arc before issuing the Move command. Then, when you clicked the Move tool, you didn't see the `Select object:` prompt. Instead, AutoCAD assumed you wanted to move the arc you had selected and went directly to the `Base point:` prompt.

Using Autoselect

Next you'll move the rest of the hole in the same direction by using the Autoselect feature.

1. Pick a point just below and to the left of the rest of the hole object. Be sure not to pick the arc you just moved. Now a window appears that you can drag across the screen as you move the cursor. If you move the cursor to the left of the last point selected, the window appears dotted (see the top image of Figure 2.17). If you move the cursor to the right of that point, it appears solid (see the bottom image of Figure 2.17).

2. Now pick a point above and to the right of the rest of the objects that make up the hole, so that they are completely enclosed by the window. Do not include the arc that you moved before. The objects are highlighted (and again, you may see small squares appear at the line's endpoints and midpoint).

3. Click the Move tool again. Just as in the last exercise, the `Base point:` prompt appears.

4. Pick any point on the screen; then enter **@.5<0**↵. The hole is reassembled.

The two different windows you've just seen—the solid one and the dotted one—represent a *standard window* and a *crossing window*. If you use a standard window, anything that is completely contained within the window will be selected. If you use a crossing window, anything that crosses through the window or is contained within it will be selected. These two kinds of windows start automatically when you click any blank portion of the drawing area with a Standard cursor or Point Selection cursor—hence the name Autoselect.

The dotted window (top image) indicates a crossing window; the solid window (bottom image) indicates a standard window.

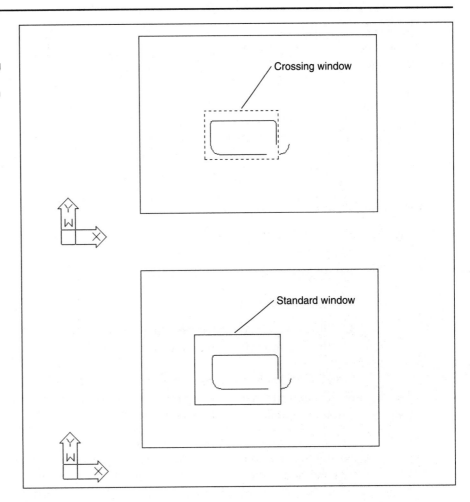

Crossing window

Standard window

Next, you'll select objects with an automatic crossing window.

1. Pick a point below and to the right of the hole. As you move the cursor to the left, the crossing (dotted) window appears.

2. Select the next point so that the window encloses the right-side arcs and one line and part of the two horizontal lines (see Figure 2.18). Three lines and two arcs are highlighted.

FIGURE 2.18:

Parts of the hole selected
by a crossing window

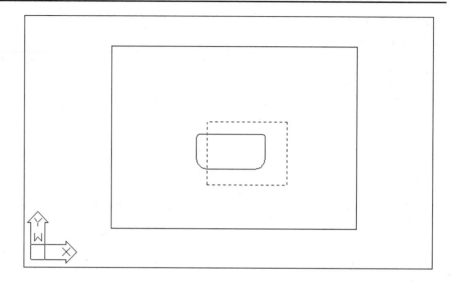

3. Click the Move tool.

4. Pick any point on the screen; then enter **@.5<180.**↵. The highlighted parts of the hole move back to their original location.

You'll find that in most cases, the Autoselect standard and crossing windows are all you need when selecting objects. They'll save you lots of time, so you'll want to get familiar with these features.

Before we continue, choose File ➢ Save to save the Base file. You won't want to save the changes you make in the next section, so saving now will store the current condition of the file on your hard disk for safekeeping.

Restrictions on Noun/Verb Object Selection

If you prefer to work with the Noun/Verb selection feature, you should know that its use is limited to the following subset of AutoCAD commands, listed here in no particular order:

Array	Mirror	Wblock	Block
Dview	Move	Erase	Explode
Change	Rotate	Chprop	Hatch
Scale	Copy	List	Stretch

For all other modifying or construction-oriented commands, the Noun/Verb selection method is inappropriate because for those commands you must select more than one set of objects. But you don't need to remember this list. You'll know if a command accepts the Noun/Verb selection method right away. Commands that don't accept the Noun/Verb selection will clear the selection and display a `Select object:` prompt.

> **NOTE**   If you want to take a break, now is a good time to do it. If you wish, you can exit AutoCAD and return to this point in the tutorial later. When you return, start Auto-CAD and open the `Base` file.

Editing with Grips

If you didn't see small squares appear on the base in the previous exercise, your version of AutoCAD may have the Grips feature turned off. Before continuing with this section, refer to the information on grips in Appendix B, "Installing and Setting Up AutoCAD."

Earlier, when you selected the base, little squares appeared at the endpoints and midpoints of the lines and arcs. These squares are called *grips*. Grips can be used to make direct changes to the shape of objects, or to quickly move and copy them.

So far, you've seen how operations in AutoCAD have a discrete beginning and ending. For example, to draw an arc, you first issue the Arc command and then go through a series of operations, including answering prompts and picking points. When you're done, you have an arc and AutoCAD is ready for the next command.

The Grips feature, on the other hand, plays by a different set of rules. Grips offer a small yet powerful set of editing functions that don't conform to the lockstep command/prompt/input routine you've seen so far. As you work through the following exercises, it will be helpful to think of the Grips feature as a subset of the standard method of operation within AutoCAD.

To practice using the Grips feature, you'll make some temporary modifications to the base drawing.

Stretching Lines Using Grips

In this exercise, you'll stretch one corner of the base by grabbing the grip points of two lines.

1. Press Esc twice to make sure AutoCAD has your attention and you're not in the middle of a command. Click a point below and to the left of the base to start a selection window.

2. Click above and to the right of the base to select it.

3. Place the cursor on the lower-left corner grip of the rectangle, *but don't press the pick button yet.* Notice that the cursor jumps to the grip point.

4. Move the cursor to another grip point. Notice again how the cursor jumps to it. When the cursor is placed on a grip, the cursor moves to the exact center of the grip point. This means, for example, that if the cursor is placed on an endpoint grip, it is on the exact endpoint of the object.

5. Move the cursor to the upper-left corner grip of the rectangle and click it. The grip becomes a solid color, and is now a *hot grip.* The prompt displays the following message:

    ```
    **STRETCH**
    Specify stretch point or [Base point/Copy/Undo/eXit]:
    ```

    This prompt tells you that Stretch mode is active. Notice the options shown in the prompt. As you move the cursor, the corner follows and the lines of the rectangle stretch (see Figure 2.19).

TIP   You can control the size and color of grips using the Selection tab of the Options dialog box (see Appendix B, "Installing and Setting Up AutoCAD").

6. Move the cursor upward and click a point. The rectangle deforms, with the corner placed at your pick point (see Figure 2.19).

NOTE   When you click the corner grip point, AutoCAD selects the overlapping grips of two lines. When you stretch the corner away from its original location, the endpoints of both lines follow.

FIGURE 2.19:

Stretching lines using hot grips. The top image shows the rectangle's corner being stretched upward. The bottom image shows the new location of the corner.

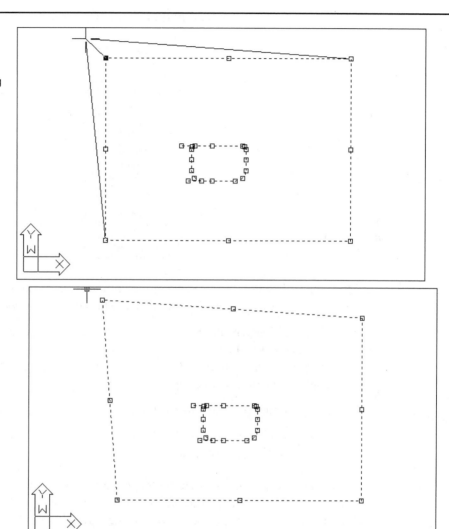

Here you saw that a command called Stretch is issued simply by clicking a grip point. As you'll see, other hot-grip commands are also available.

1. Notice that the grips are still active. Click the grip point that you moved before to make it a hot grip again.

2. Right-click the mouse. A pop-up menu of grip edit options appears.

3. Select Base Point from the menu, and then click a point to the right of the hot grip. Now as you move the cursor, the hot grip moves relative to the cursor.

4. Right-click again, and then select the Copy option from the pop-up menu and enter **@1<-30**↵. Instead of moving the hot grip and changing the lines, copies of the two lines are made, with their endpoints one unit below and to the right of the first set of endpoints.

5. Pick another point just below the last. More copies are made.

6. Press ↵ or enter **x**↵ to exit Stretch mode. You can also right-click again and select Exit from the pop-up menu.

In this exercise, you saw that you could select a base point other than the hot grip. You also saw how you could specify relative coordinates to move or copy a hot grip. Finally, with grips selected on an object, right-clicking the mouse opened a pop-up menu showing grip edit options.

Moving and Rotating with Grips

As you've just seen, the Grips feature offers an alternate method of editing your drawings. You've already seen how you can stretch endpoints, but there is much more that you can do with grips. The next exercise demonstrates some other options. You'll start by undoing the modifications you made in the last exercise.

1. Click the Undo tool in the Standard toolbar, or type **u**↵. The copies of the stretched lines disappear.

2. Press ↵ again. The deformed base snaps back to its original form.

TIP   Pressing ↵ at the command prompt causes AutoCAD to repeat the last command entered—in this case, Undo.

3. Select the entire base by first clicking a blank area below and to the right of the base.

4. Move the cursor to a location above and to the left of the rectangular portion of the base, and click. Because you went from right to left, you created a crossing window. Recall that this selects anything enclosed in or crossing through the window.

5. Click the lower-left grip of the base to turn it into a hot grip. As you move your cursor, the corner stretches.

6. Right-click, then at the grip edit pop-up menu, select Move. The Command window shows

    ```
    **MOVE**
    Specify move point or [Base point/Copy/Undo/eXit]:
    ```

    Now as you move the cursor, the entire base moves with it.

7. Position the base near the center of the screen and click there. The base moves to the center of the screen. Notice that the command prompt returns, yet the base remains highlighted, telling you that it is still selected for the next operation.

8. Click the lower-left grip again, and right-click the mouse. This time, select Rotate from the pop-up menu. The Command window shows

    ```
    **ROTATE**
    Specify rotation angle or [Base point/Copy/Undo/Reference/eXit]:
    ```

    As you move the cursor, the base rotates about the grip point.

9. Position the cursor so that the base rotates approximately 90° (see Figure 2.20). Then, while holding down the Shift key, press the mouse/pick button. A copy of the base appears in the new rotated position, leaving the original base in place.

10. Click the Undo tool in the Standard toolbar, or type **u**↵, to eliminate the rotated copy of the base. Press the Esc key until the object highlighting and grips are gone.

11. Select the entire base again using a crossing window and click the lower-left grip again.

FIGURE 2.20:

Rotating and copying the base using a hot grip. Notice that more than one object is affected by the grip edit, even though only one grip is "hot."

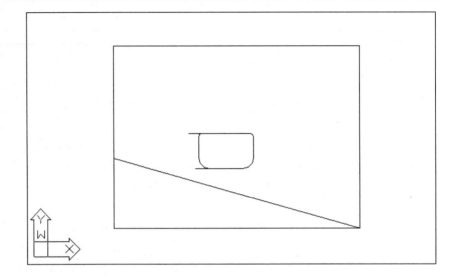

12. Right-click and select Copy. Type **@2.5<90**↵ to create another slot 2.5 units above the first slot. Your drawing should look like the one in Figure 2.20.

13. Press ↵ to exit the Grip Edit mode.

NOTE   You've seen how the Move command is duplicated in a modified way as a hot-grip command. Other hot grip commands—Stretch, Rotate, Scale, and Mirror—also have similar counterparts in the standard set of AutoCAD commands.

After you've completed any operation using grips, the objects are still highlighted with their grips still active. To clear the grip selection, press Esc twice.

In this exercise, you saw how hot grip options appeared in a pop-up list. The Properties option on this menu allows you to make adjustments to an object's properties.

Additionally, many of these grip edit options are available by pressing the spacebar or ↵ while a grip is selected. With each press, the next option becomes active. The commands then repeat if you continue to press ↵. The Shift key acts as a shortcut to the Copy option. You only have to use it once; then each time you click a point, a copy is made.

A Quick Summary of the Grips Feature

The exercises in this chapter using hot grips include only a few of the grips options. You'll get a chance to use other hot-grip commands in Chapter 8, "Using Dimensions." Meanwhile, here is a summary of grips.

- Clicking endpoint grips causes those endpoints to stretch.

- Clicking midpoint grips of lines causes the entire line to move.

- If two objects meet end to end and you click their overlapping grips, both grips are selected simultaneously.

- You can select multiple grips by holding down the Shift key and clicking the desired grips.

- When a hot grip is selected, the Stretch, Move, Rotate, Scale, and Mirror commands are available to you by right-clicking the mouse.

- Alternatively, you can cycle through the Stretch, Move, Rotate, Scale, and Mirror commands by pressing ⏎ while a hot grip is selected.

- All of the hot-grip commands allow you to make copies of the selected objects by using the Copy option or by holding down the Shift key while selecting points.

- All of the hot-grip commands allow you to select a base point other than the originally selected hot grip.

Getting Help

Eventually, you'll find yourself somewhere without documentation and you'll have a question about an AutoCAD feature. AutoCAD provides an online Help facility that will give you information on nearly any topic related to AutoCAD. Here's how to find help.

1. Click Help in the menu bar and choose AutoCAD Help. A Help Topics window appears.

2. If it isn't already selected, click the Contents tab. This window shows a table of contents. The Index and Find tabs offer assistance in finding specific

topics. The Using AutoCAD Help tab provides tips on using the Help files efficiently.

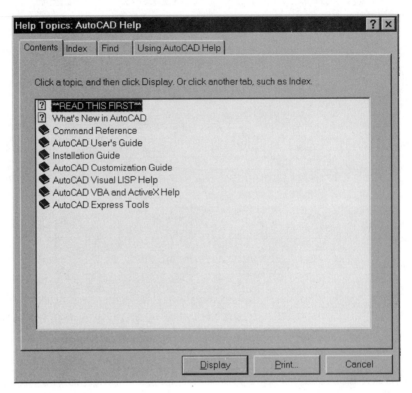

NOTE   You can also press F1 to open the AutoCAD Help window.

---

3. Scan down the screen until you see the topic named Command Reference, and double-click it. The list expands to show more topics.

4. Double-click the item labeled Commands. The Help window expands to show a list of command names.

5. At the top of the list is a set of alphabet buttons. In the main window, you'll see a list of commands, beginning with 3D. You can click the alphabetical button to go to a listing of commands that start with a specific letter. For now, scroll down the list and click the word Copy shown in green. A detailed description of the Copy command appears.

6. Click the Help Topics button at the top of the window. The Help Topics dialog box appears.

7. Click the Find tab. If this is the first time you've selected the Find tab, you'll see the Find Setup wizard. This dialog box offers options for the search database that Find uses to locate specific words.

8. Accept the default option by pressing the Next and then Finish buttons at the bottom of the dialog boxes. Find options appear with a list of topics in alphabetical order. You can enter a word to search for in the drop-down list at the top of the dialog box, or you can choose a topic in the list box.

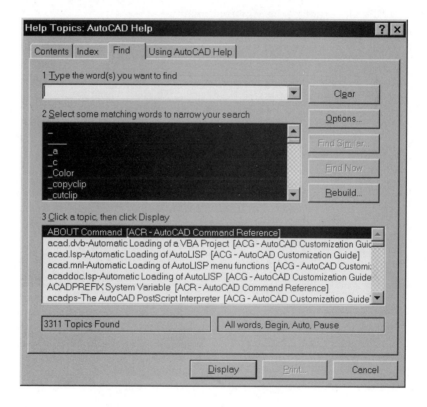

9. Type the word **change**. The list box immediately goes to the word *Change* in the list.

10. Click the word *CHANGE* in capital letters. Notice that the list box at the bottom of the dialog box changes to show some options.

11. Double-click Change Command [ACR] in the list. A description of the Change command appears.

AutoCAD also provides *context-sensitive help* to give you information related to the command you're currently using. To see how this works, try the following:

1. Close or minimize the Help window and return to the AutoCAD window.

2. Click the Move tool in the Modify toolbar to start the Move command.

3. Press the F1 function key, click the Help tool in the Standard toolbar, or choose Help ➢ AutoCAD Help. The Help window appears, with a description of the Move command.

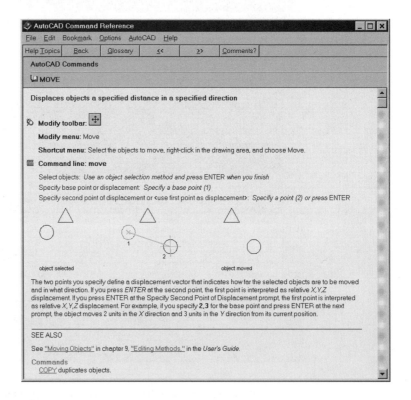

4. Click the Close button or press Esc.

5. Press Esc to exit the Move command.

If you're already familiar with the basics of AutoCAD, you may want to install the AutoCAD Learning Assistant. This tool offers quick tips and brief tutorials on a wide variety of topics, including working in collaborative groups and making the most of the Windows environment. The Learning Assistant is on its own CD as part of the AutoCAD 2000 package.

Additional Sources for Help

The Help Topics tool is the main online source for reference material, but you can also find answers to your questions through the other options found in the Help menu. Here is a brief description of the other Help menu options.

What's New   Gives you an overview of the new features found in AutoCAD 2000. If you're an experienced AutoCAD user and just want to know about the new features in AutoCAD 2000, this is a good place to start.

Learning Assistant   Offers new users some quick tutorials that show you how to combine AutoCAD commands to accomplish a task.

Support Assistant   Offers a series of answers to frequently asked questions. This unique support tool can be updated through the Autodesk Web site.

Autodesk on the Web   Offers a listing of popular pages on the Autodesk Web site. You can find news about updates, locate downloadable tools and upgrades, and find the latest AutoCAD plug-ins for specific jobs you're trying to tackle.

Connect to AutoCAD 2000 Website   Starts your default Web browser and connects you to the Autodesk Web site.

Displaying Data in a Text Window

Some commands produce information that requires a Text window. This frequently happens when you're trying to get information about your drawing. The following exercise shows how you can get an enlarged view of messages in a Text window.

1. Click the List tool (on the Distance tool's flyout, the icon that looks like a piece of paper). Alternatively, Choose Tools ➤ Inquiry ➤ List from the pull-down menus. This tool offers information about objects in your drawing.

2. At the Select objects: prompt, click one of the arcs and press ↵. Information about the arc is displayed in the AutoCAD Text window (see Figure 2.21). Toward the bottom is the list of the arc's properties. Don't worry if you don't understand this listing. As you work through this book, you'll learn what the different properties of an object mean.

    The Text window not only shows you information about objects, it also displays a history of the command activity for your AutoCAD session. This can be helpful for remembering data you may have entered earlier in a session, or to help recall an object's property that you've listed earlier. The scroll bar to the right of the Text window lets you scroll to earlier events. You can even set the number of lines AutoCAD will retain in this Text window using the Options dialog box, or you can have AutoCAD record the Text window information in a text file.

3. Press F2. The AutoCAD Text window closes.

FIGURE 2.21:

The AutoCAD text screen showing the data displayed by the List tool

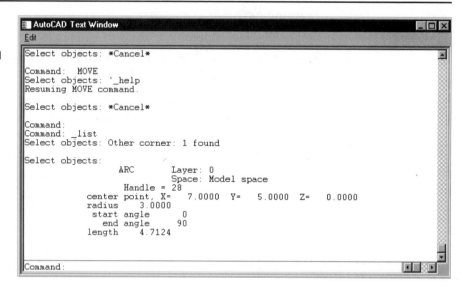

```
AutoCAD Text Window                                                    _ □ ×
Edit
Select objects: *Cancel*

Command:  MOVE
Select objects: '_help
Resuming MOVE command.

Select objects: *Cancel*

Command:
Command: _list
Select objects: Other corner: 1 found

Select objects:
                    ARC        Layer: 0
                               Space: Model space
                    Handle = 28
            center point, X=   7.0000  Y=   5.0000  Z=   0.0000
            radius    3.0000
            start angle        0
              end angle       90
            length    4.7124


Command:
```

> **TIP**
>
> The F2 function key offers a quick way to switch between the drawing editor and the Text window.

4. Now you're done with the base drawing, so choose File ➤ Close.

5. At the Save Changes dialog box, click the No button. (You've already saved this file in the condition you want it, so you don't need to save it again.)

If You Want to Experiment...

Try drawing the latch shown in Figure 2.22.

FIGURE 2.22:

Try drawing this latch.
Dimensions are provided
for your reference.

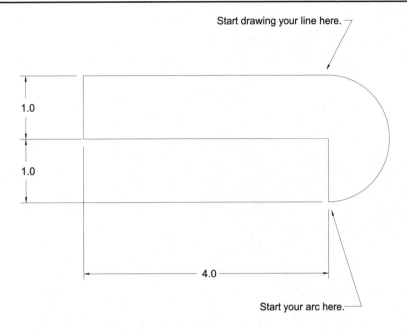

Start drawing your line here.

1.0

1.0

4.0

Start your arc here.

1. Start AutoCAD, open a new file, and name it Latch.

2. When you get to the drawing editor, use the Line command to draw the straight portions of the latch. Start a line as indicated in the figure; then enter relative coordinates from the keyboard. For example, for the first line segment, enter **@4<180**↵ to draw a line segment four units long from right to left.

3. Draw an arc for the curved part. To do this, click the Arc tool in the Draw toolbar.

4. Use the Endpoint Osnap to pick the endpoint indicated in the figure to start your arc.

5. Type **e**↵ to issue the End option of the Arc command.

6. Using the Endpoint Osnap again, click the endpoint above where you started your line. A rubber-banding line and a temporary arc appear.

7. Type **d**↵ to issue the Direction option of the Arc command.

8. Position your cursor so that the ghosted arc looks like the one in the figure, and then press the mouse button to draw the arc.

CHAPTER

THREE

Learning the Tools of the Trade

■ Setting Up a Work Area

■ Designating the Measurement System

■ Understanding Scale in AutoCAD

■ Using Snaps, Grids, and the Coordinate Readout

■ Enlarging and Reducing Your View of the Drawing

■ Trimming and Making Parallel Copies of Objects

■ Laying Out a Drawing with Lines

So far, we have covered the most basic information you need to understand the workings of AutoCAD. Now you'll put your knowledge to work. In this tutorial, which begins here and continues through Chapter 12, "Mastering 3D Solids," you'll draw a number of parts. The tutorial illustrates how to use AutoCAD commands and will give you a solid understanding of the basic AutoCAD package. Knowing these fundamentals will allow you to exploit AutoCAD to its fullest potential, regardless of the kinds of drawings you intend to create or the enhancement products you may use in the future.

In this chapter, you'll start drawing a top view of the machined bracket shown in Figure 3.1. In the process, you'll learn how to use AutoCAD's basic tools.

FIGURE 3.1:

The bracket with dimensions

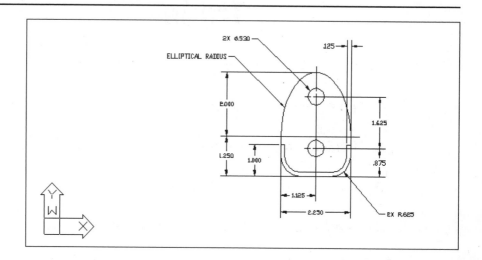

Setting Up a Work Area

Before beginning most drawings, you'll want to set up your work area. To do this you must determine the *measurement system* you want to use. The default screen work area is roughly 16"×9" at full scale, given a decimal measurement system where 1 unit equals 1 inch. These are appropriate settings for your drawing, so

you don't have to do very much setting up. However, you'll often be doing drawings of various sizes and scales. For example, you may want to create a drawing
in a measurement system where you can specify feet, inches, and fractions of
inches at 1"=1' scale, and print the drawing on an 8½"×11" sheet of paper. In this
section, you'll learn how to set up a drawing the way you want.

Specifying Units

Start by creating a new file called DT0100.

1. Start up AutoCAD; then choose File ➤ New.

2. In the Create New Drawing dialog box, click the Start from Scratch button
    and select English from the Select Default Setting list.

3. Click OK to open the new file.

4. Choose File ➤ Save As.

5. At the Save Drawing As dialog box, enter **DTO100** for the filename.

6. Check to make sure that you're saving the drawing in the Samples subdirec-
    tory or the directory you've chosen to store your exercise files, then click Save.

NOTE   You could start drawing in the AutoCAD window immediately after starting up
AutoCAD and then save the file later under the name DT0100. However, use File ➤
New for this exercise in case you're using a system that has an altered default
setup for new files.

The first thing you'll want to tell AutoCAD is the *unit style* or *type* you intend to
use. So far, you've been using the default, which is decimal. You can think of decimal as either inches or millimeters. In this style, whole units represent inches or
millimeters, and decimal units are decimal inches or millimeters. If you want to
be able to enter distances in feet, then you must change the unit style to one that
accepts feet as input. This is done through the Drawing Units dialog box.

If you're a civil engineer, you'll want to know that the Engineering unit style allows you to enter feet and decimal feet for distances. For example, the equivalent of 12'-6" would be 12.5'. AutoCAD 2000 allows keyboard entry of decimal feet, which will be converted to feet-inches-decimals of an inch. For example, 12.125' will be displayed as 12'-1.5".

1. To set the units for your drawing session, choose Format ➤ Units or type **units** ↵. The Drawing Units dialog box appears. Let's look at a few of the options available.

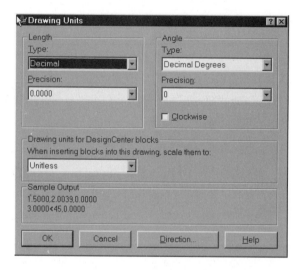

TIP   The Drawing Units settings can also be controlled by using several system variables. To set the unit style, you can type **'lunits**↵ at the command prompt. (The apostrophe lets you enter this command while in the middle of other commands.) At the New value for Lunits <2>: prompt, enter **2** for Decimal.

2. Notice the unit styles listed in the Length group. If Decimal isn't shown, click the down-pointing arrow until Decimal is visible and select it.

3. Click the down-pointing arrow in the Precision drop-down list at the bottom of the Length group. Notice the options available. The default is four-place decimal precision. AutoCAD uses this value to display settings' values in the coordinate readout on the status bar and in dialog boxes. For now, you can set the smallest unit AutoCAD will display in this drawing to three decimal places by selecting the three-decimal option.

NOTE   Units do not control the precision AutoCAD is using to create and maintain geometry or the accuracy of the database. The AutoCAD database is always accurate to slightly more than 16 decimal places.

4.   Close the drop-down list and then click the Direction button at the bottom of the dialog box. The Direction Control dialog box appears. This dialog box lets you set the direction for the 0° angle and the direction for positive degrees. For now, don't change these settings—you'll read more about them in the next section.

5.   Click the OK button.

6.   Click OK in the Drawing Units dialog box to return to the drawing.

NOTE   Remember that the status bar displays a description of the tool or pull-down menu option, including the command name. If you prefer to enter commands through the keyboard, look at the tool description in the status bar during the exercises. The command name will be listed last. You can type in the command name to issue the command instead of clicking the tool during any of these exercises. Command names are also useful when you want to create your own custom macros. You'll get a chance to create some macros in Chapter 19, "Introduction to Customization."

You picked Decimal measurement units for this tutorial, but your own work may require a different unit style. To switch to another units setting, open the Drawing dialog box and select a different option from the Type drop-down list in the length group. Table 3.1 shows examples of how the distance 15.5 is entered in each of the unit styles.

TABLE 3.1: Measurement Systems Available in AutoCAD

Measurement System	AutoCAD's Display of Measurement
Scientific	1.55E+01 (inches)
Decimal	15.5000 (inches)
Engineering	1'-3.5" (input as 1' 3.5")
Architectural	1'-31/2" (input as 1' 3-1/2")
Metric	15.5000 (mm, cm, or meters)
Fractional	151/2" (input as 15-1/2")

In the previous exercise, you needed to change only one setting. Let's take a look at the other Drawing Units settings in more detail. As you read, you may want to refer to the illustration of the Drawing Units dialog box, which appears earlier in this section.

Fine-Tuning the Measurement System

Most of the time, you'll be concerned only with the unit and angle settings of the Drawing Units dialog box. But as you saw from the last exercise, you can control many other settings related to the input and display of units.

TIP To find the distance between two points, click Distance in the Standard toolbar or type **dist**⏎ and then click the two points. If you find that this command doesn't give you an accurate distance measurement, try using Osnaps to pick the exact points on the objects you're measuring. Osnaps are discussed in Chapter 2, "Creating Your First Drawing."

The Precision drop-down list in the Length group lets you specify the smallest unit value that you want AutoCAD to display in the status line and in the prompts. If you choose a measurement system that uses fractions, the Precision list will include fractional units. This setting can also be controlled with the Luprec system variable.

The Angle group lets you set the style for displaying angles. You have a choice of five angle types: Decimal Degrees, Degrees/Minutes/Seconds, Grads, Radians, and Surveyor's units. In the Angle group's Precision drop-down list, you can determine the degree of accuracy you want AutoCAD to display for angles. These settings can also be controlled with the Aunits and Auprec system variables.

NOTE   You can find out more about system variables in Appendix D, "System Variables."

The Direction Control dialog box lets you set the direction of the 0° base angle. The default base angle (and the one used throughout this book) is a direction from left to right. You can also tell AutoCAD which direction is positive, either clockwise or counterclockwise. In this book, we use the default, which is counterclockwise. These settings can also be controlled with the Angbase and Angdir system variables.

The Drawing Units for DesignCenter Blocks setting in the Drawing Units dialog box lets you control how external files are scaled when they are imported into your current drawing. This setting provides an automatic scale translation for imported files. For example, if you know that you will be importing engineering drawings that use decimal feet as the unit of measure, you can set the Drawing Units for DesignCenter Blocks value to feet. Then the imported engineering drawings will be imported at a scale that matches your drawing. This setting also can be controlled with the Insunits system variable.

Things to Watch Out for When Entering Distances

When you're using Engineering, Architectural, or Fractional units, there are two points you should be aware of:

- Hyphens are used only to distinguish fractions from whole inches.

- You cannot use spaces while giving a dimension. For example, you can specify eight feet, four and one-half inches as 8'4-½" or 8'4.5, but not as 8'4 ½".

These idiosyncrasies are a source of confusion to many engineers and architects new to AutoCAD. This is because the program often displays dimensions with fractions in the standard format but does not allow you to enter dimensions that way.

Continued on next page

When inputting distances and angles in unusual situations, here are some tips:

- When entering distances in inches and feet, you can omit the inch (") sign. If you're using the Engineering unit style, you can enter decimal feet and forgo the inch sign entirely.

- You can enter fractional distances and angles in any format you like, regardless of the current unit system. For example, you can enter a distance as @1/2<1.5708r even if your current unit system is set for decimal units and decimal degrees (1.5708r is the radian approximation of 90°).

- If you have your angle units set to degrees, grads, or radians, you do not need to specify g, r, or d after the angle. You *do* have to specify g, r, or d, however, if you want to use these units when they are not the current default angle system.

- If your current angle system is set to something other than degrees, but you want to input angles in degrees, you can use a double less-than symbol (<<) in place of the single less-than symbol (<) to override the current angle system of measure. The << also assumes the base angle of 0° to be a direction from left to right and the positive direction to be counterclockwise.

- If your current angle system uses a different base angle and direction (other than left to right for 0° and a counterclockwise direction for positive angles), and you want to specify an angle in the standard base direction, you can use a triple less-than symbol (<<<) to indicate the angle.

- You can specify a denominator of any size when specifying fractions. However, you should be aware that the value you've set for the maximum number of digits to the right of decimal points (as the Precision setting in the Length group of the Drawing Units dialog box) will restrict the actual fractional value AutoCAD will use. For example, if your units are set for a maximum of four digits of decimals and you give a fractional value of 5/32, AutoCAD will round it off to 0.1562 if you're using decimals readout or 3/16 if you're using fractional readout.

- You're also allowed to enter decimal feet for distances in the Architectural unit style.

Setting Up the Drawing Limits

One of the big advantages to using AutoCAD is that you can draw at full scale; you aren't limited to the edges of a piece of paper the way you are in manual drawing. But you still have to consider what will happen when you want a printout of your drawing. If you're not careful, you may create a drawing that won't fit on the paper size you want at the scale you want. When you start a new drawing, you may find it useful to limit your drawing area to one that can be scaled down to fit on a standard sheet size. Although this is not absolutely necessary with AutoCAD, the limits settings will give you a frame of reference between your work in AutoCAD and the final printed output.

In order to set up the drawing work area, you need to understand how standard sheet sizes translate into full-scale drawing sizes. Table 3.2 lists widths and heights of drawing areas in inches, according to scales and final printout sizes. The scales are listed in the far-left column; the output sheet sizes are listed across the top.

TABLE 3.2: Work Area in Drawing Units by Scale and Plotted Sheet Size

Scale	81/2"×11"	11"×17"	17"×22"	22"×34"	34"×44"	28"×40"
50"=1"	0.17×0.22	0.22×0.34	0.34×0.44	0.44×0.68	0.68×0.88	0.56×0.80
20"=1"	0.42×0.55	0.55×0.85	0.85×1.1	1.1×1.7	1.7×2.2	1.4×2.0
10"=1"	0.85×1.1	1.1×1.7	1.7×2.2	2.2×3.4	3.4×4.4	2.8×4.0
4"=1"	2.12×2.75	2.75×4.25	4.25×5.5	5.5×8.5	8.5×11	7.0×10.0
2"=1"	4.25×5.5	5.5×8.5	8.5×11	11×17	17×22	14×20
3/4"=1"	11×14	14×22	22×29	29×45	45×58	37×53
1/2"=1"	17×22	22×34	34×44	44×68	68×88	56×80
3/8"=1"	22×29	29×45	45×58	58×90	90×117	74×106
1/4"=1"	34×44	44×68	68×88	88×136	136×176	112×160
1/8"=1"	68×88	88×136	136×176	176×272	272×352	224×320
1"=1'	102×132	132×204	204×264	264×408	408×528	336×480
3/4"=1'	136×176	176×272	272×352	352×544	544×704	448×640
1/2"=1'	204×264	264×408	408×528	528×816	816×1056	672×960
1/4"=1'	408×528	528×816	816×1056	1056×1632	1632×2112	1344×1920
1/8"=1'	816×1056	1056×1632	1632×2112	2112×3264	3264×4224	2688×3840
1/16"=1'	1632×2112	2112×3264	3264×4224	4224×6528	6528×8448	5376×7680
1/32"=1'	3264×4224	4224×6528	6528×8448	8448×13056	13056×16896	10752×15360
1"=10'	1020×1320	1320×2040	2040×2640	2640×4080	4080×5280	3360×4800
1"=20'	2040×2640	2640×4080	4080×5280	5280×8160	8160×10560	6720×9600

Continued on next page

TABLE 3.2 CONTINUED: Work Area in Drawing Units by Scale and Plotted Sheet Size

Scale	81/2"×11"	11"×17"	17"×22"	22"×34"	34"×44"	28"×40"
1"=30'	3060×3960	3960×6120	6120×7920	7920×12240	12240×15840	10080×14400
1"=40'	4080×5280	5280×8160	8160×10560	10560×16320	16320×21120	13440×19200
1"=50'	5100×6600	6600×10200	10200×13200	13200×20400	20400×26400	16800×24000
1"=60'	6120×7920	7920×12240	12240×15840	15840×24480	24480×31680	20160×28800

Let's take an example: To find the area needed in AutoCAD for your bracket if it is shown sitting on a 6'×3' table with the long side of the table oriented horizontally, look across from the scale ¹⁄₈"=1" to the column that reads 8¹⁄₂"×11" at the top. You'll find the value 68×88. This means the table's drawing area of 72"×36" needs to fit within an area 68"×88" in AutoCAD in order to fit a printout of a ¹⁄₈"=1" scale drawing on an 8¹⁄₂"×11" sheet of paper. You'll want the drawing area to be oriented horizontally so that the 11" will be in the X axis and the 8.5" will be in the Y axis.

Now that you know the area you need, you can use the Limits command to set up the area.

1. Choose Format ➤ Drawing Limits.

2. At the ON/OFF/<Lower left corner> 0.000,0.000>: prompt, specify the lower-left corner of your work area. Press ↵ to accept the default.

3. At the Upper right corner <12.000,9.000>: prompt, specify the upper-right corner of your work area. (The default is shown in brackets.) Enter ↵ to accept the default.

4. Next, choose View ➤ Zoom ➤ All. You can also click the Zoom All tool from the Zoom Window flyout on the Standard toolbar or type z↵a↵. Though it appears that nothing has changed, your drawing area is now set to a size that will allow you to draw your bracket at full scale.

TIP You can toggle through the different coordinate readout modes by pressing F6 or by double-clicking the coordinate readout of the status bar. For more on the coordinate readout modes, see Chapter 1, "This Is AutoCAD," and "Using the Coordinate Readout as Your Scale," later in this chapter.

5. Move the cursor to the upper-right corner of the drawing area and watch the coordinate readout. You'll see that now the upper-right corner has a Y coordinate of approximately 9.000. The X coordinate will vary depending on the proportion of your AutoCAD window.

In step 5 above, the coordinate readout shows your drawing area, but there are no visual clues to tell you where you are or what distances you're dealing with. To help you get your bearings, you can use the Grid mode, which you'll learn about in the next section. But first, let's take a closer look at scale factors and how they work.

Understanding Scale Factors

When you draft objects manually, you work on the final drawing directly with pen or pencil. With a CAD program, you're a few steps removed from the actual finished product. Because of this, you need to have a deeper understanding of your drawing scale and how it is derived. In particular, you'll want to understand *scale factors*.

For example, one of the more common uses of scale factors is in translating text size in your CAD drawing to the final plotted text size. When you draw objects manually, you simply draw your notes at the size you want. In a CAD drawing, you need to translate the desired final text size to the drawing scale.

When you start adding text and dimensions to your drawing, you'll have to specify a text height. (See Chapter 7, "Adding Text to Drawings" and Chapter 8, "Using Dimensions" for details.) The scale factor will help you determine the appropriate text height for a particular drawing scale. For example, you may want your text to appear ⅛" high in your final plot. But if you drew your text to ⅛" in your drawing, it would be multiplied by the plot scale when plotted. If your drawing were to be plotted at ⅛"=1" scale, the text would be ¹⁄₆₄" high. If the plot scale were ¼"=1", your text would be ½" high. The text has to be scaled to a size that, when scaled at plot time, will appear ⅛" high. So for a ¼"=1" scale drawing, you would multiply the ⅛" text height by a scale factor of 4 to get 0.5". Your text should be 0.5" high in the CAD drawing in order to appear ⅛" high in the final plot.

All the drawing sizes in Table 3.2 were derived by using scale factors. Table 3.3 shows scale factors as they relate to standard drawing scales. These scale factors are the values by which you multiply the desired final printout size to get the equivalent full-scale size. For example, if you have a sheet size of 11"×17", and you want to know the equivalent full-scale size for a ¼"=1"-scale drawing, you multiply the sheet measurements by 4. In this way, 11" becomes 44" (4×11) and 17" becomes 68". Your work area must be 44"×68" if you intend to have a final output of 11"×17" at ¼"=1".

TABLE 3.3: Work Area in Metric Units (Millimeters) by Scale and Plotted Sheet Size

Scale	A0 or F 841mm×1189mm (33.11"×46.81")	A or D 594mm×841mm (23.39"×33.11")	A2 or C 420mm×594mm (16.54"×3.39")	A3 or B 297mm×420mm (11.70"×16.54")	A4 or A 210mm×297mm (8.27"×11.70")
1:2	1682mm×2378mm	1188mm×1682mm	840mm×1188mm	594mm×840mm	420mm×594mm
1:5	4205mm×5945mm	2970mm×4205mm	2100mm×2970mm	1485mm×2100mm	1050mm×1485mm
1:10	8410mm×11890mm	5940mm×8410mm	4200mm×5940mm	2970mm×4200mm	2100mm×2970mm

TIP

If you get the message ****Outside limits**, it means you've selected a point outside the area defined by the limits of your drawing *and* the Limits command's Limits-Checking feature is on. (Some third-party programs may use the Limits-Checking feature.) If you must select a point outside the limits, issue the Limits command and then enter **off.↵** at the **ON/OFF <Lower left corner>…** prompt to turn off the Limits-Checking feature.

TIP

The scale factor for fractional inch scales is derived by multiplying the denominator of the scale by 12, and then dividing by the numerator. For example, the scale factor for 1/4"=1'-0" is (4×12)/1, or 48/1. For whole-foot scales like 1"=10', multiply the feet side of the equation by 12. Metric scales require simple decimal conversions.

If you're using the metric system, the drawing scale can be used directly as the scale factor. For example, a drawing scale of 1:10 would have a scale factor of 10; a drawing scale of 1:50 would have a scale factor of 50; and so on.

You'll be using scale factors to specify text height and dimension settings, so understanding them now will pay off later.

Using the AutoCAD Modes as Drafting Tools

After you've set up your work area, you can begin the top view of the bracket. We'll use this example to show you some of AutoCAD's drafting settings. These tools might be compared to a background grid (the *Grid mode*), a scale (the *coordinate readout*), and a T square and triangle (the *Ortho mode*). These drawing modes can be indispensable tools when used properly. The Drafting Settings dialog box helps you visualize the modes in an organized manner and simplifies their management.

Using the Grid Mode as a Background Grid

Using the Grid mode is like having a grid under your drawing to help you with layout. In AutoCAD, the Grid mode also lets you see the limits of your drawing and helps you visually determine the distances you're working with in any given view. In this section, you'll learn how to control the grid's appearance. The F7 key toggles the Grid mode on and off; you can also single-click the Grid button in the status bar. Start by setting the grid spacing.

1. Choose Tools ➤ Drafting Settings or type **rm**↵ to display the Drafting Settings dialog box, showing all the mode settings. You'll see three tabs: Snap and Grid, Polar Tracking, and Object Snap.

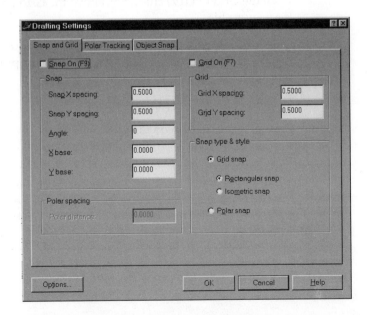

2. Let's start with the Grid group. Notice that the Grid X Spacing input box contains a value of 0.5000.

3. Double-click the Grid X Spacing input box to highlight the entry. You can now type in a new value for this setting.

TIP You can use the Gridunit system variable to set the grid spacing. Enter **'gridunit↵**, and at the **New value for GRIDUNIT <0.000,0.000>:** prompt, enter **.125,.125↵**. Note that the Gridunit value must be entered as an X,Y coordinate.

---

4. Enter **.125** and press the Tab key. Notice that the Grid Y Spacing input box automatically changes to 0.125 (the .125 you entered remains .125 until the next time you open this dialog box). AutoCAD assumes you want the X and Y grid spacing to be the same unless you specifically ask for a different Y setting.

TIP If you want to change an entry in an input box, you can double-click it to highlight the whole entry, and then replace the entry by simply typing in a new one. There are two ways to change a part of the entry. You can use your mouse to highlight the part you want to change, or you can click the input box and then use the cursor keys to move the vertical bar cursor to the exact character you want to change. You can use the Delete or Backspace keys to delete characters.

---

5. Click the Grid On (F7) check box above the Grid group. This setting makes the grid visible.

6. Click OK. The grid now appears as an array of dots with a 0.125″ unit spacing in your drawing area. They will not print or plot with your drawing.

7. Press F7, or click GRID in the status bar (you can also hold down the Ctrl key and press **g**). The grid disappears.

8. Press F7 again to turn on the grid.

TIP If your view is such that the grid spacing appears quite small, AutoCAD will not display the grid in order to preserve the readability of the drawing. If this situation occurs, you'll see the message **Grid too dense to display** in the Command window.

Using the Snap Mode

The *Snap mode* has no equivalent in hand drafting. This mode forces the cursor to step a specific distance. It is useful if you want to maintain accuracy while entering distances with the cursor. The F9 key toggles the Snap mode on and off. Or, just like the Grid mode, there is a Snap button in the status bar that you can single-click. Follow these steps to access the Snap mode.

1. Choose Tools ➤ Drafting Settings or type **rm**↵. The Drafting Settings dialog box appears.

2. In the Snap group of the dialog box, double-click the Snap X Spacing input box, enter **.125**, and press the Tab key. Like the Grid setting, AutoCAD assumes you want the X and Y snap spacing to be the same, unless you specify a different Y setting.

3. Click the Snap On (F9) check box so a check mark appears.

4. Click OK and start moving the cursor around. Notice how the cursor seems to move in "steps" rather than in a smooth motion. Also notice that in the status bar SNAP is solid black and the button appears depressed, indicating that the Snap mode is on. If this is difficult to see, type **zoom**↵**2x**↵ to zoom in closer. When you're done, type **zoom**↵**πp**↵ to return to your previous view. Zoom will be discussed a little later in this chapter under "Getting a Closer Look."

5. Press F9 or click SNAP in the status bar (you can also hold down the Ctrl key and press **b**); then move the cursor slowly around the drawing area. The Snap mode is now off.

6. Press F9 again to turn on the Snap mode.

TIP   You can use the Snapunit system variable to set the snap spacing. Enter '**snapunit**↵, then at the New value for SNAPUNIT <0.000,0.000>: prompt, enter **.125, .125**↵. Note that the Snapunit value must be entered as an X,Y coordinate.

Take a moment to look at the Drafting Settings dialog box. The other options in the Snap group allow you to rotate the cursor to an angle other than its current 0–90° (Angle), set the snap origin point (X Base and Y Base), and set the horizontal snap spacing to a value different from the vertical spacing (Snap X Spacing and Snap Y Spacing). You can also adjust other settings, such as the grid/snap orientation that allows isometric-style drawings. The Snap Type & Style group is divided into the Grid Snap options for Rectangular Snap and Isometric snap styles, and the Polar Snap type. The Polar Snap option allows you to set a snap distance for the Polar Snap feature. When you choose this option, you can enter a value in the Polar Distance input box. Now when Snap is set, the cursor will snap along polar-alignment angles relative to the starting polar-tracking point, at the Snap setting interval. This could be quite useful when acquiring a point from an existing object.

TIP   Check the Polar Angle Settings entry in the Polar Tracking tab of the Drafting Settings dialog box for the desired angle setting before you start to use Polar Snap. Better yet, keep the increment angle set to the angle increments you use most often (start out with 15°).

Using Grid and Snap Together

You can set the grid spacing to be the same as the snap setting, allowing you to see every snap point. Let's take a look at how the Grid and Snap modes work together.

1. Open the Drafting Settings dialog box.

2. Double-click the Grid X Spacing input box in the Grid group, enter **0**, and press the Tab key.

3. Click OK. Now the grid spacing has changed to reflect the 0.125 snap spacing. Move the cursor and watch it snap to the grid points.

4. Open the Drafting Settings dialog box again.

5. Double-click the Snap X Spacing input box in the Snap group and enter **.05↵**. (Did you notice that pressing ↵ closed the dialog box?)

6. The grid automatically changes to conform to the new snap setting. When the grid spacing is set to 0, the grid then aligns with the snap points. At this density, the grid is overwhelming.

7. Open the Drafting Settings dialog box again.

8. Double-click the Grid X Spacing input box in the Grid group and enter **.125↵**.

9. The grid spacing is now at 0.125 again, which is a more reasonable spacing for the current drawing scale.

With the snap spacing set to 0.05, it is difficult to tell if the Snap mode is turned on based on the behavior of the cursor, but the coordinate readout in the status bar gives you a clue. As you move your cursor, the coordinates appear as rounded numbers with no fractional distances less than the snap spacing. Next, you'll look at other ways the coordinate readout helps you.

Using the Coordinate Readout as Your Scale

As you move the cursor over the drawing area, the coordinate readout dynamically displays its position in absolute Cartesian coordinates. This allows you to find a position on your drawing by locating it in reference to the drawing origin—0,0—which is in the lower-left corner of the sheet. You can also set the coordinate readout to display relative coordinates. Throughout these exercises, coordinates will be provided to enable you to select points using the dynamic coordinate readout.

1. Click the Line tool on the Draw toolbar or type l↵. You could also select Draw ➤ Line from the menu bar.

2. Using your coordinate readout for guidance, start your line at the coordinate 2.250,2.000,0.000.

3. Press F6 until you see the relative polar coordinates appear in the coordinate readout at the bottom of the AutoCAD window. You can also single-click the coordinate readout. Polar coordinates allow you to see your current location in reference to the last point selected. This is helpful when you're using a command that requires distance and direction input.

4. Move the cursor until the coordinate readout lists 2.250<0,0.000, and pick this point. As you move the cursor around, the rubber-banding line follows it at any angle.

5. You can also force the line to be orthogonal. Press F8, or single-click ORTHO in the status bar (you can also hold down the Ctrl key and press **o** to toggle on the Ortho mode), and move the cursor around. Now the rubber-banding line will move only vertically or horizontally.

NOTE The Ortho mode is analogous to the T-square and triangle. Note that the ORTHO indicator appears depressed in the status bar to tell you that the Ortho mode is on.

6. Move the cursor down until the coordinate readout lists 1.250< 270, 0.000 and click this point.

7. Continue drawing the other two sides of the rectangle by using the coordinate readout. You should have a drawing that looks like Figure 3.2.

FIGURE 3.2:

The rectangle

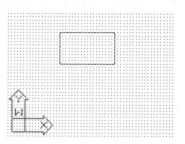

In steps 2, 4, and 6, the coordinate readout showed some extra values. The 0.000 that you see at the end of the coordinate readout listing indicates the Z value of the coordinate. This extra coordinate is significant only when you're doing 3D modeling, so for the time being, you can ignore it.

NOTE If you'd like to know more about the additional Z coordinate listing in the coordinate readout, see Chapter 13, "Using 3D Surfaces."

By using the Snap mode in conjunction with the coordinate readout, you can measure distances as you draw lines. This is similar to the way you would draw using a scale. Be aware that the smallest distance the coordinate readout will register depends on the area you've displayed in your drawing area. For example, if you're displaying an area the size of a football field, the smallest distance you can indicate with your cursor may be 6". On the other hand, if your view shows an area of only one square inch, you can indicate distances as small as 1/1000" using your cursor.

While this exercise tells you to use the Line tool to draw the bracket, you could also use the Rectangle tool. The Rectangle tool creates what is known as a *polyline*, which is a set of line or arc segments that acts like a single object. You'll learn more about polylines in Chapter 10, "Drawing Curves and Solid Fills."

Exploring the Drawing Process

In this section, you'll look at some of the more common commands and use them to complete a simple drawing. As you draw, watch the prompts and notice how your responses affect them. Also note how you use existing drawing elements as reference points.

While drawing with AutoCAD, you create geometric forms to determine the basic shapes of objects, and then modify the shapes to fill in detail. In essence, you alternately lay out, create, and edit objects to build your drawing.

AutoCAD offers many drawing tools. These include lines, arcs, circles, text, polylines, points, 3D faces, ellipses, spline curves, solids, and others. All drawings are built on these objects. You're familiar with lines and arcs; these, along with circles, are the most commonly used objects. As you progress through this book, you'll be introduced to the other objects and how they are used.

Locating an Object in Reference to Others

To continue drawing the bracket, you'll use an ellipse.

1. Click the Ellipse tool in the Draw toolbar or type **el↵**. You can also choose Draw ➤ Ellipse ➤ Axis, End.

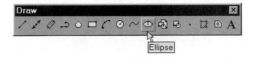

2. At the Arc/Center/<Axis endpoint 1>: prompt, pick the midpoint of the bottom horizontal line of the rectangle. Do this by bringing up the Osnap pop-up menu and selecting Midpoint; then move the cursor toward the bottom line. (Remember, to bring up the Osnap menu, press Shift and click the

right mouse button.) When you see the Midpoint Osnap marker appear on the line, press the left mouse button.

3. At the `Specify other endpoint of axis:` prompt, move the cursor up until the coordinate readout lists 4.000<90,0.000.

4. Pick this as the second axis endpoint.

5. At the `<Specify distance to other axis or [Rotation]:` prompt, move the cursor horizontally from the center of the ellipse until the coordinate readout lists 1.125< 0,0.000.

6. Pick this as the axis distance defining the width of the ellipse. Your drawing should look like Figure 3.3.

FIGURE 3.3:

The ellipse added to the rectangle

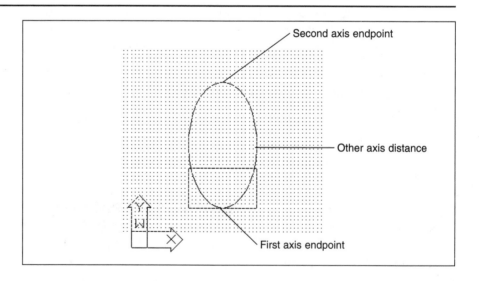

As you work with AutoCAD, you'll eventually run into NURBS. NURBS stands for Non-Uniform Rational B-Splines—a fancy term meaning that curved objects are based on accurate mathematical models. You'll learn more about polylines and NURBS curves in Chapter 10, "Drawing Curves and Solid Fills."

Getting a Closer Look

During the drawing process, you'll want to enlarge areas of a drawing to edit its objects more easily. In Chapter 1, "This Is AutoCAD," you already saw how the Zoom command is used for this purpose.

1. Click the Zoom Window tool on the Standard toolbar or type **z↵w↵**. You can also Choose View ➤ Zoom ➤ Window.

2. At the Specify first corner: prompt, pick a point below and to the left of your drawing at coordinate 1.500,0.625,0.000.

3. At the Specify opposite corner: prompt, pick a point above and to the right of the drawing at coordinate 5.125,4.875,0.000 so that the bracket is completely enclosed by the view window. You can also use the Zoom Realtime tool in conjunction with the Pan Realtime tool. The bracket enlarges to fill more of the screen (see Figure 3.4).

FIGURE 3.4:

A close-up of the details of the drawing

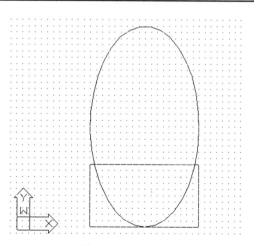

| TIP | To issue the Zoom Realtime tool from the keyboard, type **z↵↵**. |

Modifying an Object

Now let's see how editing commands are used to construct an object. To define the bottom area of the bracket, let's create copies of the three lines to define the thickness of the bottom area 0.125 units closer toward the center.

1. Click the Copy Object tool in the Modify toolbar or type **cp↵**. You can also select Modify ➤ Copy from the menu bar.

TIP   You can also use the grip edit tools to make the copy. See Chapter 2, "Creating Your First Drawing," for more on grip editing.

2. At the `Select object:` prompt, pick the bottom horizontal line. The line is highlighted. Press ↵ to confirm your selection.

3. At the `<Base point or displacement>/Multiple:` prompt, pick a base point near the line.

4. Type **@.125<90↵** to tell AutoCAD to use a relative coordinate location of .125 units distance and 90° direction from the base point.

5. Continue copying the other two sides of the rectangle by using the coordinate location method or the coordinate readout method. You should have a drawing that looks like Figure 3.5.

You'll have noticed that the Copy command acts exactly like the Move command that you used in Chapter 2, "Creating Your First Drawing." However, the Copy command does not alter the position of the objects you select.

Trimming an Object

Now you must move the ellipse and delete the part of it that is not needed. You'll use the Trim command to trim off parts of the ellipse.

1. Click the Move tool in the Modify toolbar.

FIGURE 3.5:

The copied lines

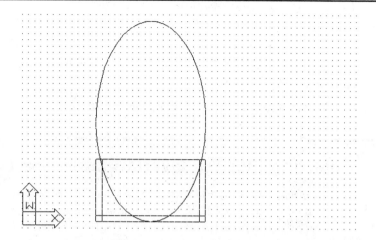

2. At the `Select objects:` prompt, click anywhere on the ellipse, then press ↵ to complete the selection set and go on to the next part of the command.

3. At the `Specify base point or displacement:` prompt, pick a point near the line.

4. At the `Specify second point of displacement <use first point as displacement>:` prompt, type **@.75<270**↵.

5. Turn off the Snap mode by pressing F9 or clicking SNAP in the status bar. The Snap mode may be a hindrance at this point in your editing session because it may keep you from picking the points you want. This mode forces the cursor to move to points at a given interval, so you'll have difficulty selecting a point that doesn't fall exactly at one of those intervals.

6. Click the Trim tool in the Modify toolbar.

You'll see the following prompt:

```
Select cutting edges: (Projmode=UCS, Edgemode=None)
Select objects:
```

7. Click the top line of the rectangle—the one that crosses through the ellipse—and press ↵ to finish your selection.

8. At the `Select object to trim or [Project/Edge/Undo]:` prompt, pick the bottom portion of the ellipse below the line. This trims the ellipse back to the line.

9. Press ↵ to exit the Trim command.

Selecting Close or Overlapping Objects

At times, you'll want to select an object that is in close proximity to or lying underneath another object, and AutoCAD won't obey your mouse click. It's frustrating when you click the object you want to select, but AutoCAD selects the one next to it instead. To help you make your selections in these situations, AutoCAD provides object-selection cycling. To use this feature, hold down the Ctrl key while simultaneously clicking the object you want to select. If the wrong object is highlighted, press the left mouse button again (you do not need to hold down the Ctrl key for the second time), and the next object in close proximity will be highlighted. If several objects are overlapping or close together, just continue to press the left mouse button until the correct object is highlighted. When the object you want is finally highlighted, press ↵ and continue with further selections.

In step 6 of the previous exercise, the Trim command produced two messages in the prompt. The first message, `Select cutting edges…`, told you that you must first select objects to define the edge to which you wish to trim an object. In step 8, you were again prompted to select objects, this time to select the objects to trim. Trim is one of a handful of AutoCAD commands that asks you to select two sets of objects: The first set defines a boundary, and the second is the set of objects you want to edit. The two sets of objects are not mutually exclusive. You can, for example, select the cutting-edge objects as objects to trim. The next exercise shows how this works.

1. Start the Trim tool again by clicking it in the Modify toolbar.

2. At the `Select cutting edges… Select objects:` prompt, click the bottom and left lines that you copied earlier. You're going to trim these lines to create a neat corner.

3. Press ↵ to finish your selection and move to the next step.

4. At the Select object to trim or [Project/Edge/Undo]: prompt, click the portion of the horizontal line that extends beyond the vertical line. The line trims back. Do the same with the portion of the line that extends below the horizontal line.

NOTE   These Trim options—Project, Edge, and Undo—are described in "The Trim Options" section later in this chapter.

5. Press ↵ to finish your selection and move to the next step. Press ↵ to begin the command again, but you'll not explicitly select new objects.

6. At the Select object to trim or [Project/Edge/Undo]: prompt, press ↵. This null-entry option sets all possible objects to be cutting edges, so you can trim a selected object back to the first object AutoCAD finds that can be used as a cutting edge.

7. Click the end of the horizontal line that extends beyond the right vertical line. The right side of the line trims back to meet the vertical line. Click the end of the right vertical line. Your drawing should look like Figure 3.6.

8. Press ↵ to exit the Trim command.

FIGURE 3.6:

Trimming the ellipse and the line

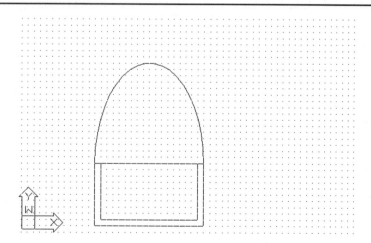

Here you saw how the objects were used both as trim objects and as the objects to be trimmed. Next you'll move the top horizontal line of the rectangle down 0.25

units and use it to trim the inside vertical lines (and the horizontal line between the inside vertical lines) to become the top of the bracket ledge.

1. Select the Move tool from the Modify toolbar.

2. At the `Select objects:` prompt, select the top horizontal line and press ↵ twice to complete the selection set.

3. At the `Base point or displacement:` prompt, pick a point near the line and at the `Second point of displacement:` prompt, type **@.25<270**↵. Your drawing should look like the top of Figure 3.7.

FIGURE 3.7:

Trimming the top of the bracket ledge

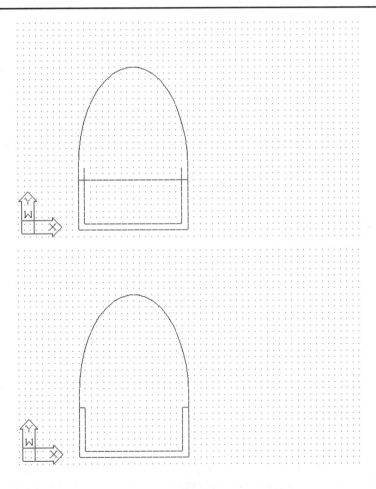

4. Select the Trim tool from the Modify toolbar.

5. At the `Select cutting edges:` prompt, press ↵ to use the Autoselect method.

6. Select the horizontal line anywhere between the two interior vertical lines to trim between them. Select the end of the left vertical line to trim back to the horizontal line. Do the same with the right vertical line. Your drawing should look like Figure 3.7.

7. Press ↵ to exit the Trim command.

Next you'll draw one mounting hole and make a copy of it for the other hole.

1. Select the Circle command from the Draw toolbar.

2. At the `Specify center point for circle or [3P/2P/Ttr (tan tan radius)]:` prompt, press Shift+right-click to select the From Osnap option. At the `Base point:` prompt, press Shift+right-click to select the Midpoint Osnap option. Pick a point along the bottom horizontal line.

3. At the `<Offset>:` prompt, type **@.875<90**↵ to start the circle 0.875 units from the bottom middle of the part.

4. At the `Diameter/<Radius>:` prompt, type **d**↵ for diameter and at the `Diameter:` prompt, type **.530**↵.

5. Select the Copy tool from the Modify toolbar. At the `Select objects:` prompt, type **l**↵ to use the last object created, and press ↵ to complete the selection set.

6. At the `Specify base point or displacement, or [Multiple]:` prompt, select a point near the circle. At the `Specify second point of displacement or <use first point as displacement>:` prompt, type **@1.625<90**↵.

Your drawing looks nearly complete, except for the arcs called *fillets* at the bottom of the part. It should look like Figure 3.8. Let's talk about the Trim options, and then fillet those corners.

FIGURE 3.8:

The bracket with holes added

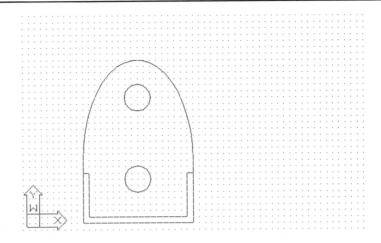

The Trim Options

AutoCAD offers four options for the Trim command: Edge, Project, Undo, and ↵ for edge inference. As described in the following list, these options give you a higher degree of control over how objects are trimmed:

> **Edge [E]**   Allows you to trim an object to an apparent intersection, even if the cutting-edge object does not intersect the object to be trimmed (see the top of Figure 3.9). Edge offers two options: Extend and No Extend. These can also be set using the Edgemode system variable.
>
> **Project [P]**   Useful when working on 3D drawings. It controls how Auto-CAD trims objects that are not coplanar. Project offers three options: None, UCS, and View. None causes Trim to ignore objects that are on different planes, so that only coplanar objects will be trimmed. If you choose UCS, the Trim command trims objects based on a top view of the current UCS and then disregards whether the objects are coplanar or not (see the middle of Figure 3.9). View is similar to UCS but uses the current view's line of sight to determine how non-coplanar objects are trimmed (see the bottom of Figure 3.9).
>
> **Undo [U]**   Causes the last trimmed object to revert to its original length.
>
> **Enter [↵]**   Causes AutoCAD to use as a cutting edge any object that intersects the object you've selected to trim at the `Select object to trim or [Project/Edge/Undo]:` prompt. This process is called *edge inference*.

FIGURE 3.9:

The Trim tool's options

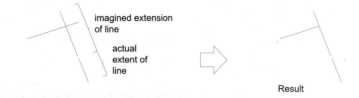

imagined extension
of line

actual
extent of
line

Result

With the Extend option, objects will trim even if the trim object
doesn't actually intersect with the object to be trimmed.

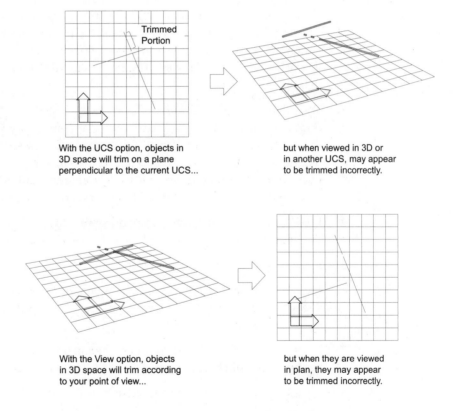

Trimmed
Portion

With the UCS option, objects in
3D space will trim on a plane
perpendicular to the current UCS...

but when viewed in 3D or
in another UCS, may appear
to be trimmed incorrectly.

With the View option, objects
in 3D space will trim according
to your point of view...

but when they are viewed
in plan, they may appear
to be trimmed incorrectly.

The inside of the detail still has some sharp corners. To round out these corners, you can use the versatile Fillet command (on the Modify toolbar). Fillet allows
you to join lines and arcs end to end, and it can add a radius where they join, so
there is a smooth transition from arc to arc or line to line. Fillet can also join two

lines that do not intersect, and it can trim two crossing lines back to their point of intersection.

1. Click the Fillet tool in the Modify toolbar or type **f**↵. You can also choose Modify ➤ Fillet from the menu bar.

2. At the `Current settings: Mode = TRIM, Radius = 0.50 Select first object or [Polyline/Radius/Trim]:` prompt, you have the option of entering **r**↵, which allows you to adjust the fillet radius to a new value. Watch the default <in these brackets> value for the current radius setting, as it may be the value you want. The last fillet radius you used becomes the default value for the next fillet.

3. The fillet radius is 0.625 minus the ledge thickness of 0.125. You'll need to create two fillets, one with a 0.625 radius and another with a 0.5 radius. At the `Specify fillet radius <0.500>:` prompt, press ↵. By accepting the default, you're telling AutoCAD that you want an 0.5 unit radius for your fillet. (If the default does not show 0.500, enter **r**↵ and **.5**↵. The Fillet command will stop; you must enter ↵ to restart and create a fillet.)

4. Press ↵ to invoke the Fillet command again; this time, pick two inside corner lines. The fillet arc joins and trims the two lines.

5. Press ↵ again and fillet the other inside corner.

6. Press ↵ to invoke the Fillet command again. At the `Polyline/Radius/Trim/ <Select first object>:` prompt, enter **r**↵. At the `Specify fillet radius <0.500>:` prompt, enter **.625**↵.

7. Press ↵ to invoke the Fillet command again; this time, pick two outside corner lines. The fillet arc joins and trims the two lines.

8. Press ↵ again to fillet the other outside corner. Your drawing should look like Figure 3.10.

9. Save the DT0100 file.

FIGURE 3.10:

The bracket with corners filleted

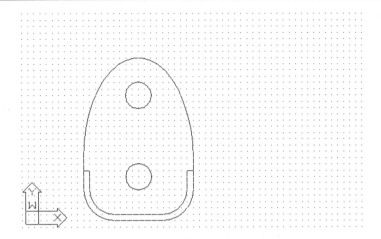

You've just seen one way to construct these details. However, there are many ways to construct objects. For example, you could have just created an ellipse arc, and you could have used the Grips feature to move the endpoints of the line to meet at the intersection. As you become familiar with AutoCAD, you'll start to develop your own ways of working, using the tools best suited to your style.

Planning and Laying Out a Drawing

As a designer, you'll often want to take a different look at a design. Let's create another bracket that will look like the drawing in Figure 3.11. You'll note that much of the original design is still the same. We'll use the parts that are the same and only create the new geometry. Let's look at how to draw the new version. This will help you get a feel for the kind of planning you can do to use AutoCAD effectively. You'll also get a chance to use some of the keyboard shortcuts built into AutoCAD. First, though, go back to the previous view of your drawing, and use the Zoom command to make room to work.

1. Return to your previous view (the one shown in Figure 3.10). A quick way to do this is to click the Zoom Previous tool on the Standard toolbar or choose View ➢ Zoom ➢ Previous. Your view will return to the one you had before the last Zoom command (Figure 3.12).

FIGURE 3.11:

The alternate bracket
design

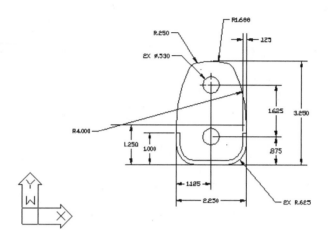

FIGURE 3.12:

The view of the bracket
after using the Zoom Previous tool. You can also obtain
this view using the Zoom All
tool from the Zoom Window
flyout.

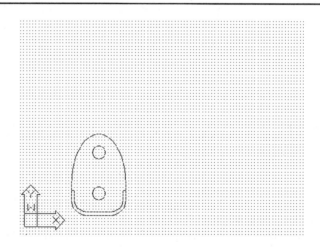

You'll begin the new design by making a copy of the current design.

2. Turn on Snap mode by pressing F9.

3. Type **cp**↵. At the Select objects: prompt, window the entire bracket or type **all**↵. You can see that the entire bracket is highlighted. Press ↵ again to complete the selection set.

4. At the Specify base point or displacement, or [Multiple]: prompt, pick a point near the object.

5. At the Specify second point of displacement or <use first point as displacement>: prompt, check the status bar to see that ORTHO is on (click the box if it is not on), and drag the cursor to the right. Type **4** and then right-click to create a copy of the bracket four units to the right of the original.

Instead of right-clicking during the direct-distance entry method, you can press ↵ or the spacebar. The difference between the two designs is the replacement of the ellipse with arcs. You'll now draw one of the large arcs. The arc begins tangent to the side, one unit up from the bottom of the bracket. You do not know exactly how long to draw the arc, so you'll draw it longer than it needs to be and trim it back.

1. In the Draw toolbar, click the Arc tool or type **a**↵. (See Figure 3.13 for other arc options available from the menu bar.) This figure shows each pull-down menu option name with a graphic above it depicting the arc, and numbers indicating the sequence of points to select. For example, if you want to know how the Draw ➤ Arc ➤ Start, Center, End option works, you can look to the graphic at the bottom-right corner of the figure. It shows the point-selection sequence for drawing an arc using that option: 1 for the start point, 2 for the center point, then 3 for the end of the arc.

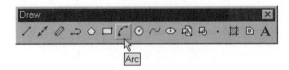

2. At the Specify start point of arc or [CEnter]: prompt, press Shift+right-click and select the Endpoint Osnap. Then pick the endpoint of the right side of the ellipse.

3. At the Specify second point of arc or [CEnter/ENd]: prompt, type **ce**↵ and at the Center: prompt, type **@4<180**↵.

4. At the Specify end point of arc or [Angle/chord Length]: prompt, type **a**↵. At the Specify included angle: prompt, type **40**↵ to draw an arc beginning at and tangent to the end of the ellipse, and ending at 40°.

5. Repeat this procedure for the other arc. Type **a**↵, and at the ARC Specify start point of arc or [CEnter]: prompt, press Shift+right-click and select the Endpoint Osnap. Then select the left endpoint of the ellipse.

6. At the Specify second point of arc or [CEnter/ENd]: prompt, type **ce** ↵ and at the Center: prompt, type **@4<0**↵.

FIGURE 3.13:

In the Draw ➤ Arc cascading menu, there are some additional options for drawing arcs. These options provide "canned" responses to the Arc command so that you select only the appropriate points as indicated by the pull-down menu option name.

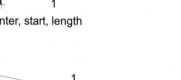

Center, start, length

Center, start, angle

Start, end, Dir

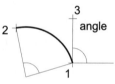

Center, start, end

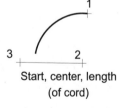

Start, end, radius

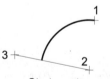

Start, end, angle

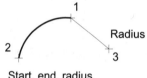

Start, center, length
(of cord)

Start, center, angle

3-point

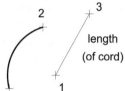

Start, center, end

7. At the `Specify end point of arc or [Angle/chord Length]:` prompt, type **a**↵. At the `Included angle:` prompt, type **–40**↵ to draw an arc beginning at and tangent to the end of the ellipse and ending at 40°. The minus sign (–) is required because AutoCAD draws arcs in the counterclockwise direction unless you change this direction in the Drawing Units dialog box. The minus sign tells AutoCAD to draw the arc in the clockwise direction. Your drawing should look like Figure 3.14.

You can use the top of the ellipse to locate a circle. The circle, after it is trimmed, will be the new arc at the top of the bracket. You'll need to zoom in a bit to see what is happening next and to select the right points.

1. Type **z**↵ to zoom.

2. When you see the Zoom options, type **e**↵ to invoke the Extents option.

3. Type **c**↵ and at the `CIRCLE Specify center point for circle or [3P/2P/Ttr (tan tan radius)]:` prompt, type **2p**↵ to draw a circle using two points on the diameter.

4. At the `Specify first end point of circle's diameter:` prompt, type **qua**↵ to use the Quadrant object snap.

5. At the `of:` prompt, select a point near the top of the ellipse. You'll see the quadrant marker appear when you're near the top of the ellipse.

FIGURE 3.14:

The bracket with both arcs

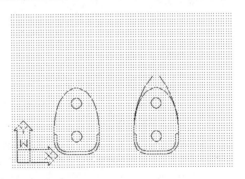

6. At the `Specify second end point of circle's diameter:` prompt, type **@3.376<270**↵. The radius is 1.688, so the diameter is 3.376 and the direction is straight down. Your drawing should look like Figure 3.15.

FIGURE 3.15:

The second bracket with
both arcs and the circle

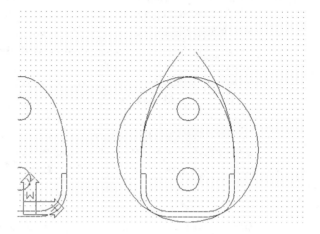

7.   Save the DT0100 file.

The ellipse has served its purpose of providing object-snap points for the new arcs. Now you can erase the ellipse.

1.   Type **e**↵ to begin the Erase command.

2.   Select the ellipse and press ↵ to complete the command.

Trim the two arcs and the circle to each other.

1.   Type **tr**↵ to begin the Trim command; select the two arcs and the circle as cutting edges.

2.   Pick the circle outside of the arcs. The circle is trimmed back to the arcs.

3.   Pick the upper ends of the arcs one at a time. The arcs are trimmed back to what was the circle.

TIP   Some of the keyboard shortcuts for tools or commands you've used in this chapter are CO (Copy), E (Erase), EL (Ellipse), F (Fillet), M (Move), O (Offset), and TR (Trim). Remember that keyboard shortcuts, like keyboard commands, can only be entered when the command prompt is visible in the Command window.

Making a Preliminary Sketch

The following exercise will show you how planning ahead can make AutoCAD work more efficiently. When drawing a complex object, you'll often have to do some layout before you do the actual drawing. This is similar to drawing an accurate pencil sketch using construction lines that you later trace over to produce a finished drawing. The advantage of doing this layout in AutoCAD is that your drawing doesn't lose any accuracy between the sketch and the final product. Also, AutoCAD allows you to use the geometry of your sketch to aid you in drawing. While planning your drawing, think about what you want to draw, and then decide what drawing elements will help you create that object.

You'll use the Offset command to establish reference lines to help you draw a circle tangent to the vertical and horizontal arcs.

Setting Up a Layout

The Offset tool in the Modify toolbar allows you to make parallel copies of a set of objects, such as the arc's detail. When an arc is offset, the radius is either increased or decreased by the offset distance. The centers of the old arc and the new arc are the same point, and the angle subtended by the old arc and the new arc remains the same. When a line is offset, the new line is parallel to and the same length as the original. In Figure 3.16 you can see examples of objects when they are offset both inside and outside of the original. The Offset tool is different from the Copy command; Offset allows only one object to be copied at a time, but it can remember the distance you specify. The Offset option does not work with all types of objects. Only lines, Xlines, rays, arcs, circles, ellipses, splines, and 2D polylines can be offset.

FIGURE 3.16:

Examples of the Offset command on a line, a circle, an arc, a polyline, a polyline rectangle, an ellipse, and a spline.

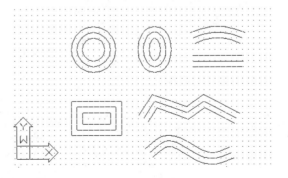

Let's continue with your drawing of the bracket.

1. Click the Offset tool in the Modify toolbar or type **offset**⏎. You can also select Modify ➢ Offset from the menu bar.

2. At the Specify Offset distance or [Through] <Through>: prompt, enter **.25**⏎. This enters the distance of 0.25 as the offset distance, which is the radius of the rounded portion of the bracket.

3. At the Select object to offset or <exit>: prompt, click the right-side arc.

4. At the Specify point on side to offset? prompt, pick a point near and to the left of the right-side arc. A copy of the arc appears. You don't have to be exact about where you pick the side to offset; AutoCAD only wants to know on which side of the object you want to make the offset copy.

5. The prompt Select an object to offset or <exit>: appears again. Click the top arc and pick a point below the arc to create a new arc. You'll produce a drawing that looks like Figure 3.17.

6. When you're done, exit the Offset command by pressing ⏎.

FIGURE 3.17:

The layout with offset arcs

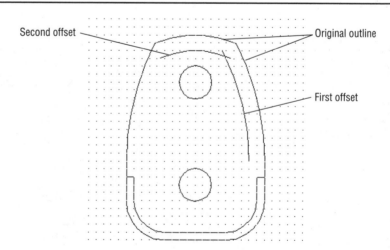

Using the Layout

You've created the intersection of two arcs that you'll use to locate the center of a circle tangent to both arcs. You'll draw the circle, then trim it and both arcs to create the rounded outside edge.

1. Choose Draw ➤ Circle ➤ Center ➤ Radius or type **c**↵. At the CIRCLE Specify center point for circle or [3P/2P/Ttr (tan tan radius)]: prompt, type **int**↵ to use the Intersection Osnap.

2. Move the cursor near the intersection of the two inside arcs until you see the intersection marker. Pick this point. At the Specify radius of circle or [Diameter]: prompt, enter **.25**↵ (the radius of the rounded edge is 0.25 units).

3. Type **z**↵↵ to use the Zoom default, which is Realtime. However, for every rule there is an exception, and if you just pick a point in the edit area or type a 2D point, you can then drag the Zoom window. When you pick or type the other corner, AutoCAD will zoom into that window. Drag the cursor close around the circle and arcs. Your view should look like Figure 3.18.

FIGURE 3.18:

Using the Zoom Window tool for a closer look at the upper corner of the new bracket

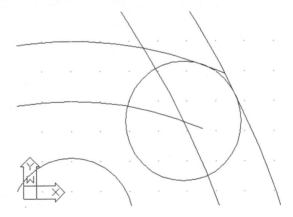

4. Type **e**↵ for Erase, pick the two inside arcs that we have used for our layout, and press ↵.

5. Type **tr**↵ for Trim and pick both arcs and the circle at the prompt: Current settings: Projection=UCS Edge=None

   Select cutting edges ...
   Select objects:

6. See Figure 3.19 to pick the portions of the arcs and circle to be trimmed. First, pick the circle near the bottom; second, pick the horizontal arc near but not on the end; and third, pick a point near the end of the vertical arc.

Putting On the Finishing Touches

The process of using the Offset tool to create guidelines is very common in Auto-CAD. Many times you'll have to choose between using the Offset tool or the Copy tool. As you'll learn in Chapter 9, "Advanced Productivity Tools," Grips also give you copy options. You may be asking yourself why there are so many options to do the same thing. Part of the answer is that this is legacy data. These commands reflect how commands were designed in prior releases. Third-party add-on software, as well as customization software, still depends on some of these commands to function as they functioned in the past. The other part of the answer is that different people work in different ways. A new or infrequent user will access the commands differently from a full-time user, and there are many specialties in engineering—each with a slightly different way of creating documents. Millions of people around the globe use AutoCAD to do all sorts of work, and Autodesk has tried to accommodate as many of these differences as possible.

FIGURE 3.19:

Pick these points to trim
the arcs and the circle

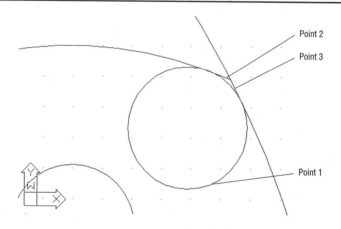

You'll now use the Fillet tool to do the same job we just completed with all of the layout and trimming. One way of solving a problem with AutoCAD isn't necessarily better than any other, but some methods may be faster than others.

1. Type **f**↵ and at the Current settings: Mode = TRIM, Radius = 0.625 Select first object or [Polyline/Radius/Trim]: prompt, type **r**↵.

2. At the Enter fillet radius <0.625>: prompt, type **.25**↵ for the radius of the rounded edge.

3. Press ↵ again to issue the Fillet command. At the (TRIM mode) Current settings: Mode = TRIM, Radius = 0.250 Select first object or [Polyline/Radius/Trim]: prompt, pick the top and the left-side arcs. AutoCAD draws the new arc tangent to both existing arcs at a radius of 0.25 units. Your drawing is now complete and should look like Figure 3.20.

4. Save your drawing and exit AutoCAD.

FIGURE 3.20:

The complete bracket after filleting

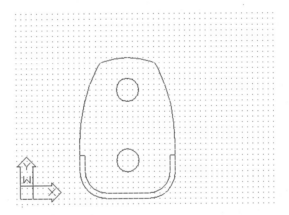

Using AutoCAD's Automatic Save Feature

As you work with AutoCAD, you may notice that it periodically saves your work for you. Your file is saved not as its current filename, but as a file called Auto.SV$. The default time interval between automatic saves is 120 minutes. You can change this interval by doing the following:

1. Enter **savetime**↵ at the command prompt.

2. At the New value for SAVETIME < 120 >: prompt, enter the desired interval in minutes. Or, to disable the Automatic Save feature entirely, enter 0 at the prompt.

If You Want to Experiment...

As you draw, you'll notice that you alternate between creating objects, and then copying and editing them. This is where the difference between hand-drafting and CAD really begins to show.

Try drawing the part shown in Figure 3.21. The figure shows you what to do, step by step. Notice how you're applying the concepts of layout and editing to this drawing.

FIGURE 3.21:

Drawing a section view of a wide-flange beam. Notice how objects are alternately created and edited instead of simply drawing each line segment of the wide flange.

1. Draw a box 7 units wide by 8 units high using the Line command.

2. Draw a vertical line through the center of the box.

3. Offset the top and bottom lines of the box a distance of 0.7 units. Offset the center at 0.35 units.

4. Break the sides of the box between the two offset lines.

5. Trim the top and bottom offset lines between the center three vertical lines.

6. Set the fillet radius to 0.4; then fillet the vertical offset lines with horizontal offset lines.

7. Erase the center vertical line. You have finished the wide-flange beam.

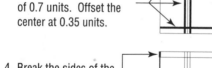
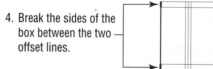

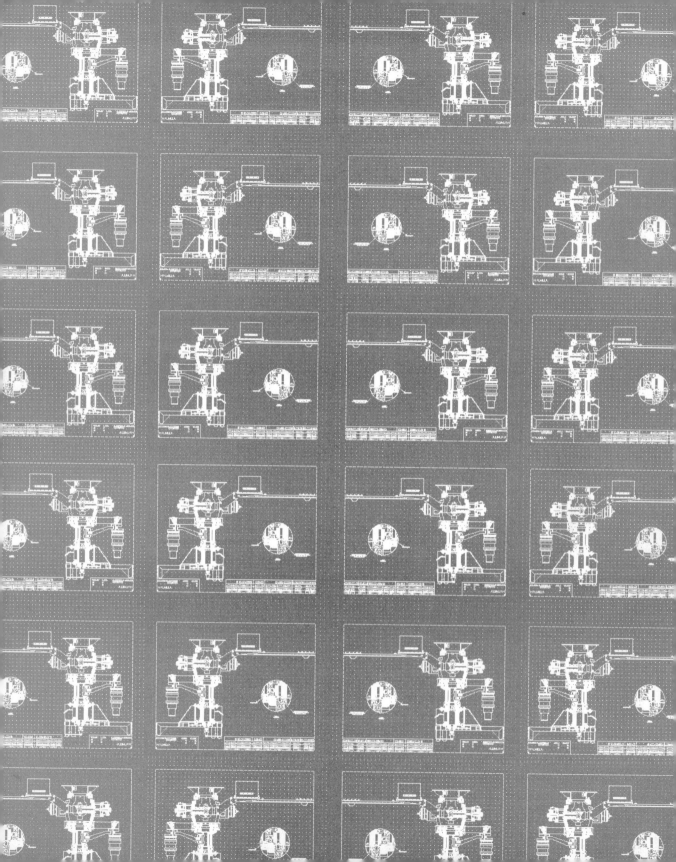

CHAPTER

FOUR

4

Organizing Your Work

■ Creating Symbols Using Blocks

■ Placing a Block

■ Restoring Erased Objects

■ Creating Layers and Assigning Objects to Them

■ Controlling Color

■ Controlling Layers

■ Using Linetypes

■ Inserting Symbols with Drag-and-Drop

Drawing the two brackets in Chapter 3, "Learning the Tools of the Trade," may have taken what seemed an inordinate amount of time. As you continue to use AutoCAD, you'll learn to draw objects more quickly. You'll also need to draw fewer of them because you can save drawings as symbols to be used like rubber stamps, duplicating drawing information instantaneously wherever it is needed. This will save you a lot of time when you're composing drawings.

To make effective use of AutoCAD, you should begin a *symbols library* of drawings you use frequently. A mechanical designer might have a library of symbols for fasteners, cams, valves, or other kinds of parts for his or her application. An electrical engineer might have a symbols library of capacitors, resistors, switches, and the like. A circuit designer will have yet another unique set of frequently used symbols.

In Chapter 3, you drew two objects. In this chapter, you'll see how to create symbols from those drawings. You'll also learn about layers and how you can use them to organize information.

Creating a Symbol

To save a drawing as a symbol, use the Block tool. In word processing, the term *block* refers to a group of words or sentences selected for moving, saving, or deleting. A block of text can be copied elsewhere within the same file, to other files, or to a separate file on disk for future use. AutoCAD uses blocks in a similar fashion. Within a file, you can turn parts of your drawing into blocks that can be saved and recalled at any time. You can also use entire existing files as blocks. Let's try it.

1. Start AutoCAD and open the existing DT0100 file. Use the one you created in Chapter 3, or open 4-DT0100.dwg on the companion CD-ROM. The drawing appears just as you left it in the last session.

2. In the Draw toolbar, click the Make Block tool (or type **bmake.↵**).

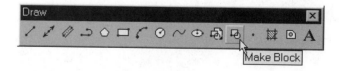

The Block Definition dialog box appears.

3. Type **bkt_v1** into the Name input box.

4. In the Base Point group of the dialog box, click the Pick Point, which is located beneath the Base Point group label. This option enables you to select a base point for the block using your cursor. (The insertion base point of a block is a reference point on the graphics of the block; you'll use this reference point to position the block later when you need to insert the block into the drawing.) When you've selected this button, the dialog box will temporarily disappear.

NOTE   Notice that the Block Definition dialog box gives you the option to specify the X, Y, and Z coordinates for the base point, instead of selecting a point.

5. Using the Midpoint Osnap, pick the midpoint of the bottom of the left bracket (the bracket that you made with the ellipse) as the base point. If you cannot see all of the bracket on the left, you can use the scroll bar at the bottom of the edit area to scroll the part into view. As you click the arrow at

either end of the scroll bar, your view changes. To set the Osnap, press Shift and right-click to pop up the menu and pick the Midpoint option. All you need to do is point near the midpoint of a line to display the Midpoint Osnap marker, then click your mouse. Once you've selected a point, the Block Definition dialog box will reappear.

6. Next, select the actual objects that you want as part of the block. Click the Select Objects button. Once again, the dialog box will momentarily disappear. You now see the familiar object-selection prompt in the Command window. Click a point below and to the left of the left bracket. Then window the entire left bracket; it will be highlighted.

WARNING   Make sure you use the Select Objects option in the Block Definition dialog box to choose the objects you want to turn into a block. AutoCAD will let you create a block that contains no objects. This can cause some confusion and frustration, even for an experienced user.

---

7. Press ↵ to confirm your selection. The dialog box appears again. Select the Retain radio button. This option retains the objects you selected as separate objects after the block is created. The other options in the Block Definition dialog box are discussed after this exercise.

8. After you're satisfied with your choices, click OK. The first bracket drawing is now a block with the name bkt_v1.

9. Repeat the process for the second bracket on the right, but this time use the center of the top circle of the bracket as the insertion base point and give the block the name bkt_v2. Use the Center Osnap to set the base point at the center of the top circle in the bracket on the right.

NOTE   You can press ↵ or right-click the mouse to start the Block tool again.

When you turn an object into a block, it is stored within the drawing file, ready to be recalled at any time. The block remains part of the drawing file, even when you end the editing session. When you open the file again, the block will be available for your use. A block acts like a single object, even though it is really made up of several objects.

The Block Definition dialog box offers several options for defining blocks. In the previous exercise, you used the Pick Point option to indicate a base point. Alternatively, you can enter X, Y, and Z coordinates to define the base point. Also, rather than using the Select Objects button to select the objects to include in the block, you can use the Quick Select button (to the right of the Select Objects button) to filter out objects based on their properties. After you select objects, you can retain them as separate objects (as in the exercise), convert them to a block so that they act as a single object, or delete them (the objects will be erased when the block is created).

The Preview Icon section in the Block Definition dialog box lets you choose whether or not to store a preview image with your block. When you select Create Icon from Block Geometry, AutoCAD creates a thumbnail picture of the block, which helps you locate and identify it later.

The Insert Units option lets you choose the units that AutoCAD uses when the block is inserted. The default is not to specify units (Unitless). The drop-down list offers Inches, Feet, Miles, Millimeters, and some far-out units such as Astronomical Units, Light Years, and Parsecs. If both the block you're creating and the drawing you eventually insert it into have the same units assigned, the block will automatically scale itself to the drawing.

In the Description box at the bottom of the Block Definition dialog box, you can enter a description or keyword to associate with the block. This is helpful if you need to find a specific block in a set of drawings.

Restoring Objects Removed by the Block Command or Block Tool

In prior versions of AutoCAD, the Block command was the only command available to create blocks. This command-line version of the Block tool is still available to those users who are more comfortable entering commands via the keyboard. If you wish to use the older Block command, invoke it by entering **-block**.

When you use the Block command, the objects you turn into a block will automatically disappear. This is also what happens when you choose Delete in AutoCAD 2000's Block Definition dialog box. If you want to restore those source objects, use the Oops command. Type **oops.⌡** at the command prompt, and the objects will reappear in their former condition and location, not as a block. You can use the Oops command in any situation where you want to restore an object you accidentally erased.

Inserting a Symbol

A block can be recalled at any time and as many times as you want, as long as you're in the file where it was created.

> **TIP**
>
> AutoCAD 2000 includes a new tool named the AutoCAD Design Center (ADC). With ADC, you can get information about blocks, layers, and other details, without opening drawings. We'll look at ADC in Chapter 19, "Introduction to Customization."

In the following exercise, you'll draw a support rail for the bracket.

1. First, delete the bracket drawings. Click the Erase tool in the Modify toolbar; then enter **all**↵↵. This erases the entire visible contents of the drawing. (It has no effect on the blocks you created previously.)
2. Turn off Grid and Snap. Look in the status bar at the bottom of the screen. If the words are black (instead of gray), double-click the box until the words become gray.
3. Draw a rectangle 60 × 5. Orient the rectangle so that the long sides go from left to right and the lower-left corner is at coordinate 2.000,3.000. If you draw the rectangle using the Rectangle tool from the Draw toolbar, make sure you explode it using the Explode tool from the Modify toolbar. This step is important for later exercises. Use Zoom All (z↵a↵) to see all of the rectangle. Your drawing should now look like Figure 4.1.

FIGURE 4.1:

The outline of the support rail (bar)

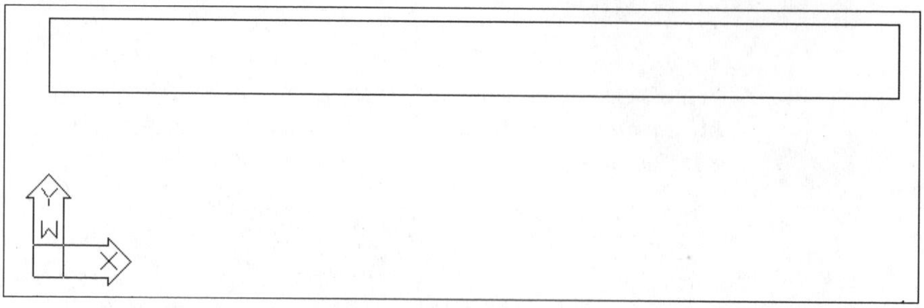

TIP   If you're in a hurry, enter **-insert**⏎ at the command prompt, enter **bkt_v1**⏎, and then go to step 7.

4. In the Draw toolbar, click the Insert Block tool or type **i**⏎.

The Insert dialog box appears.

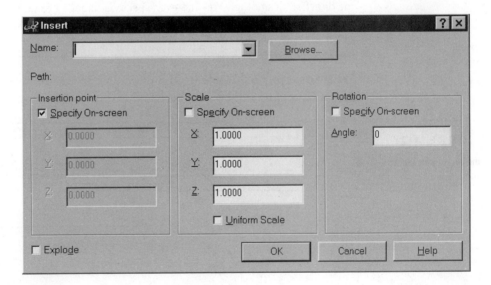

5. Click the Name drop-down list at the top of the dialog box. A list of available blocks in the current drawing appears.

6. Select the block named BKT_V1. Then check all three Specify On-Screen boxes. This allows you to place the block where you wish, at the scale (relative to the original dimensions) and rotation angle (relative to the insertion point) you specify.

7. Click OK. Now you should see a preview image of the bracket attached to the cursor. The point you picked for the bracket's insertion base point is now on the cursor intersection.

8. At the Specify Insertion point: prompt, pick a point above the left end of the bar. Once you've picked the insertion point, notice that as you move your cursor, the preview image of the bracket appears distorted.

9. At the X scale factor <1> / Corner / XYZ: prompt, press ↵ to accept the default, 1. At the Y scale factor <use X scale factor>: prompt, press ↵ to accept the default. This means you're accepting that the Y scale equals the X scale, which in turn equals 1.

NOTE   The X scale factor and Y scale factor prompts let you stretch the block in one direction or another. You can even specify a negative value to mirror the block. The default on these prompts is always 1, which inserts the block or file at the same size as it was created.

---

10. At the Rotation angle <0>: prompt, press ↵ to accept the default of 0. You should have a drawing that looks like the image of Figure 4.2.

FIGURE 4.2:

The bar with the bracket inserted

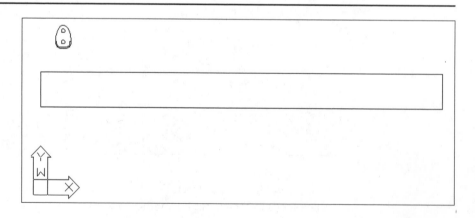

As you moved the cursor in step 8, the bracket became distorted. This demonstrated how the X and Y scale factors can affect the item being inserted. Also, in step 10, you saw the bracket rotate as you moved the cursor. You can pick a point to fix the block in place, or you can enter a rotation value. The default 0° angle inserts the block or file with the orientation at which it was created.

> **NOTE**   Checking the Explode check box at the bottom of the Insert dialog box causes the image that is inserted to be individual objects. If you want the new objects to be slightly different from the block, check Explode so that you can make changes.

Using an Existing Drawing as a Symbol

Now you need a bolt head to bolt the bracket to the bar. We've created a bolt head symbol in another drawing and saved it on the CD. This is just an AutoCAD drawing exactly like the ones that you've been creating and saving. You can bring the bolt head into this drawing file and use it as a block.

1. In the Draw toolbar, click the Insert Block tool or type **i**.

2. In the Insert dialog box, click the Browse button to the right of the Name drop-down list. The Select Drawing File dialog box appears.

3. Locate the `Chapter_4_bolt` file in the `Hardware` subdirectory of your CD and double-click it. You can insert a file as a symbol from any drive on your system or your network.

4. When you return to the Insert dialog box, click OK. As you move the cursor around, you'll notice the bolt appears centered on the cursor intersection, as shown in Figure 4.3.

FIGURE 4.3:

The bolt head

5. Pick a point to the right of the bracket and above the bar to insert the bolt (see Figure 4.4).

6. If you use the default setting for the X scale of the inserted block, the bolt will be inserted at the size it was drawn. However, as mentioned earlier, you can specify a smaller or larger size for an inserted object.

7. Press ↵ three times to accept the default Y = X and the rotation angle of 0°.

FIGURE 4.4:

The bar with the bracket and bolt head drawing being inserted

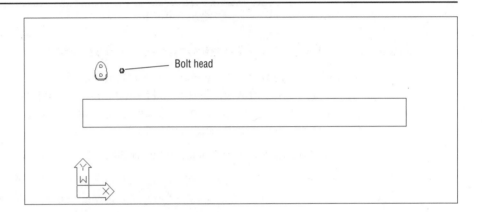

The bracket is to be mounted to the bar with the bolt. You must draw a circle representing a hole in the bar and place it three units from the left of the bar and three units from the bottom of the bar. You'll need to use object snaps to do this. In this exercise, you'll set up some of the Osnap tools to be available automatically whenever AutoCAD expects a point selection.

1. Choose Tools ➤ Object Snap Settings or type **os**↵. This opens the Drafting Settings dialog box.

2. Make sure the Object Snap tab is selected, and the Object Snap On (F3) check box is checked. Then click the check boxes labeled Endpoint, Midpoint, and Intersection so that an X appears in each box; next, click OK.

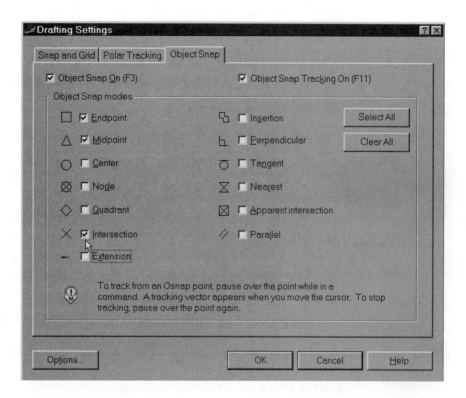

NOTE
Take a look at the graphic symbols next to each of the Osnap options in the Drafting Settings dialog box. These are the Osnap markers that appear in your drawing as you select Osnap points. Each Osnap option has its own marker symbol.

Understanding the AutoSnap Settings Options

Clicking the Options button in the Object Snap tab of the Drafting Settings dialog box presents you with the Options dialog box, which includes a set of choices pertaining to the AutoSnap settings. AutoSnap looks at the location of your cursor during Osnap selections and locates the Osnap point nearest your cursor. AutoSnap then displays a graphic called a *marker* showing you the Osnap point it has found. If it is the one you want, simply click your mouse to select it.

Continued on next page

The AutoSnap settings allow you to control the following features:

Marker turns the graphic Osnap marker on or off.

Magnet causes the cursor to "jump" to the nearest appropriate Osnap point(s).

Display AutoSnap tool tip turns the Osnap tool tip on or off. Pausing your pickbox over an object will show the name of the Osnap available.

AutoSnap Marker color controls the color of the graphic marker.

Display AutoSnap aperture box turns the old-style Osnap cursor box on or off.

AutoSnap Marker size controls the size of the graphic marker.

If you prefer the old method of using the Osnaps feature, you can turn off Marker, Magnet, and AutoSnap Tool Tip, and then turn on Display AutoSnap Aperture Box. The Osnaps feature will then work as it did prior to Release 14.

If you have problems seeing the graphic marker, you may want to change its color using the AutoSnap marker color control. When you're satisfied with your choices, select OK to exit this dialog box. You'll be returned to the Drafting Settings dialog box. When you're satisfied with your choices, select OK to exit back to the drawing.

You've just set up the Endpoint, Midpoint, and Intersection Osnaps to be on by default. This is called a *Running Osnap*: AutoCAD will automatically select the nearest Osnap point without your intervention. Now let's see how Running Osnaps work.

TIP   When you see an Osnap marker on an object, you can have AutoCAD move to the next Osnap point on the object by pressing the Tab key. If you have several Running Osnap modes turned on (such as Endpoint, Midpoint, and Intersection), pressing the Tab key will cycle through those Osnap points on the object. This feature can be especially useful in a crowded area of a drawing.

The Osnap Options

In the previous exercise, you made several of the Osnap settings automatic so that they were available without your having to select them from the Osnap pop-up menu. Another way to invoke the Osnap options is by typing in their keyboard equivalents while selecting points.

Continued on next page

Here is a summary of all the available Osnap options, including their keyboard shortcuts. You've already used many of these options in this chapter and in Chapter 3, "Learning the Tools of the Trade." Pay special attention here to those options you haven't yet used in the exercises but may prove useful in your style of work. The full name of each option is followed by its keyboard shortcut name in brackets. To use these options, you can enter either the full name or its shortcut at any point prompt. You can also pick these options from the pop-up menu obtained by pressing Shift and clicking the right mouse button, or from the Object Snap toolbar.

Apparent Intersection [app] selects the apparent intersection of two objects. This is useful when you want to select the intersection of two objects that do not actually intersect on the same plane. You'll be prompted to select the two objects.

Center [cen] selects the center of an arc or circle. You must click the arc or circle itself, not its apparent center.

Endpoint [end] selects all of the endpoints of lines, polylines, arcs, curves, and 3D face vertices.

From [fro] selects a point relative to a picked point. For example, you can select a point that is two units to the left and four units to the right of a circle's center.

Insert [ins] selects the insertion point of text, blocks, Xrefs, and overlays.

Intersection [int] selects the intersection of objects.

Midpoint [mid] selects the midpoint of a line or arc. In the case of a polyline, it selects the midpoint of the polyline segment.

Nearest [nea] selects a point on an object nearest the pick point.

Node [nod] selects a point object.

None [non] temporarily turns off Running Osnaps.

Parallel [par] snaps to an extension in parallel with an object. AutoCAD is able to track along a line parallel to the endpoint of an object. When you move the cursor over the endpoint of an object, the endpoint is marked and the cursor snaps to the parallel alignment path to that object. The alignment path is calculated from the current *from point* of the command.

Perpendicular [per] selects a position on an object that is perpendicular to the last point selected. Normally, this option is not valid for the first point selected in a string of points.

Quadrant [qua] selects the nearest cardinal (north, south, east, or west) point on an arc, ellipse, or circle.

Continued on next page

Quick [qui] improves the speed at which AutoCAD selects geometry, by sacrificing accuracy. Use Quick in conjunction with one of the other Osnap options. For example, to speed up the selection of an intersection, you would enter **quick,int⏎** at a point prompt, and then select the intersection of two objects.

Tangent [tan] selects a point on an arc or circle that represents the tangent from the last point selected. Similar to the Perpendicular option, Tangent is not valid for the first point in a string of points.

In this next exercise, we'll create a circle and then use the Move command with the displacement method to relocate the circle.

1. Choose View ➢ Zoom ➢ Window and window the area at the left end of the bar. Your drawing should look like Figure 4.5.

2. Choose the Circle tool in the Draw toolbar. At the `Specify center point for circle or [3P/2P/Ttr (tan tan radius)]:` prompt, move the cursor toward the lower-left corner of the bar. The square endpoint marker appears. If you continue to move the cursor, you'll see the endpoint and midpoint markers appear as you approach these points on any objects the crosshairs pass over. Return to the lower-left corner of the bar, and when the endpoint marker appears, click to accept this point.

FIGURE 4.5:

The enlarged bar

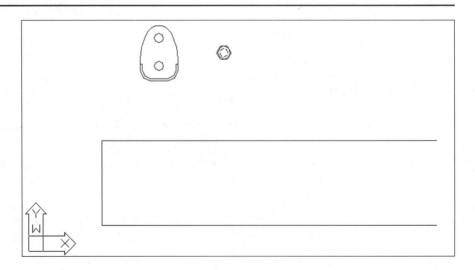

3. At the Specify radius of circle or [Diameter]: prompt, type **d**↵ (so that you input a diameter instead of a radius), and enter **.5**↵ at the Specify diameter of circle: prompt.

4. Click Move in the Modify toolbar or type **m**↵.

5. At the Select objects: prompt, click the circle.

6. At the Specify base point or displacement: prompt, enter **3,3**↵↵ to move the circle three units positive X and three units positive Y. Using the ↵ (Enter) key in this way takes advantage of the *displacement method* of copying or moving objects. Your drawing should look like Figure 4.6.

TIP

In this exercise, we used the displacement method of moving objects. This method also works with the Copy command. The displacement method allows you to tell AutoCAD 2000 exactly how far and in what direction to move the selected objects, without first setting a base point. The movement is specified as a pair or coordinates, first the X movement and then the Y. (For 3D drawings, you'll be able to add the Z movement as well.) It is important to note that the direction will automatically be in the positive direction. If you wish to move in the negative direction add a – (minus sign) to the input. So, at the Specify base point or displacement: prompt, you could enter 3,3 and then ↵ ↵ as we did above in step 6. AutoCAD 2000 will add 3.0000000000000 to the current X coordinate of the selected objects, and 3.0000000000000000 to the Y coordinate. The selected objects will be moved to this new location.

FIGURE 4.6:

The circle positioned on the bar

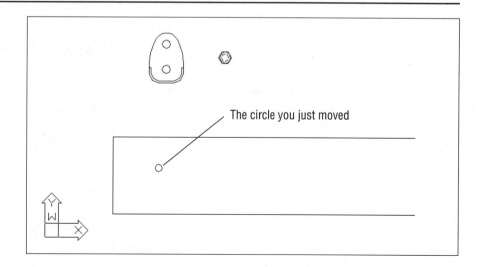

The circle you just moved

Now you'll move the bracket into position on the circle.

1. Click the Move tool in the Modify toolbar. At the `Select objects:` prompt, pick anywhere on the bracket block. Notice that the entire block is highlighted, not just the single object that you selected. The block is now a single object. Press ↵ to complete the selection set. Without exiting from the Move command, click the Object Snap Settings button in the Object Snap toolbar or select Tools ➤ Object Snap Settings from the pull-down menu. Next, click the check box for Center Under Object Snap Modes, and then click OK to close the dialog box.

2. At the `Specify base point or displacement:` prompt, move the cursor to the top circle until the round Osnap marker appears. Click to accept this location.

3. At the `Specify second point of displacement:` prompt, move the cursor, which drags the bracket. Notice how the Magnet function and the Osnap markers are working. Move the cursor until the center marker of the new circle appears, and click to accept this object snap. The bracket moves into position on the bar.

4. Use the same procedure to move the bolt head into position on the top hole of the bracket, except that now the Running Osnap for the center marker is already set.

5. Click the Copy tool in the Modify toolbar. At the `Select objects:` prompt, select the bracket. Do not use a window because this would also acquire the bolt and circle. At the `Select objects: 1 found Select objects:` prompt, press ↵.

6. At the `specify base point or displacement, or [Multiple]:` prompt, pick a point near the bracket.

7. At the `Specify second point of displacement or <use first point as displacement>:` prompt, type @9<0↵ to create a copy nine units to the right of the original. Your drawing should look like Figure 4.7.

FIGURE 4.7:

The bar and bracket assembly thus far

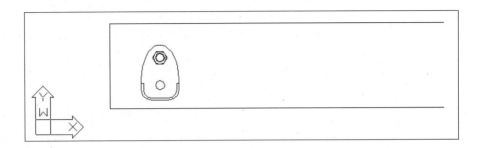

Unblocking and Modifying a Block

To *modify* a block, you break it down into its components, edit them, and then turn them back into a block. This is called *redefining* a block. If you redefine a block that has been inserted into a drawing, each occurrence of that block within the current drawing file will change to reflect the new block definition. You can use this block redefinition feature to make rapid changes to a design.

TIP   If the Regenauto setting is turned off, you'll have to issue a Regen command to see changes made to redefined blocks. See Chapter 6, "Enhancing Your Drawing Skills," for more on Regenauto.

To separate a block into its components, use the Explode command.

1. Click Explode in the Modify toolbar. You can also type **x↵** to start the Explode command.

2. Click the right-hand bracket and press ↵ to confirm your selection.

Now you can edit the individual objects that make up the bracket. In this case, you'll move the lower hole in the bracket up 0.25 units. Then you'll turn the individual objects back into a block, which will update the other copy of the block on the left.

TIP   You can simultaneously insert and explode a block by clicking the Explode check box in the lower-left corner of the Insert dialog box.

---

1. From the Modify toolbar, select Move and at the `Select objects:` prompt, pick the lower circle on the exploded block; press ↵ to complete the selection set. At the `Specify base point or displacement:` prompt, pick near the circle and at the `Specify second point of displacement or <use first point as displacement>:` prompt, type **@.25<90**↵.

2. In the Draw toolbar, select Make Block or type **b**↵.

3. In the Block Definition dialog box, enter **bkt_v1** for the block name.

4. Click the Pick Point button and pick the center of the upper circle.

5. Click the Select Objects button and select the components of the exploded bracket. Press ↵ when you've finished making your selection. The dialog box will indicate 15 objects selected. Be sure that the Delete option is checked.

6. Now click OK. You'll see a warning that says, `bkt_v1 is already defined. Do you want to redefine it?` You don't want to accidentally redefine an existing block. In this case, you know you want to redefine the bracket, so click the Yes button to proceed.

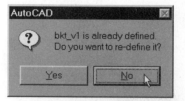

Three things happened when you redefined the block (see Figure 4.8). First, the new base point caused the other copy of the block to move. The new base point is different, but AutoCAD remembers the insertion point relative to the original base point. Second, the lower hole in the copy has moved up. You can check the alignment of the holes after you've moved the displaced copy into the correct position. Third, the objects used to describe the redefined block are deleted.

FIGURE 4.8:

The bar and blocks positioned after they were redefined

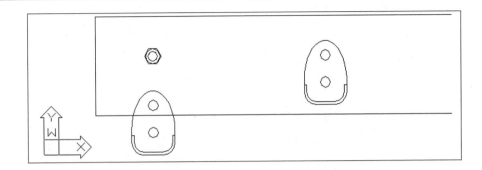

7. In the Modify toolbar, click the Move tool and use the Running Osnaps to select the top circle as the base point and the circle on the bar as the second point of displacement. Erase the old bracket still remaining in the drawing.

8. Now insert the bracket block again, using the Insert dialog box. Open the Name drop-down list and select bkt_v1. For Insertion Point, select Specify On-Screen. For Scale and Rotation, confirm that the Specify On-Screen boxes are not checked. Then click OK.

9. When prompted to Specify insertion point:, click the From Osnap option in the pop-up menu (Shift+right-click) and select the top hole on the bracket block. At the <offset>: prompt, type @9<0 to place the new block nine units to the right. Because you deselected Specify On-Screen for Scale and Rotation, AutoCAD will use the default settings in the dialog box and complete the block insertion.

In step 6, you received a warning that you were about to redefine the existing block. But you had inserted the bracket as a file, not as a block. Whenever you insert a drawing file using the Insert Block tool, the drawing automatically becomes a block in the current drawing. When you redefine a block, however, you do not affect the drawing file you imported. AutoCAD only changes the block within the current file. Next, you'll see how you can update an external file with a redefined block.

TIP   You may nave noticed that both the Select Objects and the Pick Point button icons contain a cursor arrow. Whenever you see this symbol, you know that by selecting that icon button, AutoCAD will temporarily close the dialog box and allow you to select objects, pick points, or perform other operations that require a clear view of the drawing area.

NOTE To remove unused blocks from your drawing, choose File ➤ Drawing Utilities ➤ Purge ➤ Blocks.We'll look at the Purge command in more detail in Chapter 5, "Editing for Productivity."

Saving a Block as a Drawing File

You've seen that, with very little effort, you can create a symbol that can be placed anywhere in a file. Suppose you want to use this symbol in other files. When you create a block using the Block command, the block exists within the current file until you specifically instruct AutoCAD to save it as a drawing file on disk. For an existing drawing that has been brought in and modified, such as the bracket, the drawing file on the disk associated with that bracket is not automatically updated. To update the bracket drawing file, you must take an extra step and use the Export option in the File menu. Let's see how this works.

1. Choose File ➤ Export or type **exp⏎**. The Export Data dialog box opens. This dialog box is a simple file dialog box.

TIP If you prefer, you can skip step 2, and then in step 3 enter the full filename, including the **.dwg** extension, as in **bkt_v1.dwg**.

2. Open the Save As Type drop-down list and select Block (*.dwg).

3. Click the File Name input box, highlight the filename already there, (the current drawing's name), and enter **bkt_v1**. (You don't need to type .dwg, as you declared that in step 2.)

4. Click the Save button. The dialog box closes.

5. At the Enter name of existing block or [= (block=output file)/* (whole drawing)] <define new drawing>: prompt, enter the name of the block you wish to save on disk as the bracket file—in this case, also bkt_v1. The first bracket block is now saved as a file.

AutoCAD gives you the option of saving a block as a file under the same name as the original block or under a different name. Usually, you'll want to use the same name, which you can do by entering an equal sign (=) instead of the block name at the prompt. You may also save the entire drawing under the name you entered in step 5 by using the asterisk (*) option.

NOTE Normally, AutoCAD will save a preview image with a file. This allows you to preview a drawing file prior to opening it. Previewed images are not included with files that are exported with the File ➤ Export option or the Wblock command, which is discussed in the next section.

Replacing Existing Files with Blocks

The Write Block (Wblock) command does the same thing as File ➤ Export, but output is limited to AutoCAD **.dwg** files. (Veteran AutoCAD users will want to note that Wblock is now incorporated into the File ➤ Export option.) Let's try using the Wblock command to save the second bracket block.

1. Issue the Wblock command by typing **wblock**↵ or use the keyboard shortcut by typing **w**↵. The Write Block dialog box will open.

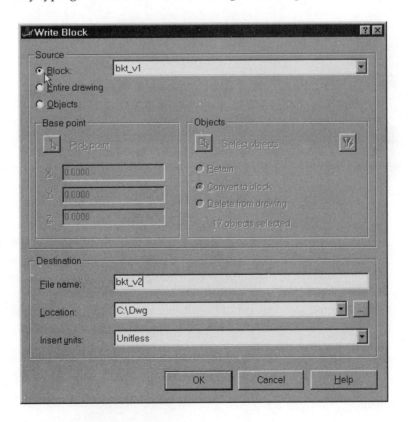

2.   In the Source group, if you click the Block radio button, you can open the accompanying drop-down list and select the bkt_v2 block in the current drawing's library. If you click the Objects radio button, you can pick a base point and select objects.

3.   In the Destination group, adjust the settings as necessary.

   • In the File Name box, indicate what to call the new drawing.

   • For Location, indicate where to file it, but don't use the default folder because you'll never think to look there when you need it.

   • Insert Units shows what units will be used when, or if, the drawing is used elsewhere. If you're not sure, select Unitless.

4.   When you're satisfied with your choices, click OK to save your new drawing.

In this exercise, you typed the Wblock command at the command prompt instead of using File ➤ Export. Either way you access the command, the results are the same.

Other Uses for Blocks

So far, you've used the Block tool to create symbols and the Export and Wblock commands to save those symbols to disk. As you can see, symbols can be created and saved at any time while you're drawing. You've made the bracket symbols into drawing files that you can see when you check the contents of your current directory.

TIP   Blocks provide a way to keep your drawings compact. Blocks take up less space in the drawing's database than the equivalent objects.

However, creating symbols is not the only use for Insert Block, Block, Export, and Wblock. You can use them in any situation that requires grouping objects (although you may prefer to use the more flexible Object Group command discussed in the next section). Export and Wblock also allow you to save a part of a drawing to disk. You'll see instances of these other uses of Block, Export, and Wblock in Chapter 5, "Editing for Productivity" through Chapter 8, "Using Dimensions."

Block, Export, and Wblock are extremely versatile and, if used judiciously, can boost your productivity and simplify your work. Planning your drawings helps you to determine which elements will work best as blocks and to recognize situations where other methods of organization would be more suitable.

An Alternative to Blocks

Another way to create symbols is to create *shapes*. Shapes are special objects made up of lines, arcs, and circles. They can regenerate faster than blocks, and they take up less file space. Unfortunately, shapes are considerably more difficult to create and less flexible to use than blocks.

You create shapes by using a coding system developed by Autodesk. The codes define the sizes and orientations of lines, arcs, and circles. First, sketch your shape, convert it into the code, and then copy that code into a DOS text file. We won't get into detail on this subject, so if you want to know more about shapes, see your *AutoCAD Customization Manual*.

Still another way of working with symbols is to use AutoCAD's external reference capabilities. Externally-referenced files, otherwise known as *Xrefs*, are files inserted into a drawing in a way similar to blocks—the difference is that Xrefs do not actually become part of the drawing's database. Instead, they are loaded along with the current file at start-up time. It is as if AutoCAD opened several drawings at once—the main file and the Xrefs associated with that file.

When you open a drawing containing attached Xrefs, AutoCAD checks the individual files and loads in any changes made since you last edited the drawing. You don't have to update the Xrefs as you must for blocks. For example, if you used the Xref option in the Reference toolbar to insert the bolt drawing, and you later made changes to the bolt, the next time you opened the bracket file, you would see the new version of the bolt. (The Reference toolbar is discussed in Chapter 9, "Advanced Productivity Tools.")

Xrefs are especially useful in workgroup environments, where several people are working on the same project. One person might be updating several files that have been inserted into a variety of other files. Before Xrefs were available, everyone in the workgroup had to be notified of the changes and had to update all of the affected blocks in every drawing that contained them. With Xrefs, the updating is automatic. There are many other features unique to these files, discussed in more detail in Chapter 9.

Grouping Objects

Blocks are an extremely useful tool, but for some situations, they are too restrictive. At times, you'll want to group together objects so that they're connected but can still be edited individually. A group can be copied, then each object can be edited without the group losing its identity. The following exercises demonstrate how this works.

1. Make a copy of the right bracket block nine units to the right of the current one and then explode the block.

2. Use the Zoom command to enlarge just the view of the new bracket, as shown in Figure 4.9.

3. At the Command: prompt, enter **group**⌐, or use the keyboard shortcut **g**⌐. The Object Grouping dialog box appears.

4. Type **bkt_1_mod**. As you type, your entry appears in the Group Name input box.

5. Click New in the Create Group area, near the center of the dialog box. The dialog box temporarily disappears to allow you to select objects for your new group.

FIGURE 4.9:

The bracket after the zoom

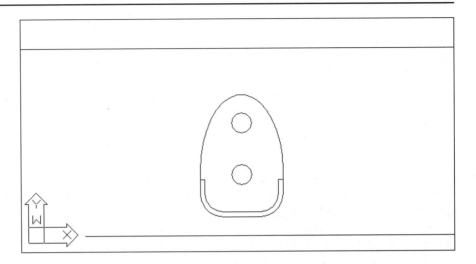

6. At the `Select objects for grouping:` prompt, window the entire bracket and press ↵. The Object Grouping dialog box returns. Notice that the name BKT_1_MOD appears in the Group Name list box at the top of the dialog box.

7. Click OK. You've just created a group.

Now, whenever you want to select the bracket, you can click any part of it and the entire group will be selected. At the same time, you'll still be able to modify individual parts of the group—the holes, ellipses, and so on—without losing the grouping of objects.

Modifying Members of a Group

Next, you'll remove a notch from the left side of the ellipse. Figure 4.10 is a sketch of the proposed modification.

This exercise shows you how to complete your drawing to reflect the design requirements of the sketch.

1. Draw a rectangle in the area of the bracket. The rectangle should be approximately two units wide and three units high.

2. Move the rectangle so that the lower-right corner is at the center of the lower bracket mounting hole.

3. Move the rectangle again from the current position to **@-0.5,0.75**, as shown in Figure 4.11.

FIGURE 4.10:

A sketch of the modified bracket

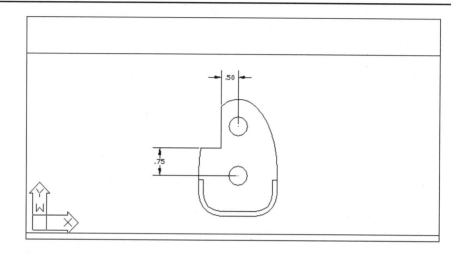

FIGURE 4.11:

The rectangle placed and ready to be used to edit the group

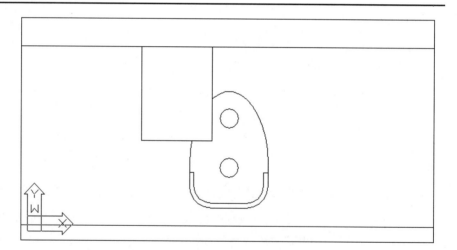

Now that you've got a copy of the bracket as a group and a sketch of the notch, you need to edit the bracket group. If you had used a block for the bracket, you would have had to explode the bracket before you could edit it. Groups, however, let you make changes without undoing their grouping.

1. Press Ctrl+a. This temporarily turns off groupings. You'll see the message `<Group off>` in the Command window. With Group On, selecting any

object in a group selects all of the objects, similar to selecting a block. But with Group off, any individual object can be selected and edited.

2. Using the Trim tool, remove the appropriate parts of the rectangle and the ellipse, as shown in Figure 4.12.

3. Press Ctrl+a again and you'll see the message <Group on> in the Command window.

4. To check your bracket, click one of the objects to see if all of its components are highlighted together. The new parts of the notch are not part of the bracket group. The ellipse arc is now two ellipse arcs, and only one is part of the group.

FIGURE 4.12:

The bracket with the notch removed

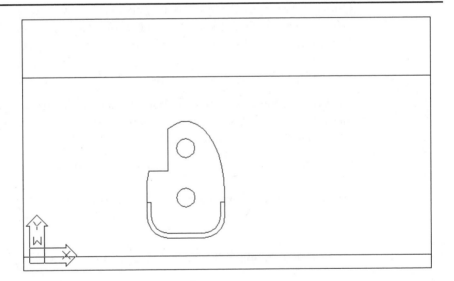

5. To add the notch and ellipse objects to the group, issue the Group command again by pressing **g**↵.

6. At the top of the dialog box, pick the bracket group and click the Add button in the Change Group area.

7. The dialog box disappears, and the prompt says Select objects to add to group. Pick the two lines that make up the notch, and press ↵ to complete the selection set. If you wish to check your work, with Group Name BKT_1_MOD

selected, click the Highlight button. The dialog box should disappear, and all the objects in the named group will be highlighted. Click Continue to return to the dialog box. When you're satisfied with your work, click OK.

8. Save the current drawing.

TIP   You can also use the Pickstyle system variable to control groupings. See Appendix D, "System Variables," for more on Pickstyle.

Working with the Object Grouping Dialog Box

Each group has a unique name, and you can also write a brief description of the group in the Object Grouping dialog box. When you copy a group, AutoCAD assigns an arbitrary name to the newly created group. Copies of groups are considered unnamed, but these still can be listed in the Object Grouping dialog box by clicking the Unnamed check box. You can use the Rename button in the Object Grouping dialog box to rename unnamed groups appropriately.

Objects within a group are not bound solely to that group. One object can be a member of several groups, and you can have nested groups, which is another advantage that groups have over blocks.

Here are descriptions of the options available in the Object Grouping dialog box.

The Group Identification Area   The options in this area allow you to identify your groups, using unique elements that help you remember what each group is for.

Group Name   This input box lets you create a new group by naming it first.

Description   This input box lets you include a brief description of the group.

Find Name   This button lets you find the name of an existing group by temporarily closing the dialog box so you can click a group.

Highlight   This button highlights a group that has been selected from the group list, which helps you locate a group in a crowded drawing.

Include Unnamed   This check box determines whether unnamed groups are included in the Group Name list. Check this box to display the names of copies of groups for processing by this dialog box.

The Create Group Area   This is where you control how a group is created.

New   This button lets you create a new group. It temporarily closes the dialog box so you can select objects for grouping. To use this button, you must have either entered a group name or checked the Unnamed check box.

Selectable   This check box lets you control whether the group you'll create is selectable or not. See the description of the Selectable button in "The Change Group Area" section next.

Unnamed   This check box lets you create a new group without naming it.

The Change Group Area   These buttons are available only when a group name is highlighted in the Group Name list at the top of the dialog box.

Remove   Lets you remove objects from a group.

Add   Lets you add objects to a group. While you're using this option, grouping is temporarily turned off to allow you to select objects from other groups.

Rename   Lets you rename a group.

Reorder   Lets you change the order of objects in a group. Objects are numbered in the order they are added to the group. This numbering system may cause problems later if you process your drawing using some third-party programs.

Description   Lets you modify the description of a group.

Explode   Separates a group into its individual components. Unlike blocks, this doesn't change the properties of the affected objects, but merely removes the group links.

Selectable   Turns individual groupings on and off. When a group is selectable, it is selectable as a group. When a group is not selectable, the individual objects in a group can be selected, but not the group.

Organizing Information with Layers

Another tool for organization is the layer. *Layers* are like overlays on which you keep various types of information (see Figure 4.13). Each layer is perfectly registered to the original so that if you're doing a layout of an assembly and wish to create another layer to draw a part, there will be no loss in accuracy or clarity. It's also a good idea to keep notes and reference information for the drawing, as well as the drawing's dimensions, on separate layers. As your drawing becomes more complex, the various layers can be turned on and off to allow easier display and modification. If you're making working drawings, you'll want to use layers to help you with plotting. A color can be assigned to each layer, as well as a linetype. Most plotters, whether they are of the modern raster type or older pen plotters, use the pen color to assign line weight (width). If you've created a green layer called Dimension, placed all of your dimensions on the Dimension layer, and assigned a 0.015-wide pen to the color green, the plotter will draw all of the green objects—your dimensions—with a 0.015-wide pen. In Chapter 14, "Printing and Plotting," we'll look at other ways to control line weight.

FIGURE 4.13:

A comparison of layers and overlays

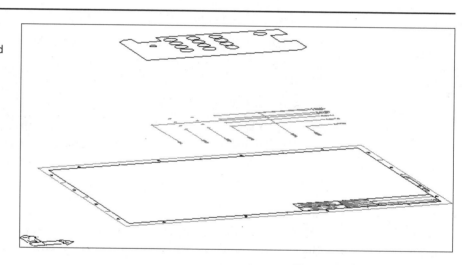

AutoCAD allows an unlimited number of layers, and you can name each layer anything you want.

Creating and Assigning Layers

To continue with your bar and bracket drawing, you'll create some new layers:

1. Open the DT0100 file you saved earlier in this chapter. If you didn't create one, use the file 4b-dto100.dwg from the companion CD-ROM.

2. To display the Layer Properties Manager dialog box, click the Layers tool in the Object Properties toolbar; choose Format ➤ Layer from the menu bar; or type **la**⏎ to use the keyboard shortcut.

The Layer Properties Manager dialog box shows you the status of your layers at a glance. Right now, you only have one layer, but as your work expands, so will the number of layers. You'll then find this dialog box indispensable.

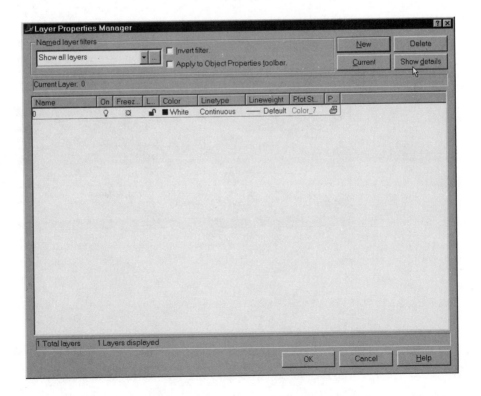

Click the New button in the upper-right corner of the dialog box. A new layer named Layer1 appears in the list box. Notice that the name is highlighted.

This tells you that by typing in a new name, you can change the default name to something better suited to your needs.

4. Type **hidden**. As you type, your entry replaces the Layer1 name in the list box. You can use up to 256 characters to name layers. Spaces are allowed, but not the following characters: < > / \ " : * | = and '.

NOTE   If you're going to use the drawing as an external reference, the name of the file will be added to the layer name when the drawing is attached. If the resultant layer name is greater than 256 characters, you'll be asked to use a substitute as an *alias* for the layer name.

---

5. Click the Show Details button (a toggle that changes to Hide Details) near the upper-right corner of the dialog box. Additional layer options will appear in the Details area.

6. With the hidden layer name highlighted, click the downward-pointing arrow to the right of the Color drop-down list. You'll see a listing of colors that you can assign to the Hidden layer. At first glance, this list may seem a bit limited, but you actually have a choice of 256 colors.

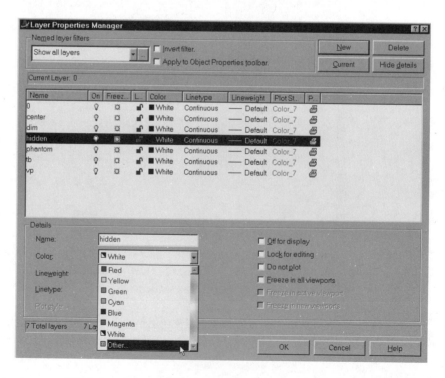

NOTE   Although it isn't readily apparent, all of the colors in the Color drop-down list, except for the first seven, are designated by numbers. So when you select a color after the seventh color, the color's number—rather than its name—appears in the Color input box at the bottom of the dialog box. Some users may experience confusion if their system doesn't support 256 colors, often due to a video card with insufficient video RAM. When you display the drawing on a system with 2MB of video RAM (or more) and set to at least 256 colors, the image will be corrected.

7. Select Other at the bottom of the list to see the full palette.

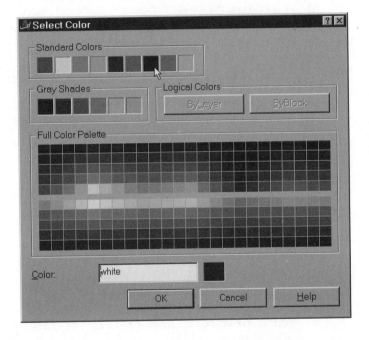

8. In the Standard Colors group, select the red square and then click OK. Notice that the color swatch in the Hidden layer listing is now red. You could have chosen red from the Colors drop-down list, but you selected red from the Select Color dialog box in order to be aware that many other colors are available.

9. When the Layer Properties Manager dialog box returns, click OK to close it.

TIP   Just as you can adjust Windows Explorer, you can also adjust the width of each column in the list of layers in the Layers Properties Manager dialog box by clicking and dragging either side of the column-head buttons. You can also sort the layer list based on a property by clicking the property name at the top of the list. Similar to other Windows list boxes, you can Shift+click names to select a block of layer names or Ctrl+click individual names to select multiple names that do not appear together. These features will become helpful as your list of layers expands.

Controlling Layers through the Layer Command

You've seen how the Layer Properties Manager dialog box makes it easy to view and edit layer information, and how layer colors can be easily selected from this dialog box. But layers can also be controlled through the command prompt.

1. Press Esc to make sure any current command is canceled.

2. At the command prompt, enter **–layer.**⏎. Make sure you include the minus sign (–) in front of the word *layer*. The following prompt appears:

   ```
   [?/Make/Set/New/ON/OFF/Color/Ltype/LWeight
   /Plot/Freeze/Thaw/LOck/Unlock]:
   ```

   You'll learn about many of the options in this prompt as you work through this chapter.

3. Enter **n**⏎ to select the New option.

4. At the `Enter name list for new layer(s):` prompt, enter **dim**⏎. The ?/Make/Set... prompt appears again.

5. Enter **c**⏎.

6. At the `Enter color name or color (1-255):` prompt, enter **green**⏎. Or you can enter **3**⏎, the numeric equivalent of the color green in AutoCAD.

7. At the `Enter name list of layer(s) for color 3 (green) <0>:` prompt, enter **dim**⏎. The ?/Make/Set... prompt appears again.

8. Press ⏎ to exit the Layer command.

Each method of controlling layers has its own advantages. The Layer Properties Manager dialog box offers information about your layers at a glance. On the other hand, the Layer command offers a quick way to control and create layers if you're in a hurry. Also, if you intend to write custom macros, you'll want to know how to use the Layer command as opposed to the dialog box, because dialog boxes cannot be controlled through custom toolbar buttons or scripts.

TIP Another advantage to using the keyboard commands is that you can recall previously entered keystrokes by using the ↑ (up arrow) and ↓ (down arrow) cursor keys. For example, to recall the layer named Dim you entered in step 4 in the previous exercise, press the ↓ key until **DIM** appears in the prompt. This can save time when you're performing repetitive operations such as creating multiple layers. This feature does not work with tools selected from the toolbars.

Assigning Objects to Layers

When you create an object with any of the Draw tools, that object is automatically assigned to the current layer. Until now, only one layer has existed, Layer 0, which contains all the objects you've drawn so far. Because it was the only layer, it was automatically the current layer. Now that you've created some new layers, you can reassign objects to them using the Properties tool on the Object Properties toolbar. The drawing you've been working on is a channel with the flanges facing away from you. Let's draw the lines representing the flanges using the Offset tool and then assign them to the Hidden layer.

In Chapter 3, "Learning the Tools of the Trade," we worked with the Offset tool. Let's use it now to create the lines we need:

1. Click the Offset tool in the Modify toolbar. At the `Specify offset distance or <Through>` prompt, type **.19**↵. Pick the top horizontal line of the bar at the `Object to offset or <exit>:` prompt, and then pick any point below the line at the side to offset. A new line appears. The command is still working, so pick the bottom line at the `Object to offset or <exit>:` prompt and choose a point above the line for the side to offset.

2. Click Properties in the Standard toolbar.

3. The Properties dialog box will appear. A drop-down list at the top of the dialog box will say No Selection. Click the two new lines representing the flanges, and this will change to Line (2) to indicate that two lines are selected.

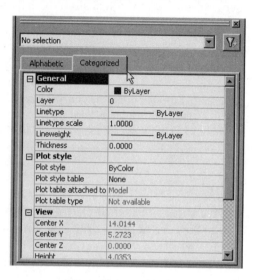

5. Click the Layer item. A drop-down list appears. Selecting it presents you with a list of the layers in the drawing, including the ones that you just created.

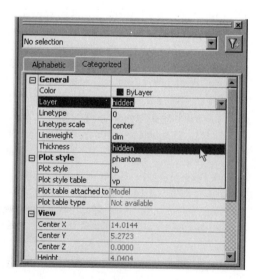

NOTE When the Properties dialog box is first opened, if no objects are selected, it shows general property settings. But if you preselect objects by turning on their grips, you'll be presented with all of the properties shared by the selected objects. Generally, this allows you to change the layer assignment, color, linetype, and thickness of an object. You'll learn about linetypes later in this chapter in the section "Assigning Linetypes to Layers."

6. Click the Hidden layer. The drop-down list will close.

NOTE You don't need to exit the Properties dialog box to see the effects of changes. You can even select objects after you've opened the dialog box, without closing it. In fact, if you have sufficient screen space, you can leave the Properties dialog box open all the time.

7. Click the X in the upper-right corner to close the dialog box.

The flanges are now assigned to the new layer, named Hidden, and the lines are changed to red. Objects on different layers are more easily distinguished from one another when you use colors to set apart the layers.

Next, you'll practice the commands you learned in this section by creating some new layers.

1. Bring up the Layer Properties Manager dialog box, (use Format ➤ Layers or click the Layers button in the Object Properties toolbar). Create a new layer called Center and give it the color yellow.

TIP You can change the name of a layer by clicking it in the Layer Properties Manager dialog box, and once it is highlighted, click it again so that a box surrounds the name. You can then rename the layer. This works in the same way as renaming a file or folder in Windows 95/98/NT. You may also rename layers, blocks, dimension styles, linetypes, text styles and views using the Rename command. You select the type of named object you wish to rename, and a list of those objects is presented in the drawing. Highlight your choice and enter the new name in the box next to the Rename To button. Be sure to click the Rename To button to complete the operation. You may rename other objects, or click OK to exit the dialog box.

NOTE   Within a block, you can change the color assignment and linetype of only the objects that are on Layer 0. See the sidebar, "Controlling Colors and Linetypes of Blocked Objects," at the end of this section.

2.   Use the Layer Properties Manager dialog box to create three more layers for phantom lines called Phantom (color cyan), for a title block called Tb (color white), and for viewports called Vp (color magenta), as shown in Table 4.1. You can open the Select Color dialog box by choosing Other from the Color drop-down list or by clicking the color swatch in the layer listing.

TABLE 4.1:   Create These Layers and Set Their Colors as Indicated

Layer Name	Layer Color (number)
Vp	Magenta (6)
Tb	White (7)
Phantom	Cyan (4)

TIP   As you create these new layers, try this little trick. Open the Layer Properties Manager dialog box, and select New. In the Name box, enter **Vp,Tb,Phantom⏎.** Notice that the new layers are made. They are now ready for color settings.

Controlling Colors and Linetypes of Blocked Objects

Layer 0 has special importance to blocks. When objects that are assigned to Layer 0 are used as parts of a block, those objects take on the characteristics of the layer on which the block is inserted. On the other hand, if those objects are on a layer other than 0, they will maintain their original layer characteristics even if you insert or change that block to another layer. For example, suppose the bracket is drawn on the Hidden layer, instead of on Layer 0. If you turn the bracket into a block and insert it on the Vp layer, the objects that the bracket is composed of will maintain their assignment to the Hidden layer, although the bracket block is assigned to the Vp layer.

Continued on next page

It might help to think of the Block function as a clear plastic bag that holds together the objects that make up the bracket. The objects inside the bag maintain their assignment to the Hidden layer even while the bag itself is assigned to the Vp layer.

AutoCAD also allows you to have more than one color or linetype on a layer. You can use the Properties button on the Standard toolbar to alter the color or linetype of an object on any layer. That object then maintains its assigned color and linetype—no matter what its layer assignment. Likewise, objects specifically assigned a color or linetype won't be affected by their inclusion in blocks.

Working on Layers

So far, you've created layers and assigned objects to them. However, the current layer is still 0, and every new object you draw will be on Layer 0. Here's how to change the current layer.

1. In the Object Properties toolbar, click the arrow button next to the layer name. A drop-down list opens, showing you all the layers available in the drawing.

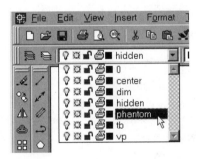

Notice the icons that appear next to the layer names; these control the status of the layer. Also notice the box directly to the left of each layer name. This shows you the color of the layer.

2. Click the Phantom layer name. The drop-down list closes, and the name Phantom appears in the toolbar's layer name box. Phantom is now the current layer.

NOTE
You can also use the Layer command to reset the current layer. To do this here, enter **–layer** (be sure to include the minus sign) at the command prompt, and at the prompt, enter **s** for *set*. At the `Enter layer name to make current` prompt, enter **phantom** and then press twice to exit the Layer command.

---

3. Zoom into the notched bracket, and draw a 5" horizontal line, starting at the center of the top mounting hole. You may use direct distance entry to draw this line. With the Running Osnap center we set earlier, select the circle, and with Ortho on, drag to the right. Enter **5.0** and press .

4. Draw a similar line from the bottom mounting hole. Your drawing should look like Figure 4.14.

FIGURE 4.14:

Bar and bracket with objects on layers

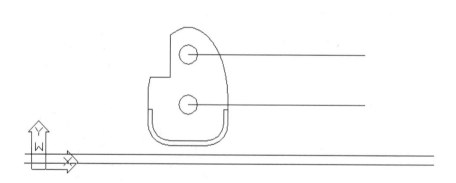

Because you assigned the color cyan to the Phantom layer, the two lines you just drew are cyan. This gives you immediate feedback about what layer you're on as you draw.

Next, you'll change the Layer property of the new line to the Center layer. But instead of using the Properties tool, as you've done in earlier exercises, you'll use a shortcut method.

1. Click the new cyan-colored line to highlight it. Notice that the layer listing in the Object Properties toolbar changes to Phantom. Whenever you select

an object to expose its grips, the layer, color, and linetype listings in the Object Properties toolbar will change to reflect those properties of the selected object.

2. Click the layer name in the Object Properties toolbar. The layer drop-down list will open.

3. Choose the Center layer name. The list closes and the line you selected changes to yellow, showing you that it is now on the Center layer. Also, notice that the color list in the Object Properties toolbar has changed to reflect the new color for the line.

4. Press the Esc key twice to clear the grip selection. Notice that the current layer displayed returns to Phantom.

In this exercise, you saw that by selecting an object with no command active, the object's properties were immediately displayed in the Object Properties toolbar in the layer, color, and linetype boxes. Using this method, you can also change an object's color and linetype independent of its layer. Using the layer drop-down list, you can change the layers of multiple objects.

Controlling Layer Visibility

When you're working on developing a new design, it's often nice to have hidden lines, dimensions, and so on available, but off.

WARNING   Words of caution before you get carried away with layer visibility: Objects on layers that are not displayed are not selected unless they are part of a group or block that is visible. If you have the Hidden layer turned off and you move the rest of the view to a better location, you'll find that the hidden lines did not move.

---

1. Open the Layer Properties Manager dialog box by clicking the Layers tool in the Object Properties toolbar.

2. Click the Hidden layer in the layer list.

3. Click the lightbulb icon in the layer list next to the Hidden layer name. You can also click the check box labeled Off for Display in the Details section of the dialog box. In either case, the lightbulb icon changes from yellow to gray to indicate that the layer is off.

4. Click OK to exit the dialog box. When you return to the drawing, the red hidden lines disappear because you've made them invisible by turning off their layer (see Figure 4.15).

You can also control layer visibility using the layer drop-down list in the Object Properties toolbar.

1. In the Object Properties toolbar, click the layer drop-down list.

2. Find the Hidden layer and notice that its lightbulb icon is gray. This tells you that the layer is off and not visible.

3. Click the lightbulb icon to make it yellow.

FIGURE 4.15:

Bar with the hidden lines turned off

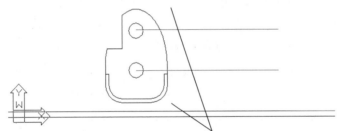

The missing red lines show that the Hidden layer is off.

TIP
By momentarily placing the cursor on an icon in the layer-control section of the Object Properties toolbar, you'll get a tool tip giving you a brief description of the icon's purpose.

4. Now click the drawing area to close the layer list, and the hidden lines reappear.

Figure 4.16 explains the role of the other icons in the Layer drop-down list.

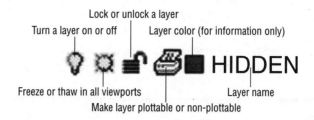

The layer drop-down list icons

Finding the Layers You Want

With only a handful of layers, it's fairly easy to find the layer you want to turn off. This becomes much more difficult, however, when the number of layers exceeds 20 or 30. The Layer Properties Manager dialog box offers some useful tools to help you find the layers you want fast.

Now suppose you have several layers whose names begin with C, such as C-flange, C-grill, and C-seal, and you want to find those layers quickly. You can click the Name button at the top of the layer list to sort the layer names in alphabetical order. Click the Name button again to reverse the order. To select those layers for processing, click the first layer name that starts with C; then scroll down the list until you find the last layer of the group and Shift+click it. All of the layers between those layers will be selected. You can also select multiple layer names by holding down the Ctrl key while clicking each name. If you make a mistake and want to deselect an item, Ctrl+click it again.

The Color and Linetype buttons at the top of the list let you control which layers appear in the list by color or linetype assignments. Other buttons will sort the list by status: On/Off, Freeze/Thawed, Locked/Unlocked, and so forth. See the "Other Layer Options" sidebar later in this chapter.

NOTE   To delete all of the objects on a layer, you can set the current layer to the one that you want to edit, and then freeze or lock all of the others. Click Erase in the Modify toolbar and then type **all↵**. The All option selects all of the objects currently visible, *and those on layers that are set to off.* If you wish to protect other objects from being erased accidentally, look at the "Other Layer Options" sidebar at the end of this section.

You can turn off a set of layers by single-clicking a lightbulb icon. You can freeze/thaw, lock/unlock, or change the color of a group of layers in a similar manner by clicking the appropriate layer property. For example, if you had clicked a color swatch of one of the selected layers, the Select Color dialog box would appear, allowing you to set the color for all of the selected layers.

Other Layer Options

You may have noticed the Freeze and Thaw buttons in the Layer Properties Manager dialog box. These options are similar to the On and Off buttons. However, Freeze not only makes layers invisible, it also tells AutoCAD to ignore the contents of those layers when you use the All response to the `Select Object` prompt. Freezing layers can also save time when you issue a command that regenerates a complex drawing. This is because Auto-CAD ignores objects on frozen layers during a regeneration. You'll get firsthand experience with Freeze and Thaw in Chapter 6, "Enhancing Your Drawing Skills."

Another pair of Layer Properties Manager options, Lock and Unlock, offers a function similar to Freeze and Thaw. If you lock a layer, you'll be able to view and snap to objects on that layer, but you won't be able to edit those objects. This feature is useful when you're working on a crowded drawing and you don't want to accidentally edit portions of it. You can lock all of the layers except those you intend to edit, and then proceed to work without fear of making accidental changes.

Taming an Unwieldy List of Layers

You may eventually end up with a fairly long list of layers. Managing such a list can become a nightmare, but AutoCAD provides help for locating and isolating only those layers you need to work with.

To use layer filters, click the Show All Layers drop-down list near the top of the Layer Properties Manager dialog box. This drop-down list contains the options described in Table 4.2. See "Using External References (Xrefs)" in Chapter 9 for information on Xref-dependent layers.

TABLE 4.2: The Filter Options

Filter Options	What It Filters
Show All Layers	All layers regardless of their status
Show All Used Layers	All layers that have objects assigned to them
Show All Xref-Dependent Layers	All layers that contain Xref objects

You can also create your own filter criteria and add it to this drop-down list by clicking on the Ellipsis button with the three dots on it.

When you click this button, the Named Layer Filters dialog box appears. Here you can create a filter list based on the layer's characteristics.

Now suppose you have a drawing whose layer names are set up to help you easily identify groups of layers associated with each part in an assembly layout, such as the following:

v1-case	v1-cover	v1-chassis	v2-case
v2-cover	v1-panel	v1-screw	

If you want to isolate only those layers that have to do with covers, regardless of their assembly, assign a filter name such as Covers, enter **??-cover*** in the Layer Name input box, click the Add button, and then click Close. Now only the layers whose names contain the letters -*cover* following the first two characters, and any combination of numbers or letters following -*cover* will appear in the list of layers. You can then easily turn off all of these layers, change their color assignment, or change other settings quickly, without having to wade through other layers you don't want to touch.

NOTE
In the **??-cover*** example, the question marks (??) tell AutoCAD that the first two characters in the layer name can be anything. The -*cover* tells AutoCAD that the layer name must contain the word -*cover* in these six places of the name. The asterisk (*) at the end tells AutoCAD that the remaining characters can be anything. The question marks and asterisk are known as *wildcard characters*. They are commonly used filtering tools for Windows and UNIX operating systems.

All of the drop-down lists let you filter layers by status—On/Off, Freeze/Thawed, Locked/Unlocked, and so forth. See the "Other Layer Options" sidebar earlier in this chapter. The other input boxes—Color, Lineweight, Linetype and Plot Style—let you control what layers appear by their various assignments. (Lineweight will be discussed in the next section and plot style will be discussed in Chapter 14, "Printing and Plotting.")

As the number of layers in a drawing grows, you'll find layer filters to be an indispensable tool. But bear in mind that the successful use of layer filters depends on a careful layer-naming convention.

Assigning Linetypes to Layers

You may have been wondering how to make the lines on the various layers, which seem to be named after linetypes, look like those linetypes. A hidden line, for example, should be made up of a dashed line, and a center line should be a line broken by a dashed line. Drafters use different linetypes to differentiate drafting lines from object lines and other kinds of information. AutoCAD lets you use both color and linetype to indicate the difference between lines. Many people work on a relatively small monitor (relative to a full-sized piece of paper) and need all the help that they can get. Color plotting is often not an option for many reasons. A company may not have a color plotter, or color plots may not be allowed by document control because the company doesn't have a large document color copier. Color is still valuable because your plotter uses color to determine line weight.

You can set a layer to have not only a color assignment but also a linetype assignment. AutoCAD comes with several linetypes, as shown in Figure 4.17. From the top of Figure 4.17 downward, we can see standard linetypes, then ISO and complex linetypes, and then a series of lines that can be used to illustrate gas and water lines in civil work, or batt insulation in a wall cavity. ISO linetypes are designed to be used with specific plotted-line widths and linetype scales. For example, if you're using a pen width of 0.5mm, the linetype scale of the drawing should be set to 0.5 as well (see Chapter 14, "Printing and Plotting," for more information on plotting and linetype scale). You can also create your own linetypes (see Chapter 21, "Integrating AutoCAD into Your Projects and Organization").

WARNING Be aware that linetypes containing text, such as the gas sample, will use the current text height and font to determine the size and appearance of the text displayed in the line. A text height of 0 (zero) will display the text properly in most cases. See Chapter 7, "Adding Text to Drawings," for more on text styles.

AutoCAD stores linetype descriptions in an external file named ACAD.LIN. You can edit this file in a word processor to create new linetypes or to modify existing ones. You'll see how this is done in Chapter 21, "Integrating AutoCAD into Your Projects and Organization."

Adding a Linetype to a Drawing.

To see how linetypes work, change the linetype of the Hidden layer to a hidden linetype.

1.   Open the Layer Property Manager dialog box.

TIP If you're in a hurry, you can simultaneously load a linetype and assign it to a layer by using the Layer command. In this exercise, you would enter **–layer**↵ at the command prompt, enter **l**↵, **hidden**↵, **hidden**, and then press ↵ to exit the Layer command.

FIGURE 4.17:

Standard AutoCAD linetypes

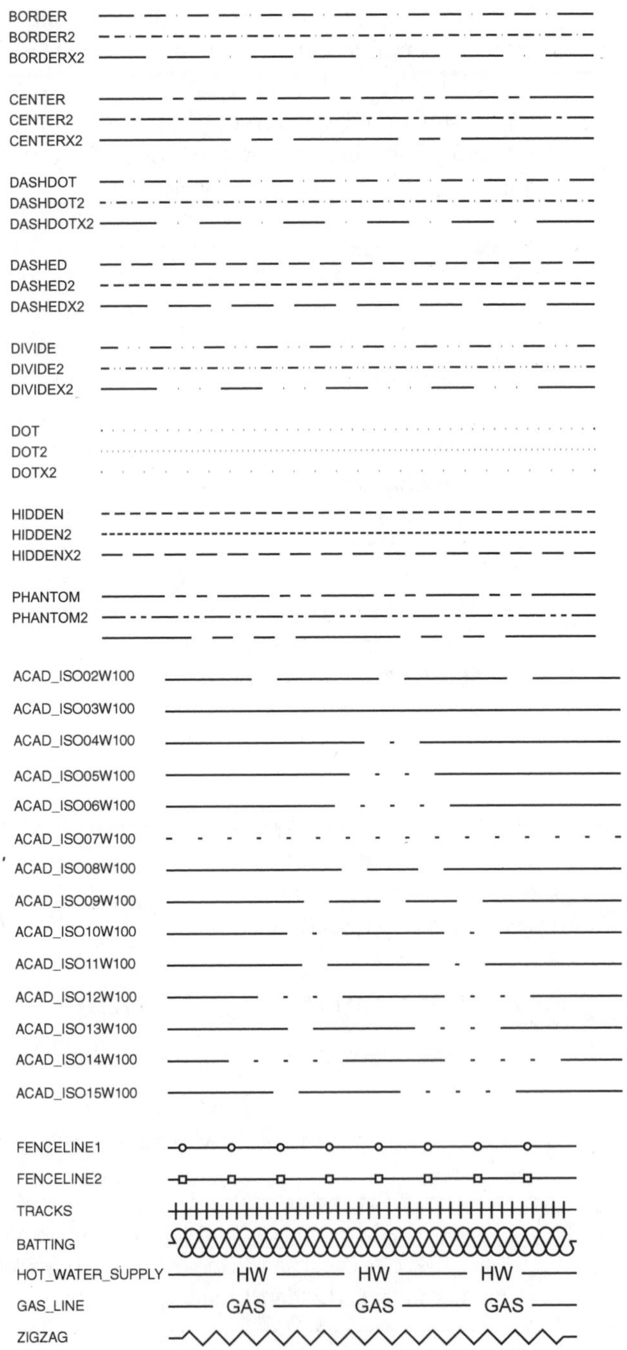

BORDER	
BORDER2	
BORDERX2	
CENTER	
CENTER2	
CENTERX2	
DASHDOT	
DASHDOT2	
DASHDOTX2	
DASHED	
DASHED2	
DASHEDX2	
DIVIDE	
DIVIDE2	
DIVIDEX2	
DOT	
DOT2	
DOTX2	
HIDDEN	
HIDDEN2	
HIDDENX2	
PHANTOM	
PHANTOM2	
ACAD_ISO02W100	
ACAD_ISO03W100	
ACAD_ISO04W100	
ACAD_ISO05W100	
ACAD_ISO06W100	
ACAD_ISO07W100	
ACAD_ISO08W100	
ACAD_ISO09W100	
ACAD_ISO10W100	
ACAD_ISO11W100	
ACAD_ISO12W100	
ACAD_ISO13W100	
ACAD_ISO14W100	
ACAD_ISO15W100	
FENCELINE1	
FENCELINE2	
TRACKS	
BATTING	
HOT_WATER_SUPPLY	HW HW HW
GAS_LINE	GAS GAS GAS
ZIGZAG	

2. Click the word `Continuous` in the Hidden layer listing (under the Linetype column).

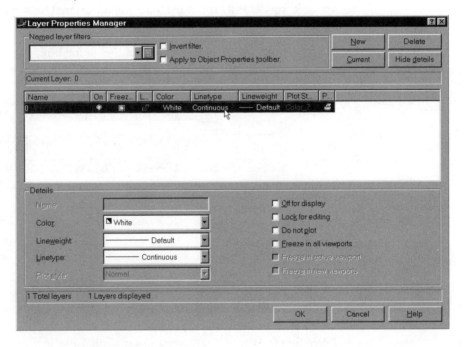

The Select Linetype dialog box appears.

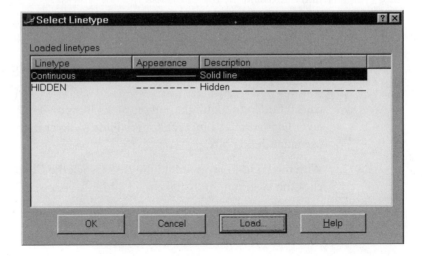

This dialog box offers a list of linetypes to choose from. In a new file, such as the drawing file you've been working on, only the Continuous linetype is available, by default. You must load any additional linetype you may want to use.

3. Click the Load button at the bottom of the dialog box. The Load or Reload Linetypes dialog box appears.

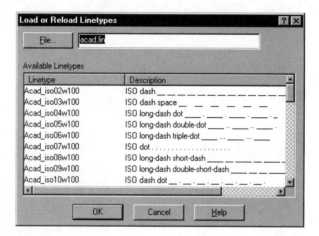

Notice that the list of linetype names is similar to the layer list. You can sort the names alphabetically or by description by clicking the Linetype or Description buttons at the top of the list.

4. In the Available Linetypes list, scroll down to locate the Hidden linetype, click it, and then click OK.

5. Notice that the Hidden linetype is now added to the linetypes available in the Select Linetype dialog box.

6. Click Hidden to highlight it, then click OK. Now Hidden appears in the layer listing under Linetype for the Hidden layer in the Layer Properties Manager dialog box.

7. With the Hidden layer still highlighted, click the Current button, (or double-click the layer name) to make the Hidden layer current.

8. Click OK to exit the Layer Properties Manager dialog box. The red lines for the flanges have become dashed lines.

Controlling Linetype Scale

Although you've designated that this line is to be a hidden line, it may appear to be solid. Zoom into a small part of the line, and you'll see that the line is indeed as you specified.

Because you're working at a scale of 1"=1", you may adjust the scale of your line-types to see them more easily on the screen. You'll have to return the linetypes to your plot scale before you print the drawing. This is accomplished in the Linetype Manager dialog box.

1. Choose Format ➢ Linetype. The Linetype Manager dialog box appears. Click the Show Details button to expand the Details area of the dialog box.

NOTE   You can also use the Ltscale system variable to set the linetype scale. Type **ltscale↵**, and at the LTSCALE Enter new linetype scale factor <1.0000>: prompt, enter **12↵**.

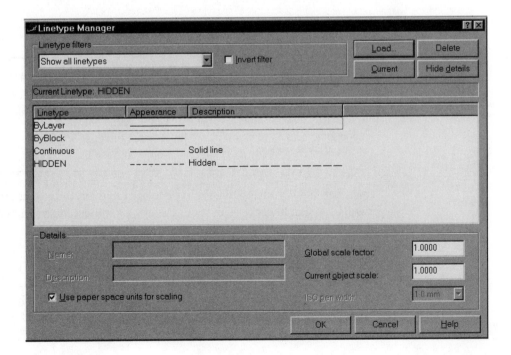

2. Double-click the Global Scale Factor input box, and then type **0.5**. This is the scale conversion factor for a 1"=0.5" scale (see Table 3.3).

3. Click OK. The drawing regenerates, and the red lines are displayed in the linetype and at the scale you designated. Your drawing looks like Figure 4.18.

4. Choose File ➤ Save to record your work up to now.

FIGURE 4.18:

The hidden linetype at a smaller scale

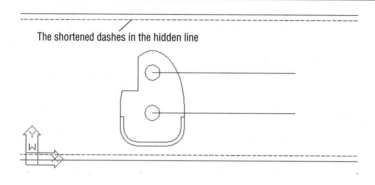

The shortened dashes in the hidden line

TIP

If you change the linetype of a layer or object but the object remains a continuous line, check the Ltscale system variable. It should be set to your drawing scale factor. If this doesn't work, set the Viewres system variable to a higher value (see Chapter 6, "Enhancing Your Drawing Skills"). Also, linetype scales act differently depending on whether you're in Model Space or Paper Space. See Chapter 9, "Advanced Productivity Tools," for more on Model and Paper Space.

Remember that if you assign a linetype to a layer, everything you draw on that layer will be of that linetype. This includes arcs, polylines, circles, dimension lines, and traces. As explained in the "Setting Individual Colors, Linetypes, and Linetype Scales" sidebar at the end of this section, you can also assign different colors and linetypes to individual objects, rather than relying on their layer assignment to define color and linetype. However, you may want to avoid assigning colors and linetypes directly to objects until you have some experience with AutoCAD and a good grasp of your drawing's organization. You should also check your company's drafting standards for any restrictions.

In the last exercise, you changed the global linetype scale setting. This affected all noncontinuous linetypes within the current drawing. You can also change the linetype scale of individual objects, using the Properties button in the Standard toolbar. Alternately, you can set a default linetype scale for all new objects, with the Current Object Scale option in the Linetype Manager dialog box.

When individual objects are assigned a linetype scale, they are still affected by the global linetype scale set by the Ltscale system variable. For example, say you assign a linetype scale of two to the hidden lines in the previous example. This scale would then be multiplied by the global linetype scale of 12, for a final linetype scale of 24.

TIP   The default linetype scale setting for individual objects can also be set using the Celtscale system variable. Once this is set, it will affect only newly created objects. You must use the Properties tool to change the linetype scale of individual existing objects.

If the objects you draw appear in a different linetype from that of the layer they are in, check the default linetype. Click the linetype drop-down list and select ByLayer. Also, check the linetype scale of the object itself, using the Properties button. A different linetype scale can make a line appear to have an assigned linetype that may not be what you expect.

The display of linetypes also depends on whether you're in Paper Space or Model Space. If your efforts to control linetype scale have no effect on your linetype's visibility, you may be in Paper Space. See Chapter 14, "Printing and Plotting," for more information on how to control linetype scale while in Paper Space.

Setting Individual Colors, Linetypes, and Linetype Scales

If you prefer, you can set up AutoCAD to assign specific colors and linetypes to objects, instead of having objects take on the color and linetype settings of the layer on which they reside. Normally, objects are given a default color and linetype called ByLayer, which means each object takes on the color or linetype of its assigned layer. (You've probably noticed the word ByLayer in the Object Properties toolbar and in various dialog boxes.)

Continued on next page

Use the Properties tool in the Standard toolbar to change the color or linetype of existing objects. This tool opens a dialog box that lets you set the properties of individual objects. For new objects, use the Color tool in the Object Properties toolbar to set the current default color to red (for example), instead of ByLayer. The Color tool opens the Select Color dialog box, where you select your color from a palette. Then everything you draw will be red, regardless of the current layer's color.

For linetypes, you can use the linetype drop-down list in the Object Properties toolbar to select a default linetype for all new objects. The list only shows linetypes that have already been loaded into the drawing, so you must have loaded a linetype before you can select it.

Another possible color and linetype assignment is ByBlock, which is also set with the Properties button. ByBlock makes everything you draw appear white, until you turn those drawing objects into a block and then insert the block on a layer with an assigned color. The objects then take on the color of that layer. This behavior is similar to that of objects drawn on Layer 0. The ByBlock linetype behaves like the ByBlock color.

Finally, if you want to set the linetype scale for each individual object, instead of relying on the global linetype scale (the Ltscale system variable), you can use the Properties tool to modify the linetype scale of individual objects. In place of using the Properties tool, you can set the Celtscale system variable to the linetype scale you want for new objects.

Remember, you should stay away from assigning colors and linetypes to individual objects until you're comfortable with AutoCAD; and even then, use color and linetype assignments carefully. Other users who work on your drawing may have difficulty understanding your drawing's organization if you assign color and linetype properties indiscriminately.

Keeping Track of Blocks and Layers

The Insert dialog box and the Layer Properties Manager dialog box let you view a list of the blocks and layers available in your drawing by listing them in a window. The Layer Properties Manager dialog box also includes information on the status of layers. However, you may forget the layer on which an object resides. The List button in the Standard toolbar enables you to get information about individual objects, as well as blocks.

1.   Click and hold the Distance tool so that a flyout appears.

2. Drag the pointer down to the List tool and select it.

TIP   If you just want to check quickly what layer an object is on, click the object. Its layer, color setting, and linetype will appear in the drop-down lists of the Object Properties toolbar. You can also click the Properties tool in the Standard toolbar, and then click the object in question. You'll get a dialog box showing you the basic properties of the object, including its layer setting.

---

3. At the object-selection prompt, click the hidden line and then press ⏎. The AutoCAD Text window appears.

4. In the Text window, you'll see a listing that shows not only the layer that the line is on, but also its space, insertion point, name, color, linetype, rotation angle, and scale.

Using the Log File Feature

Eventually, you'll want a permanent record of block and layer listings. This is especially true if you work on drawing files that are being used by others and your company doesn't have a layer standard. Here's a way to get a permanent record of the layers and blocks within a drawing. Use the Tools ➢ Options ➢ Files tab and choose Log File Location to change the name and location of the log file. The default log filename is acad.log.

If you don't turn off the log file when you're finished, a new log session begins each time you start AutoCAD. (Dashed lines separate sessions in the log file.) Here's a short exercise using the log file.

1. Minimize the Text window (click the Minimize button in the upper-right corner of the Text window, or press the F2 key).

2. Choose Tools ➤ Options or type **op**↵. The Options dialog box appears.

3. Click the Open and Save tab at the top of the dialog box.

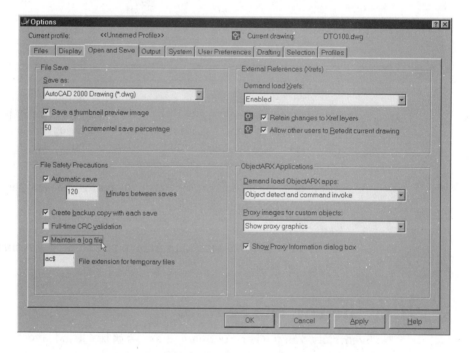

4. Click the Maintain a Log File check box in the bottom-left area of the dialog box. A check mark appears in the check box.

5. Click OK. Type **–layer**↵ (don't forget the minus sign) at the command prompt, and then **?**↵↵. The AutoCAD Text window appears, and a listing of all the layers scrolls into view.

6. Press F2 to return to the AutoCAD drawing screen. Then choose Tools ➤ Options to reopen the Options dialog box.

7. Deselect the Maintain a Log File check box, and then click OK. (If you don't do this, the file will be locked by AutoCAD, and you won't be able to open it.)

8. Select Tools ➤ Options ➤ Files tab and choose Log File Location. Click the plus sign (+) to see where the acad.log file has been placed. Use the Windows Notepad to open the AutoCAD log file, acad.log, located in the indicated directory. You'll see that the layer listing is recorded there.

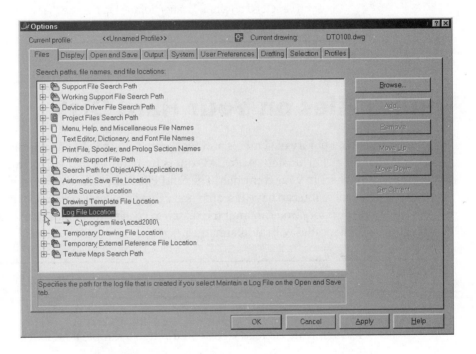

With the Log File feature, you can record virtually anything that appears in the command prompt. You can even record an entire AutoCAD session. The log file can also be helpful in constructing script files to automate tasks. (Script files will be discussed in Chapter 19, "Introduction to Customization.") To have a hard copy of the log file, just print it from an application such as Windows Notepad or your favorite word processor.

If you wish, you can arrange to keep the acad.log file in a directory other than the default AutoCAD subdirectory. As mentioned above, this setting is in the Options dialog box under the Files tab. Locate the Log File Location listing in the

Search Path, File Names, and File Locations list box. Click the plus sign (+) next to this listing. You'll see the location of the acad.log file. You can double-click the acad.log file location listing to open a Browse for Folder dialog box and specify a different location for your log file. This dialog box is a typical Windows file dialog box.

TIP	Once you've settled on a disk location for the log file, use the windows Explorer to associate the log file with the Windows Notepad or Write application.

Finding Files on Your Hard Disk

As your library of symbols and files grows, you might have difficulty keeping track of them. Fortunately, AutoCAD offers two utilities that let you quickly locate a file anywhere in your computer. The Find File utility searches your hard disk for specific files. You can have it search one drive or several, or you can limit the search to one directory. You can limit the search to specific filenames or use DOS wildcards to search for files with similar names.

NOTE	New in AutoCAD 2000, is the AutoCAD Design Center. We'll use this tool in Chapter 19, "Introduction to Customization."

The following exercise steps you through a sample Find File task.

1. Choose File ➤ Open. In the Select File dialog box, click Find File to display the Browse/Search dialog box.

TIP	Find File can also be accessed using the Find File button in any AutoCAD file dialog box, including the File option of the Insert Block tool. Find File can help you access and maintain your symbols library.

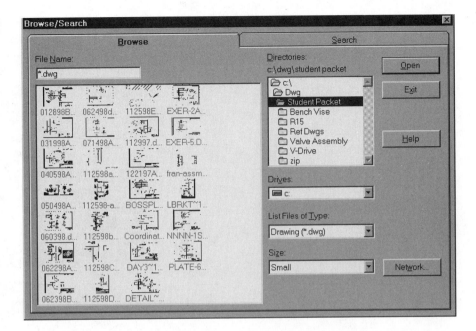

The Browse/Search dialog box has two tabs: Browse and Search. On the Browse page are all of the drawings in the current directory, displayed as thumbnail views so you can identify them easily. You can open a file by double-clicking its thumbnail view or by entering its name in the File Name input box at the top. The Size drop-down list lets you choose the size of the thumbnail views shown in the list box—small, medium, or large. You can scroll through the views using the scroll bars at the bottom of the list box.

2. Click the Search tab to open the Search page. Use the Search Pattern input box to enter the name of the file for which you wish to search. The default is *.dwg, which will cause Find File to search for all AutoCAD drawing files. Additional input boxes help you set a variety of other search criteria, such as the date stamp of the drawing, the type of drawing, and the drive and path to be searched. For now, leave these settings as they are.

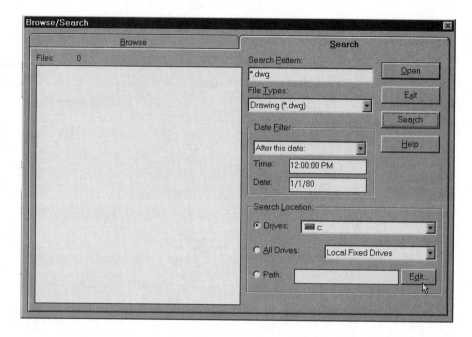

3. Click the Search button. In a few seconds, a listing of files that meet the criteria specified in the input boxes appears in the Files list on the left, along with thumbnail views of each file. You can click a filename in the list, and then click the Open button to open the file in the drawing editor.

4. When you're ready, click Exit to leave the Browse/Search dialog box, and then click Cancel to exit the Select File dialog box.

NOTE   In the Browse/Search dialog box, a drawing from a pre-Release 13 version of AutoCAD will be represented as a box with an X through it.

In this exercise, you performed a search using the default settings. These settings caused AutoCAD to search for files with the .dwg filename extension, created after 12:00 midnight on January 1, 1980, on the C:\ drive.

Here are descriptions of the items in the Browse/Search dialog box.

Search Pattern   Lets you give specific filename search criteria using wild-card characters.

File Types   Lets you select from a set of standard file types.

Date Filter   Lets you specify a cutoff time and date. Files created before the specified time and date are ignored.

Drives   Lets you specify the drives to search.

All Drives   Lets you search all of the drives on your computer.

Path   Lets you specify a path to search.

Open   Opens the file highlighted in the file list, after a search is performed.

Search   Begins the search process.

Help   Provides information on the use of Browse/Search.

Edit   Opens another dialog box, displaying a directory tree from which you can select a search path.

Inserting Symbols with Drag-and-Drop

If you prefer to manage your symbols library using Windows Explorer or another third-party file manager, you'll appreciate AutoCAD's support for *drag-and-drop*. With this feature, you can click and drag a file from Windows Explorer into the AutoCAD window. You can also drag-and-drop from the Windows Find File or Folder utilities. AutoCAD will automatically start the Insert command to insert the file. Drag-and-drop also works with a variety of other AutoCAD support files. Additionally, AutoCAD supports drag-and-drop for other types of data from applications that support Microsoft's ActiveX technology. Table 4.3 shows a list of files with which you can use drag-and-drop and the functions associated with them.

TABLE 4.3: AutoCAD Support for Drag-and-Drop

File Type	Command Issued	Function Performed When File Is Dropped
.dxf	Dxfin	Imports .dxf files
.dwg	Insert	Imports drawing files
.txt	Mtext	Imports text via Mtext

Continued on next page

TABLE 4.3 CONTINUED: AutoCAD Support for Drag-and-Drop

File Type	Command Issued	Function Performed When File Is Dropped
.lin	Linetype	Loads linetypes
.mnu, .mnx	Menu	Loads menus
.ps	Psin	Imports PostScript files
.psb, .shp, .shx	Style	Loads fonts or shapes
.scr	Script	Runs script
.lsp	(Load..)	Loads AutoLISP routine
.exe, .exp	(Xload..)	Loads ADS application

TIP   You can also drag-and-drop from folder shortcuts placed on your Desktop or even from a Web site.

If You Want to Experiment...

You might want to experiment with creating some other types of symbols. Perhaps you want to start thinking about a layering system that suits your particu-lar needs.

Open a new file called Mytemp. In it, create layers named 1 through 8 and assign each layer the color that corresponds to its number. For example, give layer 1 the color 1 (red), layer 2 the color 2 (yellow), and so on. Draw each part shown in Figure 4.19, and turn each part into a file on disk using the Export (File ➤ Export) or the Wblock command. When specifying a filename, use the name indicated for each part in the figure. For the insertion point, use the points indicated in the figure. Use the Osnap modes (see Chapter 2) to select the insertion points.

When you're finished creating the parts, exit the file using File ➤ Close, and then open a new file using File ➤ New. Set up the drawing as an engineering drawing with a scale of 1/4"=1" on an 11"×17" sheet. Create the drawing in Figure 4.20 using the Insert Block command to place your newly created parts.

FIGURE 4.19:

A typical set of symbols

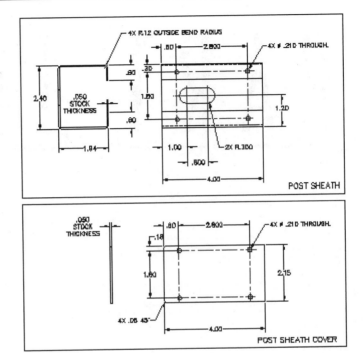

FIGURE 4.20:

Draw this part using symbols; create and insert a block for the tapped holes.

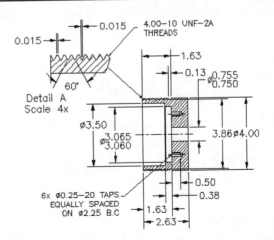

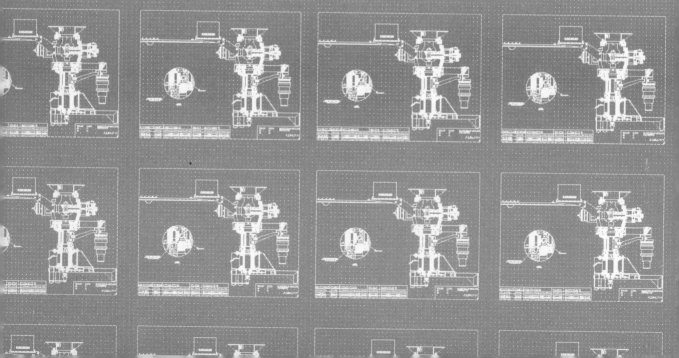

PART II

Building on the Basics

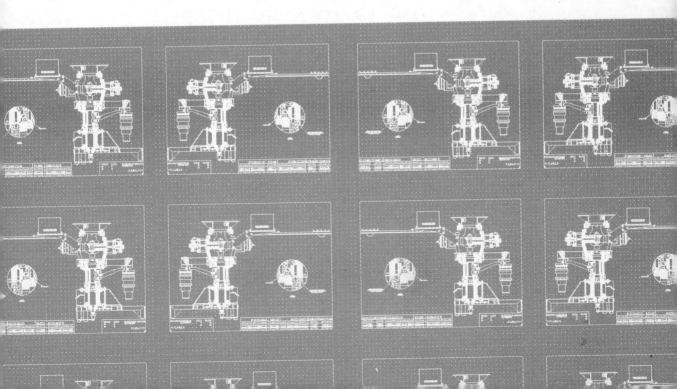

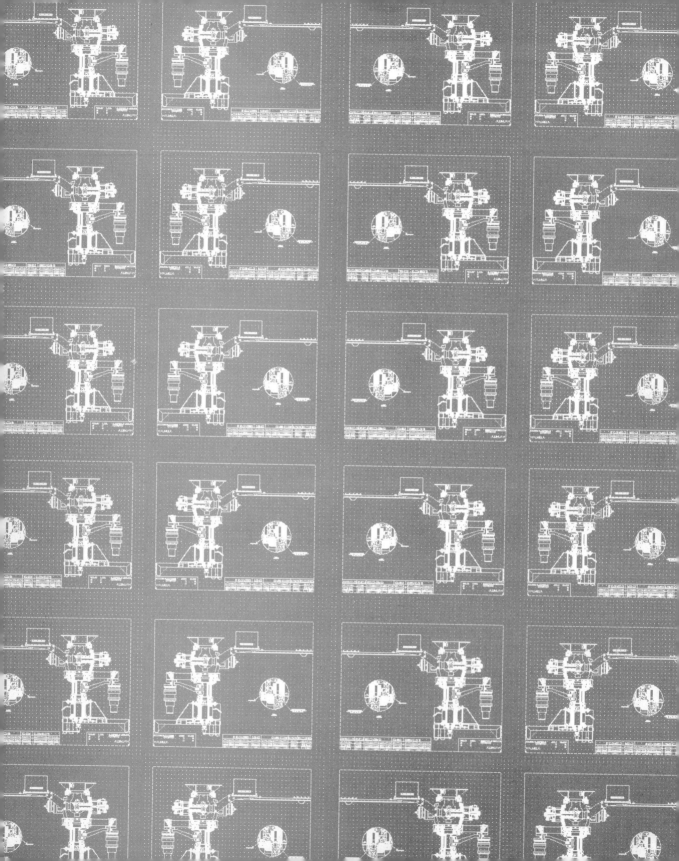

CHAPTER

FIVE

5

Editing for Productivity

■ Using Existing Drawings as Prototypes for New Drawings

■ Making Polar and Rectangular Arrays

■ Using Construction Lines

■ Marking Regular Intervals Along a Line or Arc

■ Drawing Parallel Lines

■ Removing Unused Elements Such As Blocks and Layers

■ Understanding Methods for Constructing New Drawings

In this chapter, as you add more detail to your drawing you'll explore some of the ways to exploit existing files and objects. For example, you'll use existing files as prototypes for new files, eliminating the need to set up layers, scales, and sheet sizes for similar drawings. With AutoCAD you can also duplicate objects in multiple arrays. In Chapter 3, "Learning the Tools of the Trade," and Chapter 4, "Organizing Your Work," you've already seen how to use the Osnap overrides on objects to locate points for drawing complex forms. Here, we'll look at other ways of using lines to aid your drawing.

Because you'll begin to use Zoom more in the exercises of this chapter, we'll review this command as we go along. We'll also introduce you to the Pan command—another tool to help you get around in your drawing.

You're already familiar with many of the commands. So, rather than going through every step of the drawing process, we'll sometimes ask you to copy the drawing from a figure, using notes and dimensions as guides and putting objects on the indicated layers. If you have trouble remembering a command you've already learned, just go back and review the appropriate section of the book.

Creating and Using Templates

If you're familiar with the Microsoft Office suite, you're probably familiar with *templates*. A template is a file that is already set up for a specific application. For example, you might want to have letters set up in a way that is different from reports or invoices. You can have a template for each type of document; each template is set up for the needs of that document. That way, you don't have to spend time reformatting each new document that you create.

Similarly, AutoCAD offers templates, which are drawing files that contain custom settings designed for a particular function. Out of the box, AutoCAD offers templates for ISO, ANSI, DIN, and JIS standard drawing formats. But you aren't limited to these canned templates. You can create your own templates set up for your particular style and method of drawing.

The ISO and ANSI standards are common standards for engineers in the United States. The ISO (International Standards Organization) sets international standards in all fields except electrical and electronic engineering. ANSI (American National Standards Institute) standards pertain to mechanical design and manufacturing, as well as to programming languages. DIN (German Industrial Standards) and JIS (Japanese Industrial Standards) are important to designers with projects where these standards apply.

If you find that you use a particular drawing setup frequently, you can turn one or more of your typical drawings into a template. For example, you may want to create a set of drawings with the same scale and sheet size as an existing drawing. By turning a typical drawing into a template, you can save a lot of setup time for subsequent drawings. Templates can contain partially created objects that are used over again. Even libraries of blocks (discussed in Chapter 4, "Organizing Your Work") can be stored in template drawings.

Creating a Template

The following exercise guides you through creating and using a template drawing for a version of the bar and bracket. You'll use the DT0100 bracket that you drew and a new, thinner bar. Because the new version will use the same layers, settings, scale, and sheet size as the new version drawing, you can use the DT0100 file as a prototype.

1. Start AutoCAD in the usual way.

2. Choose File ➤ Open.

3. At the Select File dialog box, locate the drawing you created in the last chapter. You can also use the file DT0100.dwg from the companion CD-ROM.

If you continued this exercise from Chapter 4, "Organizing Your Work," you may already have the drawing from that chapter open. New in AutoCAD 2000 is the ability to open more than one drawing during an editing session. If the drawing you've chosen is already open, this could be confusing. So, how do you check for currently open files? The answer is quite simple. In fact, as with any Windows-compliant software, AutoCAD provides a Window pull-down menu. There you'll see each of the files you have opened. If you wish, you may choose File ➤ Close to close the file you're currently editing, or use the Window menu to make another file current, and then close it.

4. Click the Erase button in the Modify toolbar; then type **all.⌐⌐**. This will erase all of the objects that make up the current drawing.

5. Choose File ➤ Save As. Then, at the Save Drawing As dialog box, open the Save As Type drop-down list and select Drawing Template File (*.dwt). The file list will change to display the current template files in the \Template folder.

NOTE   When you choose the Drawing Template File option in the Save Drawing As dialog box, AutoCAD automatically opens the folder containing the template files. The standard installation creates the folder named Template to contain the template files. If you wish to place your templates in a different folder, you can change the default template location using the Options dialog box (choose Tools ➤ Options). Use the Files tab and then double-click Drawing Template Drawing File Location in the list. Double-click the folder name that appears just below Drawing Template Drawing File Location. Then select a new location from the Browse for Folder dialog box that appears. Select Apply, and then exit from Options by clicking OK.

6. Double-click the File Name input box and enter the name **versions**.

7. Click Save. The Template Description dialog box appears.

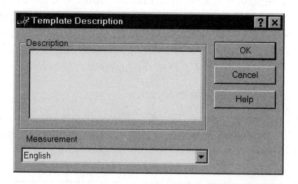

8. Enter the following description:

    Bar and bracket template for multiple versions

9. Click OK. You've just created a template.

Notice that the current drawing is now the template file that you just saved. As with other Windows programs, the File ➤ Save As option makes the saved file current. This also shows that you can edit template files just as you would regular drawing files.

WARNING Selecting AutoCAD File ➤ Save As allows you to change the drawing's name, storage location, and in this case, the file type. Unfortunately, it also leaves you in the newly renamed or relocated file—in this example, the .dwt file. As you may have already realized, this could cause confusion, so choose File ➤ Close to exit this file. If you have other open files, you'll automatically go to one of them.

Using a Template

Now let's see how a template is used. You'll use the template you just created as the basis for a new drawing you'll work on in this chapter.

1. Choose File ➤ New.

2. At the Create New Drawing dialog box, click the Use a Template button. The Select a Template list box appears, along with a file preview window.

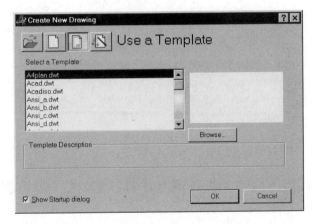

3. Click versions.dwt in the Select a Template list box. Normally, the file is displayed in the preview window, but since it is a blank file, nothing is displayed. Also notice that the description you entered earlier in the Template Description dialog box appears in the Template Description area below the list.

TIP   The Browse button in the Create a New Drawing dialog box lets you browse your hard drive to locate other files for use as a template.

4. Click OK. It may not be obvious, but your new file is set up with the same units and drawing limits as the previous drawing. It also contains the two versions of the bracket blocks.

5. Now you need to give your new file a name. Choose File ➢ Save. Because you created this as a new drawing, you'll see the Save Drawing As dialog box. Enter **bar** for the filename and select the appropriate folder in which to save your new bar file.

> **NOTE** Before you named it, the filename was the AutoCAD default of Drawing#—(# represents a sequential number assigned by AutoCAD). If you forget to give your file a name, you can easily rename the file from Windows Explorer or use the Save As option in the File menu to save your work under the correct filename.

6. Click Save to create the bar file and close the dialog box.

You've created and used your own template file. Later, when you've established a comfortable working relationship with AutoCAD, you can create a set of templates that are custom-made to your particular needs.

You don't need to create a template every time you want to reuse settings from another file. You can use an existing file as the basis or prototype for a new file without creating a template. Open the prototype file and then choose File ➢ Save As to create a new version of the file under a new name. You can then edit the new version without affecting the original file.

Copying an Object Multiple Times

Now let's explore the tools that let you quickly duplicate objects. In the next exercise, you'll begin to draw a new bar. This exercise introduces the Array command, which enables you to draw rectangular and circular arrangements of objects.

> **NOTE** An array can be in a circular pattern called a *polar array,* or in a matrix of columns and rows called a *rectangular array.*

Making Polar Arrays

You don't need a circular array to complete the current drawing, but you will draw and discard a circular array to see how this command works. Here, you'll set an undo marker so that you can back up to this point in the drawing process. Auto-CAD either temporarily or permanently keeps a log of the commands that you issue. (We'll discuss the Log feature, which is the permanent record of the log, in Chapter 18, "Getting and Exchanging Data from Drawings.") The ability to undo to a certain point provides the AutoCAD user with the opportunity to iterate a design.

1. Type **undo**↵. No toolbar tool or dialog box exists for the control of the Undo command; this must be done on the command line. At the `Enter the num-ber of operations to undo or [Auto/Control/BEgin/End/Mark/ Back]:` prompt, type **m**↵ to mark this point in the command sequence of this file. Take a look at the following sidebar, "The Undo Feature," to see what the rest of the functions can do for you.

The Undo Feature

The Undo feature allows you to undo parts of your editing session. This can be useful if you accidentally execute a command that destroys part or all of your drawing. Undo also allows you to control how much of a drawing is undone. Typing **Undo** ↵ at the command prompt will return the following response:

```
Enter the number of operations to undo or [Auto/Control/BEgin
/End/Mark/Back]:
```

Enter the number of operations to undo   Allows you to enter a number indicating how many steps or operations to remove.

Auto   Makes AutoCAD view menu macros as a single command. If Auto is on, the effect of macros issued from a menu will be undone regardless of the number of commands the macro contains.

Control   Allows you to turn off the Undo feature to save disk space or to limit the Undo feature to single commands. You're prompted for All, None, or One. All fully enables the Undo feature, None disables Undo, and One restricts the Undo feature to a single command. (This feature was very important when the dominant computer hardware was the PC-AT, but it's not so important today.)

Continued on next page

BEgin, End   Allow you to mark a group of commands to be undone together. Begin marks the beginning of a sequence of operations. All edits after that point become part of the same group. End terminates the group (important for custom macros).

Mark, Back   Allow you to experiment safely with a drawing by first marking a point in your editing session to which you can return. Once a mark has been issued, you can proceed with your experimental drawing addition. Then you can use Back to undo all of the commands back to the point at which the mark was issued.

NOTE   Many commands (such as Line, Pline, and so on) offer an Undo option. The Undo option under a main command will act like the single Undo command and will not offer the options described here.

2. Set the current layer to 0, and toggle on the Snap mode.

NOTE   Because you used the **DT0100** file as a template, the Running Osnaps for End-point, Midpoint, Center, and Intersection are already turned on and available in this new file.

3. Click the Circle tool in the Draw toolbar or type **c↵**.

4. At the Specify center point for circle or [3P/2P/Ttr (tan tan radius)]: prompt, pick a point at coordinate 14,15.

5. At the Specify radius of circle or [Diameter] <3.2609>: prompt, enter **3↵**. The circle appears.

If your circle does not appear on your screen, use Zoom Extents (type **z↵e↵**, use the Zoom tool located in the Standard toolbar, or choose View ➤ Zoom ➤ Extents). Now you're ready to use the Array command to draw a polar array. You'll first draw one line, and then use Array to create the copies.

1. Draw a four-unit long line starting from the coordinate 15.5,15 and ending to the right of that point.

2. Zoom into the circle and line to get a better view. Your drawing should look like Figure 5.1.

3. Click the Array tool in the Modify toolbar or type **ar**↵.

4. At the `Select objects:` prompt, enter **1**↵. This highlights the last object, the line you just drew.

5. Press ↵ to confirm your selection.

6. At the `Enter the type of array [Rectangular/Polar] <R>:` prompt, type **p**↵ to use the Polar Array option.

7. At the `Specify Center point of array:` prompt, pick the center of the circle. You can either use the snap point at the center of the circle at coordinate 14.000,15.000, or use the Center Osnap.

FIGURE 5.1:

A close-up of the circle and line

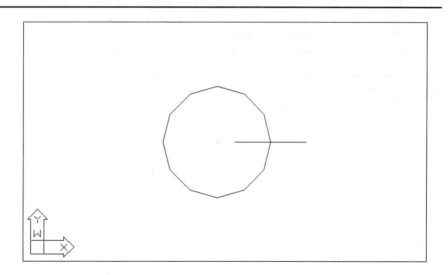

TIP   Remember to access Osnaps other than those set up as Running Osnaps, Shift+right-click, then select the Osnap from the pop-up menu.

WARNING   If you use the Center Osnap, you must place the cursor on the circle, not on the circle's center point.

8. At the `Number of items:` prompt, enter 8↵. This tells AutoCAD you want seven copies plus the original.

9. At the `Specify the angle to fill (+=ccw, -=cw) <360>:` prompt, press ↵ to accept the default. The default value of 360 tells AutoCAD to copy the objects so that they are spaced evenly over a 360° arc. (If you had instead entered 180, the lines would be evenly spaced over a 180° arc, filling only half the circle.)

NOTE If you want to copy in a clockwise (CW) direction, you must enter a minus sign (–) before the number of degrees.

---

10. At the `Rotate objects as they are copied? <Y>:` prompt, press ↵ again to accept the default. The line copies around the center of the circle, rotating as it copies. Your drawing will look like Figure 5.2.

NOTE In step 10, you could have the line maintain its horizontal orientation as it is copied around, by entering n↵. But because you want it to rotate around the array center, accept the default, which is Yes.

FIGURE 5.2:

The completed array

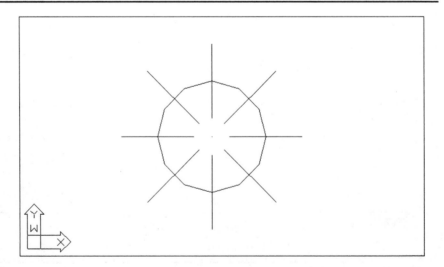

11. Now use the Undo command to return to the point before you created the polar array. Type **undo**↵. At the `Enter the number of operations to`

undo or [Auto/Control/BEgin/End/Mark/Back]: prompt, type **b**↵. You see this response on the command line: ARRAY LINE CIRCLE Mark encountered. It means that AutoCAD searched back through the drawing's history, found the mark that you placed, and undid all of the commands back to that point.

Making Rectangular Arrays

You only drew one hole in the bar in the previous exercise, but to mount the bracket properly you needed two holes per bracket. You need to draw another rectangle that represents the new, thinner bar and two concentric circles that represent a tapped mounting hole. The larger circle will be on the Hidden layer. You'll then insert the second version of the bracket to find the distance between the mounting holes. (You could also open and review the bracket drawing that you created in Chapter 3, "Learning the Tools of the Trade.") You'll finish by creating a rectangular pattern of tapped holes along the bar.

1. Set the current layer to 0, and toggle on the Snap mode.

NOTE   Because you used the previous exercise as a template, the Running Osnaps for Endpoint, Midpoint, Center and Intersection are already turned on and available in this new file.

2. Draw a rectangle 60 units long, and approximately 3.375 units high from point 2,3.

3. Draw a circle using the From Osnap at a point three units from the left end of the bar and one unit up from the bottom of the bar.

4. At the Specify radius of circle or [Diameter]: prompt, type **d**↵ to select Diameter, then enter **.422**↵. The circle represents the minor diameter of the thread.

5. From the Object Properties toolbar, choose the layer drop-down list and select the Hidden layer to make it current.

6. Using the Center Osnap, draw another circle concentric with the first, with a diameter of 0.500. The circle appears in red and consists of a dashed line.

7. Change the current layer to 0 using the layer drop-down list in the Object Properties toolbar.

8. Insert the block bkt_v2 near, but not on, the top-left end of the bar. Your drawing should look like Figure 5.3.

FIGURE 5.3:

The bar, bracket, and circle

Using the Distance Tool

You'll next use the Distance tool to find the distance between the bracket mounting holes. This information is required to create the array of mounting holes along the length of the bar.

The Distance Tool

The Distance tool measures the vector distance and angle between any two points that you select. The information appears on the command line as

```
Distance = 1.625, Angle in XY Plane = 90, Angle from XY Plane = 0
Delta X = 0.000, Delta Y = 1.625, Delta Z = 0.000
```

The current units setting determines the number of decimal places in the report. The positive or negative sign of delta values are returned by the order in which you pick the points. For example, if you pick two points from left to right and from bottom to top, the values will be positive. The angles are also determined by the order of point selection.

1. From the Standard toolbar, choose the Distance tool, or type **dist** ↵.

2. At the Specify first point: prompt, pick the lower circle in the bracket. At the Specify second point: prompt, pick the upper circle. The response looks like this:

```
Distance = 1.625, Angle in XY Plane = 90, Angle from XY Plane = 0
Delta X = 0.000, Delta Y = 1.625, Delta Z = 0.000
```

Using the Array Command

Now you're ready to use the Array command to draw the mounting holes.

1. Click the Array tool in the Modify toolbar or type **array**↵.

2. At the `Select objects:` prompt, select the two circles and press ↵. The prompt states `2 found Select objects:`. Press ↵ to finish selecting objects and begin the next step of the command. The circles do not remain highlighted.

3. At the `Select objects: Enter the type of array [Rectangular/ Polar] <R>:` prompt, type **r**↵ to use the Rectangular Array option.

4. The next prompt is `Enter the number of rows (--) <1>:`. The series of dashes indicates that a row is horizontal. You need two rows for the two mounting holes per bracket, so you should type **2**↵. The next prompt is `Enter the number of columns (|||) <1>:`. The design calls for seven brackets equally spaced on this rail; type **7**↵.

5. From the distance inquiry, you determined the distance between the bracket mounting holes is 1.625. So at the `Enter the distance between rows or specify unit cell (--):` prompt, enter **1.625**↵.

6. You'll have to do a little math to determine the input for the `Specify the distance between columns (|||):` prompt. If the hole pattern is to be centered on the 60-unit bar and the holes are three units in from each end, then you must divide the remaining 54 units by six (the number of spaces between seven evenly-spaced units). This gives a distance of nine units. Enter **9**↵. Your drawing should look like Figure 5.4.

FIGURE 5.4:

The new bar with the hole pattern

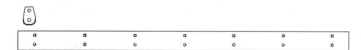

NOTE   The Array command now "remembers" whether you last used the Polar or Rectangular Array option and offers that option as the default.

AutoCAD usually draws a rectangular array from bottom to top, and from left to right. You can reverse the direction of the array by giving negative values for the distance between columns and rows.

TIP   At times you may want to do a rectangular array at an angle. To accomplish this you must first set the Snap Angle setting in Tools ➤ Drafting Settings ➤ Snap and Grid to the desired angle. Then proceed with the Array command. Another method is to set the UCS to the desired angle. See Chapter 11, "Introducing 3D," for more information.

You can also use the cursor to graphically indicate an *array cell* (see Figure 5.5). An array cell is a rectangle defining the distance between rows and columns. You may want to use this option when an object is available as a reference from which to determine column and row distances. For example, consider the spacing between the holes in the bracket (1.625); all you need is the distance between the brackets (9). Now, select the circle using a Center Osnap and then type a relative coordinate **@9,1.625** ↵. This unit cell spacing completes the array.

FIGURE 5.5:

An array cell

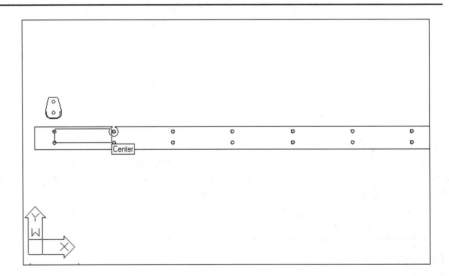

Fine-Tuning Your View

Notice that most of the bracket does not appear on your screen. To move the view over, you can use the Pan command. Pan is similar to Zoom in that it changes your view of the drawing. However, Pan does not alter the magnification of the view the way Zoom does. Rather, Pan maintains the current magnification while moving your view across the drawing, just as you would pan a camera across a landscape.

NOTE   Pan is especially helpful when you have magnified an area to do some editing, and you need to get to part of the drawing that is near your current view.

TIP   If you're not too fussy about the amount you want to zoom out, you can choose View ➤ Zoom ➤ Out to quickly enlarge your viewing area.

You've been looking at an overall view of the bar. Zoom in on the bracket and holes at the left end of the bar.

1. From the Standard toolbar, select the Zoom Window tool or type **z↵**. Pick a point slightly below and to the left of the leftmost set of holes. For the second point, pick a point slightly above and to the right of the bracket. Notice that you didn't need to enter **w↵** to specify a windowed zoom. The window option is the default.

The Zoom Flyout

There are a few tools in the Standard toolbar that offer more than one option. The Zoom tool, located on the Standard toolbar to the right of the Zoom Realtime tool, is one of these. You might have to look closely to see a small, southeast-pointing, black triangle on the tool. Only a few tools have this triangle, which indicates that a flyout toolbar can be accessed by depressing this tool.

Move your cursor to the Zoom tool and press the left mouse button, holding it down. (If you just click the tool button, you'll select the tool itself.) When you hold down the left mouse button, the flyout toolbar is displayed, and if you drag the mouse, still pressing the left button, you'll see tool tips appear to identify various tools. These are called *flyout tools*. To select a flyout tool, release the mouse button over the tool that you want.

The last flyout tool that you used will become the default tool, shown in place of the original tool on the Standard toolbar. This last-used feature is designed to make the most recently used tool easy to use again. This feature also has the disconcerting side effect of making the toolbar look different each time you use a distinct flyout tool. If you don't see Zoom Window as the default tool in the Standard toolbar, click whatever Zoom tool appears to the right of the Zoom Realtime tool. Then hold down the left mouse button

Continued on next page

until you see the flyout tools. Continue to hold down the left mouse button, drag the cursor down until you highlight the Zoom Window tool, then release the mouse button. If you're not sure which tool is which, keep pressing the left mouse button and hover the cursor over the tool button a moment, until the tool tip appears.

To activate the Pan command, follow these steps:

1. Click the Pan Realtime tool in the Standard toolbar or type **p**↵. A small, hand-shaped cursor appears in place of the AutoCAD cursor.

2. Place the hand cursor in the center of the drawing area, then click and drag it downward and to the left. The view follows the motion of your mouse.

3. Continue to drag the view until it looks similar to Figure 5.6; then let go of the mouse.

FIGURE 5.6:

The panned view of the bar

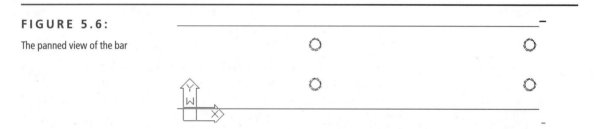

4. Now right-click the mouse. A pop-up menu appears.

NOTE The Pan/Zoom pop-up menu also appears when you right-click your mouse during the Zoom Realtime command.

5. Select Zoom from the list. The cursor changes to the Zoom Realtime cursor.

6. Now place the cursor close to the top of the screen and click and drag the cursor downward to zoom out until your view looks like Figure 5.7. You may need to click and drag the Zoom cursor a second time to achieve this view.

7. Right-click again, then choose Exit from the pop-up menu.

NOTE To exit the Pan Realtime or Zoom Realtime command without opening the pop-up menu, press the Esc key.

FIGURE 5.7:

The final view of the bar holes and bracket

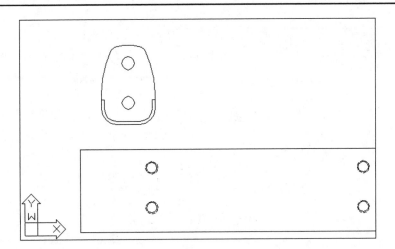

This exercise showed you how you can fine-tune your view by easily switching between Pan Realtime and Zoom Realtime. Once you get the hang of these two tools working together, you'll be able to quickly access the best view for your needs. Each of the other options in the pop-up menu—Zoom Window, Zoom Previous, and Zoom Extents—performs the same function as the tool with the same name in the Standard toolbar and View pull-down menu.

While we're on the subject of display tools, don't forget the scroll bars to the right and bottom of the AutoCAD drawing area. They work like any other Windows scroll bar, offering a simple way to move up, down, left, or right in your current view, even in the middle of a command.

TIP   If the scroll bars don't appear in AutoCAD, go to the Display tab of the Options dialog box (Tools ➤ Options) and make sure the Display Scroll Bars in Drawing Window option is checked. If you prefer to turn off the scroll bars, make sure this option is unchecked.

This would be a very good time to save your work. You could take a break before going on to the next exercise.

Making Random Multiple Copies

The Draw ➤ Array command is useful when you want to make multiple copies in a regular pattern. But what if you need to make copies in a random pattern? You have two alternatives for accomplishing this: the Copy command's Multiple option and the Grips Move option.

To use the Copy command to make random multiple copies:

1. Click the Copy Object tool in the Modify toolbar or type **cp**↵.

2. At the `Select objects:` prompt, select the objects you want to copy and press ↵ to confirm your selections.

3. At the `Specify base point or displacement, or [Multiple]:` prompt, enter **m**↵ to select the Multiple option.

4. At the `Specify base point:` prompt, select a base point as usual.

5. At the `Specify second point of displacement or <use first point as displacement>:` prompt, select a point for the copy. You will be prompted again for a second point, allowing you to make yet another copy of your object.

6. Continue to select points for more copies as desired.

7. Press ↵ to exit the Copy command when you are done.

Continued on next page

When you use the Grips feature to make random multiple copies, you get an added level of functionality because you can also rotate, mirror, and stretch copies by using the pop-up menu (right-click while a grip is selected). Of course, you must have the Grips feature turned on; it's usually on by default, but you may find yourself on a system that has it turned off for some reason.

1. Press Esc twice to make sure you're not in the middle of a command; then select the objects you want to copy. Grips will be turned on (in blue).

2. Click a grip point as your base point. (It will turn red.)

3. Right-click your mouse and select Move from the pop-up menu.

4. Right-click again and select Copy.

5. Click the location for the copy. Notice that the rubber-banding line persists and you still see the selected objects follow the cursor.

6. If desired, click other locations for more copies.

Finally, you can make square-arrayed copies using grips by following steps 1-3 above, but instead of step 4, Shift+click on a copy location. Continue to hold down the Shift key and select points. The copies will snap to the angle and distance you indicate with the first Shift+select point. Release the Shift key and you can make multiple random copies.

Developing Your Drawing

When using AutoCAD, you first create the most basic forms of your drawing; then you refine them. In this section, you'll create two views of the original bracket—top and side—that demonstrate this process in more detail.

You'll also further examine how to use existing files as blocks. In Chapter 4, "Organizing Your Work," you inserted a file into another file. There is no limit to the size or number of files you can insert. As you may have guessed already, you can also *nest* blocks; that is, you can create blocks that contain other blocks.

Importing Settings

In this exercise, you'll use the Versions template. Then you'll import the bracket, thereby importing the layers and blocks contained in the file.

As you go through this exercise, observe how the drawings begin to evolve from simple forms to complex, assembled forms.

1. Open a new file using the Versions template. If you skipped creating the template, use the template file named Versions on the companion CD-ROM.

2. Ensure that Layer 0 is the current layer. Use the Block Insert tool to insert the file DT0100. This file is also available on the companion CD-ROM.

3. The inserted block comprises both brackets. You only need the bracket on the left for this exercise. Use the Explode tool in the Modify toolbar to explode the block.

TIP
Did you notice that selecting one of the brackets exploded both of them? They were nested in the block DT0100 created when you inserted the drawing of the same name.

---

4. Erase the right-hand bracket. Use a window to surround the bracket, which activates the grips. Then select the Erase tool from the Modify toolbar.

5. Save the new file as bkt_v1.

By using the Versions template to start this drawing, you are using the layer and Running Osnaps that you've set in that template.

Using the Quick Setup Wizard

You can use Quick Setup *wizard* to start a new drawing. A wizard is an aid designed to assist users by prompting them for information. You will find wizards in many Windows 95/98/NT programs. AutoCAD 2000's Setup wizard is a combination of the File command and the New option, with settings for Limits and Units. In prior releases of AutoCAD, it was necessary to set these system variables in each new drawing or create a template file that already had these system variables set. AutoCAD would then inherit these settings at the start of a new drawing. The difference between the quick and the advanced versions of the Setup wizard is that the advanced version gives you access to more unit-setting options. The drawback to using this feature is that the wizard does not retain any of your settings, so the next time that you use the wizard you will find that the settings have reverted to the original values (the defaults) that shipped with AutoCAD.

Let's create a new drawing with the Quick Setup wizard.

1. Choose File ➤ New.

2. In the Create New Drawing dialog box, click Use a Wizard. You'll see two options in the Select a Wizard list box.

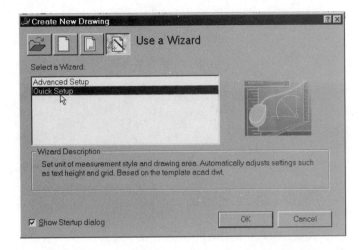

3. Choose Quick Setup and then click OK. The Quick Setup dialog box appears. Notice the two listed steps at the left side labeled Units and Area. The Decimal radio button is the default.

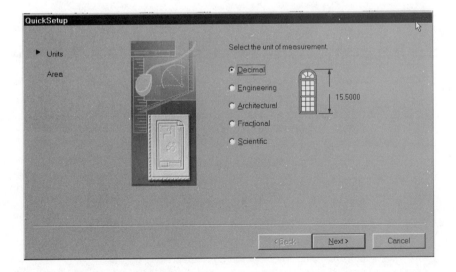

4. Click the Next button.

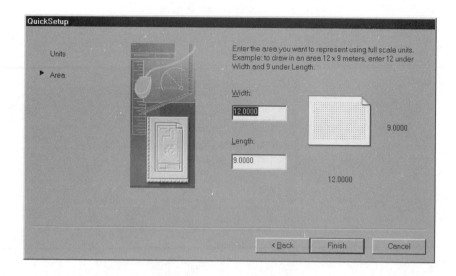

5. Enter **34** in the Width input box and **22** in the Length input box. These are the appropriate dimensions for a "D" size ANSI drawing engineering sheet. Be sure to use the Tab key or mouse to move between input boxes; pressing ↵ will dismiss the dialog box.

6. Click Finish.

Importing Settings from External Reference Files

As explained in the sidebar, "An Alternative to Blocks," in Chapter 4, you can use the External Reference (Xref) Attach option to use another file as a background or Xref file. Xref files are similar to blocks, except that they do not actually become part of the current drawing's database; neither do the settings from the cross-referenced file automatically become part of the current drawing.

If you want to import layers, linetypes, text styles, and so forth from an Xref file, you must use the Xbind command, which you'll learn more about as you work through this book. Xbind allows you to attach dimension-style settings, layers, linetypes, or text styles from a cross-referenced file to the current file.

You can also use Xbind to turn a cross-referenced file into an ordinary block, thereby importing all of the new settings contained in that file.

See Chapter 9, "Advanced Productivity Tools," for a more detailed description of how to use the Xref and Xbind commands.

Using and Editing Lines

You'll draw lines in most of your work, so it's important to know how to manipulate lines to your best advantage. In this section, you'll look at some of the more common ways to use and edit these fundamental drawing objects. The following exercises show you the process of drawing lines, rather than just how individual commands work.

Roughing In the Line Work

The bracket you inserted in the last section had only one view. In the next exercise, you'll draw the side view. The dimensions for this view are shown in Figure 5.8.

FIGURE 5.8:

The bracket with a side view and dimensions

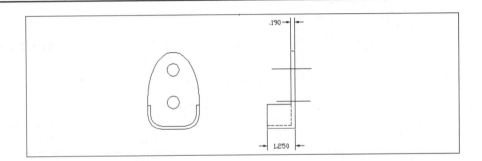

You may notice that some of the arcs in your bracket drawing are not smooth. Don't be alarmed; this is how AutoCAD displays arcs and circles in enlarged views. The arcs will be smooth when they are plotted. If you want to see them now as they actually are stored in the file, you can regenerate the drawing by typing **regen** ↵ at the command prompt. We'll look more closely at regeneration in Chapter 6, "Enhancing Your Drawing Skills."

1. Select 0 from the layer drop-down list in the Object Properties toolbar to make 0 the current layer.

2. Choose Draw ➤ Line or type **l**↵.

3. Shift+right-click to open the Osnap menu; then select From. This option lets you select a point relative to another point.

4. At the _from Base point: prompt, open the Osnap menu again and using the Endpoint Osnap, click the lower-right corner of the bracket (see Figure 5.9). Nothing appears in the drawing area yet. You'll see the prompt <Offset>:.

FIGURE 5.9:

This point is the lower-right corner of the bracket.

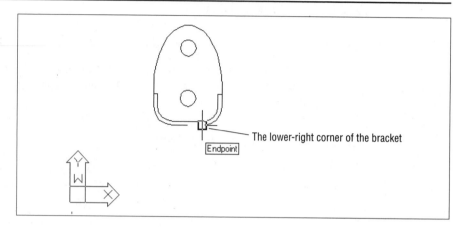

The lower-right corner of the bracket

Endpoint

5.   Type @5<0.↵. Now a line starts five units to the right of the corner of the bracket.

6.   To draw the line up, use a point filter. At the `Specify next point or [Undo]`: prompt, Shift+right-click to open the Osnap menu. Select Point Filters ➤ .Y (AutoCAD will only use the Y value of the next point that you pick).

TIP

Point filters can be very valuable in saving the experienced AutoCAD user a good deal of layout time. You can use the Shift+right-click method to pop up the Osnap menu, or at the `To or From point:` prompt, you can type a period (**.**) followed by the filter that you would like to use. You can use the X, Y, Z component or combinations XY, XZ, or YZ components of the next point that you pick. AutoCAD will then prompt you for a point in the missing direction.

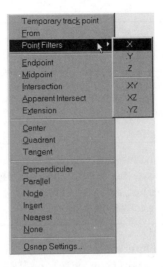

7. At the .Y of: prompt, move the cursor until you see the triangle marker for the top of the bracket and click the left mouse button. To complete the command, move the cursor to any point above the top right of the bracket and press ↵. Your drawing should look like Figure 5.10.

FIGURE 5.10:

The line and Osnap locations using a point filter

If you don't select the correct point when using a point filter, the option is discontinued and you must select the filter again. If you find the point filter too difficult to use, try to draw a line from the midpoint of the bottom of the bracket to the quadrant of the ellipse. Then move the line five units to the right.

In the preceding exercise, the From Osnap allowed you to specify a point in space relative to the lower edge of the bracket. The point filter lets you specify the second point of the line in space relative to the upper edge.

1. Use the Offset or the Copy command to make a copy of the new line 0.19 units to the left.

TIP

The Perpendicular Osnap override can also be used to draw a line perpendicular to a non-orthogonal line—one at a 45° angle, for instance.

2. Draw a line connecting these two lines at the top, and another line connecting them at the bottom.

NOTE

If you find the object snaps and point filters somewhat difficult to understand at first, you can use some of the more classic drafting procedures. You can create layout lines first, and then trim away the parts of the lines that you don't need. Use the Xline command to create some guidelines.

3. In the Draw toolbar, click the Construction Line tool or type **xl**↵. At the XLINE Specify a point or [Hor/Ver/Ang/Bisect/Offset]:Specify through point: prompt, move the cursor to the top of the ledge area until the endpoint box marker appears, and pick this point. Check the status bar to see that Ortho mode is on. Then at the Through point: prompt, move the cursor to the right and pick any point. You'll see a construction line that extends from one side of your screen to the other. Press ↵ to end the selection set.

4. Use the Endpoint Osnap to pick the left endpoint of the lower horizontal line in the side view. With Ortho on, pick a location 1.25 units to the left. To accomplish this, you may use the coordinate meter, if it is set to the correct units. Usually it isn't. So, try a method named *direct-distance entry*. While dragging the crosshairs to the left, type **1.25**↵.

5. Draw a vertical construction line from the left end of the lower horizontal line in the side view. Click the Construction Line tool in the Draw toolbar and use it to draw the vertical construction line. Your drawing should look like Figure 5.11.

FIGURE 5.11:

The construction lines added to the bracket

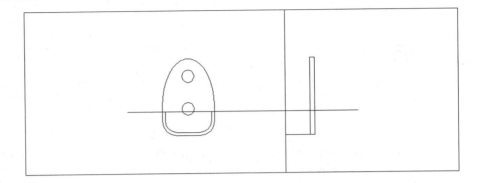

The Xline Options

There's more to the Xline command than you've seen in the exercises of this chapter. Here's a list of the Xline options and their uses.

Hor   Draws horizontal Xlines as you click points.

Ver   Draws vertical Xlines as you click points.

Angle   Draws Xlines at a specified angle as you pick points.

Bisect   Draws Xlines bisecting an angle or a location between two points.

Offset   Draws Xlines offset at a specified distance

Cleaning Up the Line Work

You've drawn some of the lines, approximating their endpoint locations. Next you'll use the Fillet command to join lines exactly end to end.

1.   Click the Fillet tool in the Modify toolbar.

2. Type **r↲0↲** to set the fillet radius to 0; then press ↲ to repeat the Fillet command.

NOTE   There is another command, the Chamfer command, that performs a similar function to Fillet. Unlike Fillet, the Chamfer command allows you to join two lines with an intermediate beveled line, rather than an arc. Chamfer can be set to join two lines at a corner in exactly the same manner as Fillet.

3. Fillet the two lines by picking the vertical line and horizontal Xline as indicated in Figure 5.12. Notice that the points you must pick lie on the portion of the line you want to keep. Your drawing will look like Figure 5.13.

FIGURE 5.12:

Pick these points to fillet the first corner.

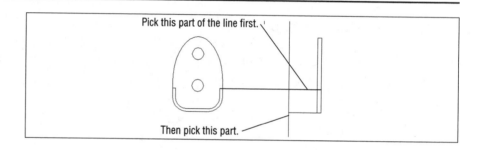

FIGURE 5.13:

Your drawing after the first fillet

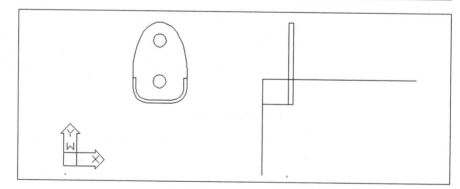

WARNING   If you're a veteran AutoCAD user, you should note that the default value for the Fillet command is now 10.0 instead of 0.

4. Fillet the bottom of the bracket with the left vertical line, as shown in Figure 5.14. Make sure the points you pick on the lines are on the side of the line you want to keep, not on the side you want trimmed.

FIGURE 5.14:

Pick these points to fillet your second corner.

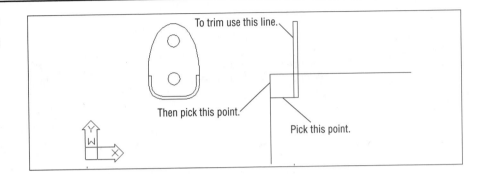

NOTE Before completing the cleanup of this view, use the List tool and select the filleted Xline. You should find that the name has been changed to ray. A ray has a beginning point (unlike an Xline) and only one end goes to the edge of the screen.

5. Use the inside vertical line to trim the rest of the construction line as shown in Figure 5.14. Your drawing should look like Figure 5.15.

FIGURE 5.15:

The construction lines after you've cleaned them up

TIP You can select two lines at once for the Fillet operation by using a crossing window. Type **c↵** at the Select first object: prompt. The two endpoints closest to the fillet location will be trimmed.

Where you select the lines will affect how the lines are joined. As you select objects for the Fillet command, the side of the line you click is the side that remains when the lines are joined. Figure 5.16 illustrates how the Fillet command works and shows what the options do.

TIP If you select two parallel lines during the Fillet command, they'll be joined with an arc.

FIGURE 5.16:

Where you click the object to select it determines what part of an object gets filleted.

These first two examples show how the pick location affects the way lines are filleted.

Pick here   Result

Pick here   Result

This example shows what happens when you set a fillet radius to greater than 0 and the Fillet Trim option is turned off.

Pick here   Result
fillet radius

This example shows how Fillet affects polylines. You can fillet polylines using the Polyline and Radius options of the Fillet command.

Pick here   Result
fillet radius

Next you'll draw the hidden lines. The left vertical line, which is the thickness of the back of the bracket below the ledge, is a hidden line. You'll break the vertical line into two segments so that one segment can be an object line and the other segment can be a hidden line.

1. Click the Break tool in the Modify toolbar, then select the vertical line anywhere along its length. (The Break tool crates a gap in a line, an arc or a circle.)

2. Type **f↵** to use the First Point option; then select the intersection of the upper horizontal line. (Use Shift+right-click to open the Osnap menu or type **int↵** and move the mouse close to the intersection until you see the X intersection marker. You could also pick one of the lines and AutoCAD will ask you to pick the intersecting line.)

3. At the `Specify second break point:` prompt, type **@↵**. The at (@) symbol forces AutoCAD to use the last point again. You use the same procedure to remove a section of line, except in that case you would pick two different points on the line or arc and AutoCAD would remove the line or arc between the points. For more details, see the following sidebar, "Using the Break Command."

4. Click the lower portion of the left vertical line to turn on the Grips feature. Then from the layer drop-down list, select the Hidden layer. Press Esc twice to clear the grips.

5. Click the layer drop-down list again and click the Hidden layer to make it current.

Using the Break Command

When you drew the side view of the bracket, you used the Break command with the F option to accurately break the line into two parts. You can also break a line without the F option—with a little less accuracy. By not using the F option, the point at which you select the object is used as the first break point. If you're in a hurry, you can dispense with the F option and simply place a gap in an approximate location. You can then use other tools later to adjust the gap.

In addition, you can use locations on other objects to select the first and second points of a break. For example, you may want to align an opening with another opening some distance away. Once you've selected the line to break, you can then use the F option and select two points on the existing opening to define the first and second break points. The break points will align in an orthogonal direction to the selected points.

Using Construction Lines as Tools

Now you'll draw the hidden line, which represents the inside of the ledge, in the side view. Start by drawing a ray.

1. Choose Draw ➤ Ray from the menu bar.

2. At the Specify start point: prompt, start the ray from the endpoint of the horizontal line that represents the inside thickness of the ledge in the front view. If the Ortho mode is on, turn it off. As you move your mouse, you'll see the ray follow its direction (see the top image of Figure 5.17).

FIGURE 5.17:

A ray used to project the ledge thickness into the side view

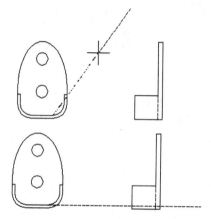

3. Type @1<0↵. The ray is fixed in a 0° angle. The Ray command persists, allowing you to add more lines.

4. Press ↵ to exit the Ray command.

> **NOTE** A *ray* is a line that starts from a point you select and continues off to some infinite distance. For this reason, you weren't required to enter a distance value in step 3.

5. Zoom in to see the detail at the bottom of the side view.

6. Trim the ray and the vertical line to look like Figure 5.18.

FIGURE 5.18:

The bracket with the ray and vertical line trimmed

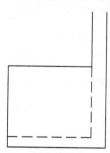

7. Choose View ➤ Zoom ➤ All to view the entire drawing. It will look like Figure 5.19.

FIGURE 5.19:

The bracket after using Zoom All

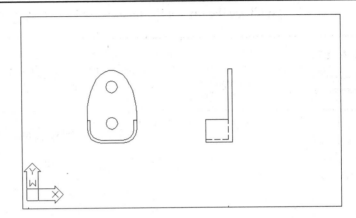

8. Save your work.

In this exercise, you used the Ray command to accurately position a construction line that became part of the drawing. This shows that you can freely use rays and construction lines (Xlines) as objects to help construct your drawing.

Finding Distances Along Arcs

You've seen how you can use lines to help locate objects and geometry in your drawing. But if you need to find distances along a curved object such as an arc, lines don't always help. Following are two ways of finding exact distances on arcs. Try these exercises when you're not working through the main tutorial.

Finding a Point a Particular Distance from Another Point

At times you'll need to find the location of a point on an arc that lies at a known distance from another point on the arc. The distance could be described as a *chord* of the arc, but how do you find the exact chord location? To find a chord along an arc, do the following:

1. Click the Arc tool in the Draw toolbar and create an arc on your drawing. At the Specify start point of arc or [CEnter]: prompt, pick a location randomly on your screen. At the Specify second point of arc or [CEnter/ENd]: prompt, pick another random location. At the Specify end point of arc: prompt, drag the crosshairs far enough to create an arc.

2. Click the Circle tool in the Draw toolbar. Use the Endpoint Osnap to click the endpoint of the arc.

3. At the end of `Specify radius of circle or [Diameter]:` prompt, enter the length of the chord distance you wish to locate along the arc.

The point where the circle intersects the arc is the endpoint of the chord distance from the endpoint of the arc (see Figure 5.20). You can then use the Intersect Osnap override to select the circle and arc intersection.

FIGURE 5.20:

Finding a chord distance along an arc using a circle

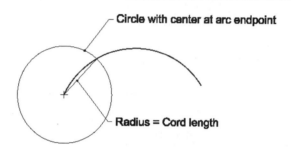

Circle with center at arc endpoint

Radius = Cord length

Finding an Exact Distance Along an Arc

To find an exact distance along an arc or curved (nonlinear) object, or to mark off specific distance increments along an arc or curve, do the following:

1. Choose Format ➤ Point Style from the menu bar to open the Point Style dialog box.

TIP   You can also set the point style by setting the Pdmode system variable to 3. See Appendix D, "System Variables," for more on Pdmode.

2. In the Point Style dialog box, click the icon that looks like an X in the top row. Also be sure that the Set Size Relative to Screen radio button is selected. Then click OK.

3. Choose Draw ➢ Point ➢ Measure from the menu bar.

TIP   Another command called Divide (Draw ➢ Point ➢ Divide) marks off a line, arc, or curve into equal divisions, as opposed to divisions of a length you specify. You would use Divide to divide an object into 12 equal segments, for example. Aside from this difference in function, Divide works in exactly the same way as Measure.

4. At the `Select object to measure:` prompt, click the arc near the end from which you wish to find the distance.

5. At the `Specify length of segment or [Block]:` prompt, enter the distance you are interested in. A series of Xs appears on the arc, marking off the specified distance along the arc. You can use the Xs' reference points with the Node Osnap override (Figure 5.21).

FIGURE 5.21:

Finding an exact distance along an arc using points and the Measure command

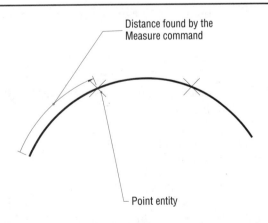

Distance found by the Measure command

Point entity

TIP   The Block option of the Measure command allows you to specify a block to be inserted at the specified segment length, in place of the Xs on the arc. You have the option of aligning the block with the arc as it is inserted. (This is similar to the polar array's Rotate Objects As They Are Copied option.)

The Measure command also works on Spline objects (Bézier curves). You'll get a more detailed look at the Measure command in Chapter 10, "Drawing Curves and Solid Fills."

As you work with AutoCAD, you'll find that constructing temporary geometry, such as the circle and points in the two previous examples, will help you solve problems in new ways. Don't hesitate to experiment. Remember that you've always got the Save and Undo commands to help you recover from mistakes.

Changing the Length of Objects

Suppose, after finding the length of an arc, you realize that you need to lengthen the arc by a specific amount. The Modify ➤ Lengthen command lets you lengthen or shorten arcs, lines, splines, and elliptical arcs. Here's how to lengthen an arc:

1. Click the Lengthen tool in the Modify toolbar.

    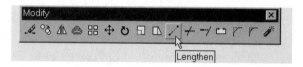

2. At the `Select an object or [DElta/Percent/Total/DYnamic]:` prompt, type **t↵**.

3. At the `Specify total length or [Angle] <1.0000)>:` prompt, enter the length you want for the arc, then press ↵.

4. At the `Select object to change or [Undo]:` prompt, click the arc you wish to change. Be sure to click the point nearest the end you want to lengthen. The arc increases in length to the size you specified. Press ↵ to end the command.

Lengthen will also shorten an object if the object is currently longer than the value you enter.

In this short example, we've demonstrated how to change an object to a specific length. You can use other criteria to change an object's length, using these options available in the Lengthen command:

> **DElta**   Lets you lengthen or shorten an object by a specific length. To specify an angle rather than a length, use the Angle sub-option.

Percent   Lets you increase or decrease the length of an object by a percentage of its current length.

Total   Lets you specify the total length or angle of an object.

DYnamic   Lets you graphically change the length of an object using your cursor.

Extending an Object to Meet Another Object

For some drawings, such as projections, you need to join an object to another object. With the Extend tool, you can extend arcs, lines, polylines, and rays. Figure 5.22 shows an example of extending objects with the Extend tool.

FIGURE 5.22:

Using the Extend tool to make objects meet

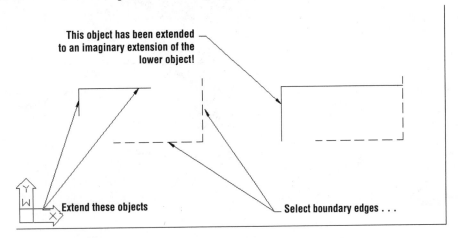

To join objects with the Extend tool, click the Extend tool in the Modify toolbar. At the `Select boundary edges...` prompt, select the objects to join, or press ø to select all objects as potential boundaries. Valid boundary objects include polylines, arcs, circles, ellipses, floating viewports, lines, rays, regions, splines, text, and Xlines. (For 2D polylines, AutoCAD ignores the width and extends objects to the center line.) Next, you'll see the `Select object to extend or [Project/Edge/Undo]:` prompt. Select the objects you want to extend on their end nearest to the objects you selected as boundaries.

You can also force a line to touch an invisible extension by using the Edge option. When you see the `Select object to extend or [Project/Edge/ Undo]:` prompt, type **eø.** You'll see the `Enter an implied edge extension mode [Extend/No extend] <No extend>:` prompt. Type **eø** again. This changes the Extend tool (and Trim tool) to look for extended boundaries (and cutting edges).

Stretching Objects

When you need to change the size of objects, either to accommodate design changes or to fix mistakes, you can use the Stretch tool. Figure 5.23 shows an example of using the Stretch tool to correct problems.

In Figure 5.23, the view shown on the left is 0.19 unit too long. It should be the same size as the view on the right. Here's how you can use the Stretch tool to fix this:

1. Click the Stretch tool on the Modify toolbar.

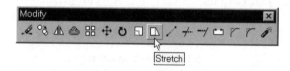

FIGURE 5.23:

Using the Stretch tool to correct problems

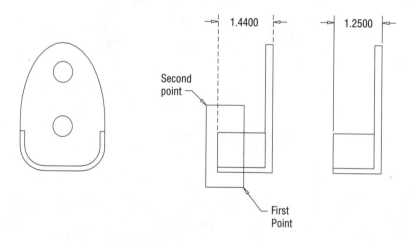

2. At the Select objects to stretch by crossing-window or crossing-polygon... prompt, pick the points labeled "first point" and "second point" in Figure 5.23. The crosshair moves from right to left, which automatically creates a crossing window.

3. When you see the prompt Select objects: Specify opposite corner: 5 found, press ↵.

4. At the `Specify base point or displacement:` prompt, type **0.19,0** and press ↵.

The view on the left should now look like the one on the right. Notice that the dimension of 1.44 changed to 1.25. You'll learn more about dimensions in Chapter 8, "Using Dimensions."

Drawing Parallel Lines

Multilines are generally not used in mechanical engineering. But civil and architectural engineers will find them convenient when working on projects. You'll first do your schematic layout using simple lines for walls. Then, as the design requirements begin to take shape, you can start to add more detailed information about the walls—for example, indicating wall materials or locations for insulation. Auto-CAD provides *multilines* (accessed through the Mline command), which are double lines that can be used to represent walls. Multilines can also be customized to display solid fills, center lines, and additional linetypes shown in Figure 5.24. You can save your custom multilines as Multiline styles, which are in turn saved in special files for easy access from other drawings.

FIGURE 5.24:

Samples of multiline styles

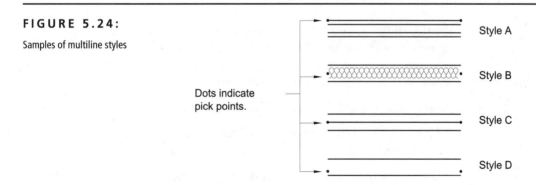

The following exercise shows how you might build information into your drawings by using multilines to indicate wall types.

1. Click the Multiline tool in the Draw toolbar or type **ml**↵. You'll see two lines in the prompt area:

```
Current settings: Justification = Top, Scale = 1.00, Style = STANDARD
Specify start point or [Justification/Scale/STyle]:
```

NOTE   The first line of the prompt gives you the current settings for Mline. You'll learn more about those settings in the section "Customizing Multilines," which follows this exercise.

---

2. At the `Specify start point or [Justification/Scale/STyle]:` prompt, type **s.↵** (for *scale*).

3. At the `Enter Mline Scale <1.00>:` prompt, press ↵.

4. Pick a point to start the double line.

5. Continue to select points to draw more double-line segments, or type **c↵** to close the series of lines.

Now let's take a look at the meaning of the Mline settings included in the prompts you saw in steps 1 and 2 above.

Justification   Controls how far off center the double lines are drawn. The default sets the double lines equidistant from the points you pick. By changing the justification value to be greater than or less than 0, you can have AutoCAD draw double lines off-center from the pick points.

Scale   Lets you set the width of the double line.

Close   Closes a sequence of double lines, similar to the Line command's Close option.

Style   Lets you select a style for multilines. You can control the number of lines in the multiline, as well as the linetypes used for each line in the multiline style, by using the Mledit command (discussed in the next section).

Customizing Multilines

In Chapter 4, "Organizing Your Work," you learned how to make a line appear dashed or dotted by using linetypes. In a similar way, you can control the appearance of multilines by using the Multiline Styles dialog box. This dialog box allows you to

- Set the number of lines that appear in the multiline

- Control the color of each line

- Control the linetype of each line

- Apply a fill between the outermost lines of a multiline

- Control whether and how ends of multilines are closed

To access the Multiline Styles dialog box, choose Format ➤ Multiline Style from the menu bar, or type **mlstyle.**⏎ at the command prompt. You'll see the Multiline Styles dialog box.

WARNING   Once you've drawn a multiline in a particular style, you cannot modify the style settings for that style in the Element Properties and Multiline Properties dialog boxes described later in this section. The Multiline Styles dialog box only allows you to set up a style before it is used in a drawing.

At the top of the dialog box, you see a group of buttons and input boxes that allow you to select the multiline style you want to work with. The Current drop-down list offers you a selection of existing styles. In the Name input box you can name a new style you're creating, or rename an existing style. The Description input box lets you attach a description to an Multiline style for easy identification. Use the Save and Add buttons to create and save multiline styles as files so they can be accessed by any AutoCAD drawing. The Load button lets you retrieve a saved style for use in the current drawing. Rename lets you change the name of a Multiline style (the default style in a new drawing is called Standard).

In the lower half of the Multiline Styles dialog box are two buttons—Element Properties and Multiline Properties—that allow you to make adjustments to the Multiline style currently indicated at the top of the dialog box. This Multiline style is also previewed in the middle of the dialog box.

Element Properties

In the Element Properties dialog box, you can control the properties of the individual elements of a newly created style, including the number of lines that appear in the Multiline, their color, and the distance they appear from your pick points. The Element Properties settings are not available for existing Multiline styles.

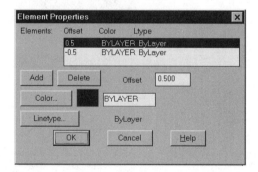

For example, click the Add button, and another line is added to your multiline. The offset distance of the new line appears in the list box. The default value for new lines is 0.0, which places the line at the center of the standard multiline. To delete a line, highlight its offset value in the list box and click the Delete button. To change the amount of offset, highlight the current value in the Offset input box and enter a new value.

To change the color and linetype of individual lines, use the Color and Linetype buttons, which open the Color and Select Linetype dialog boxes. In Figure 5.25, you see some examples of multilines and their corresponding element properties settings.

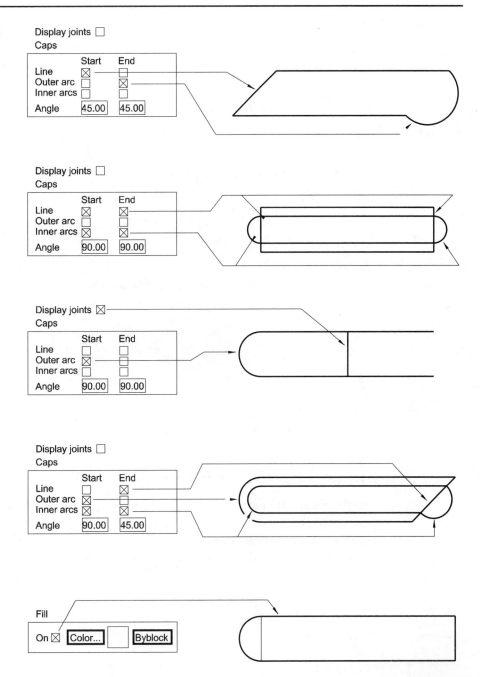

Samples of Mline styles you can create

TIP You can easily indicate insulation in a drawing by adding a third center line (offset of 0.0), and giving that center line a Batting line. This linetype draws an S-shaped pattern typically used to represent fiberglass batt insulation. To see how other patterns can be created, see the sections on linetype customization in Chapter 21, "Integrating AutoCAD into Your Projects and Organization."

Multiline Properties

The Multiline Properties dialog box lets you control how the Multiline is capped at its ends, and whether joints are displayed.

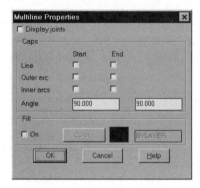

To turn on a cap, click the check box next to the type of cap you want. If you prefer, you can give your multiline style a solid fill, using the Fill check box.

Joining and Editing Multilines

Multilines are unique in their ability to combine several linetypes and colors into one entity. For this reason, you need special tools to edit them. On the Modify menu, the Edit Multiline option (and the Mledit command) has the sole purpose of allowing you to join multilines in a variety of ways, as demonstrated in Figure 5.26.

FIGURE 5.26:

The Mledit options and their meanings

CLOSED CROSS Trims one of two intersecting multilines so that they appear overlapping.	
OPEN CROSS Trims the outer lines of two intersecting multilines.	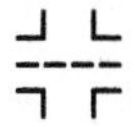
MERGED CROSS Joins two multilines into one multiline.	
CLOSED TEE Trims the leg of a tee intersection to the first line.	
OPEN TEE Joins the outer lines of a multiline tee intersection.	
MERGED TEE Joins all the lines in a multiline tee intersection.	
CORNER JOINT Joins two multilines into a corner joint.	
ADD VERTEX Adds a vertex to a multiline. The vertex can later be moved.	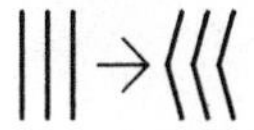
DELETE VERTEX Deletes a vertex to straighten a multiline.	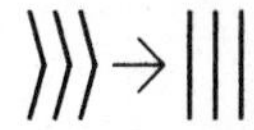
CUT SINGLE Creates an opening in a single line of a multiline.	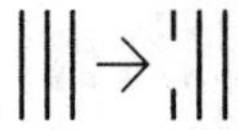
CUT ALL Creates a break across all lines in a multiline.	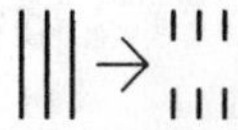
WELD Closes a break in a multiline.	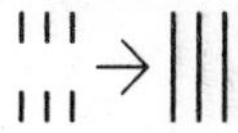

Here's how the Mledit command is used.

1. Type **mledit** ↵ at the command prompt. The Multiline Edit Tools dialog box appears (see Figure 5.27), offering a variety of ways to edit your multilines.

2. Click the graphic that best matches the edit you want to perform.

3. Select the multilines you want to join or edit.

Another option is to explode multilines and edit them with the editing tools you've used in this and previous chapters. When a multiline is exploded, it's reduced to its component lines. Linetype assignments and layers are maintained for each component.

FIGURE 5.27:

Multiline Edit Tools
dialog box

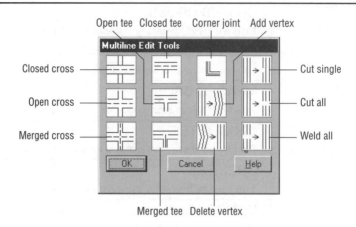

If you're doing a lot of work with multilines, you can open the Modify II toolbar. It contains a Multiline Edit tool that opens the Multiline Edit Tools dialog box. To open the Modify II toolbar, right-click any toolbar and then click Modify II in the Toolbars dialog box.

Eliminating Blocks, Layers, Linetypes, Shapes, and Styles

A prototype may contain blocks and layers you don't need in your new file. When you erase a block, it remains in the drawing file's database. The erased block is considered unused because it doesn't appear as part of the drawing. Such extra blocks can slow you down by increasing the amount of time needed to open the file. They will also increase the size of your file unnecessarily. There are two commands for eliminating unused elements from a drawing: Purge and Wblock.

Selectively Removing Unused Elements with Purge

The Purge command is used to remove unused individual blocks and shapes, as well as multiline, dimension, plot, and text styles from a drawing file. To help keep the file size down and to make layer maintenance easier, you'll want to purge your drawing of unused elements.

As you'll see in the File ➢ Drawing Utilities ➢ Purge cascading menu and at the Purge command prompt, you can purge other unused drawing elements, such as linetypes and layers, as well. Bear in mind, however, that Purge will not delete certain primary drawing elements—namely, Layer 0, the Continuous linetype, and the Standard text style.

1. Click File ➢ Open and open the DT0100 file.

2. Click File ➢ Drawing Utilities ➢ Purge ➢ Blocks.

3. At the `Enter Name(s) to purge <*>:` prompt, you can enter the name of a specific block or press ↵ to purge all of the blocks.

4. Go ahead and press ↵; you'll see the prompt: `Verify each name to be purged? [Yes/No] <Y>:`. This lets you selectively purge blocks by displaying each block name in succession.

5. Press ↵ again.

6. At the `Purge block "BKT-V2"? <N>:` prompt, enter **y**↵. This prompt will repeat for each unused block in the file. Continue to enter **y**↵ at each prompt until the Purge command is completed.

The DT0100 file is now purged of most, but not all, of the unused blocks. Now let's take a look at how to delete all of the unused elements at once.

Opening a File as Read-Only

When you open an existing file, you might have noticed the Read Only Mode check box in the Open Drawing dialog box. If you open a file with this option checked, AutoCAD will not let you save the file under its original name. You can still edit the drawing any way you please, but if you attempt to use File ➢ Save, you'll get the message `Drawing file is write-protected`. You can save your changed file under another name.

The read-only mode provides a way to protect important files from accidental corruption. It also offers another method for reusing settings and objects from existing files by letting you open a file as a prototype, and then saving the file under another name.

Removing All Unused Elements with Wblock

Purge doesn't remove nested blocks on its first pass. For example, say you had a bolt block inserted into a bracket block that was inserted into the bar drawing. To remove them using Purge, you would have to start the command again and remove the nested blocks. For this reason, Purge can be a time-consuming way to delete a large number of elements.

NOTE   Cross-referenced (Xref) files do not have to be purged because they never actually become part of the drawing's database. You must use the Xref command to detach these files.

In contrast, the File ➤ Export option (Wblock) enables you to remove all unused elements simultaneously—including blocks, nested blocks, layers, linetypes, shapes, and styles. You cannot select specific elements or types of elements to remove.

WARNING   Be careful: In a given file, there may be an unused block that you want to keep, so you might want to keep a copy of the unpurged file.

If You Want to Experiment...

Try using the techniques you learned in this chapter to create a new file. Then draw the top view of the bracket, using Xlines, rays, Osnaps, and point filters. You might also wish to create the mechanical symbols shown in Figure 5.28.

FIGURE 5.28:

Mechanical Symbols

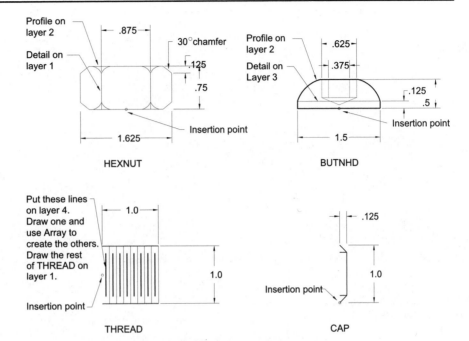

HEXNUT

BUTNHD

THREAD

CAP

Note:
Create four layers named 1, 2, 3, and 4, if they do not already exist.
Give each layer the same color as their number.
Give layer 3 the HIDDEN line type.
Don't draw dimensions, just use them for reference.

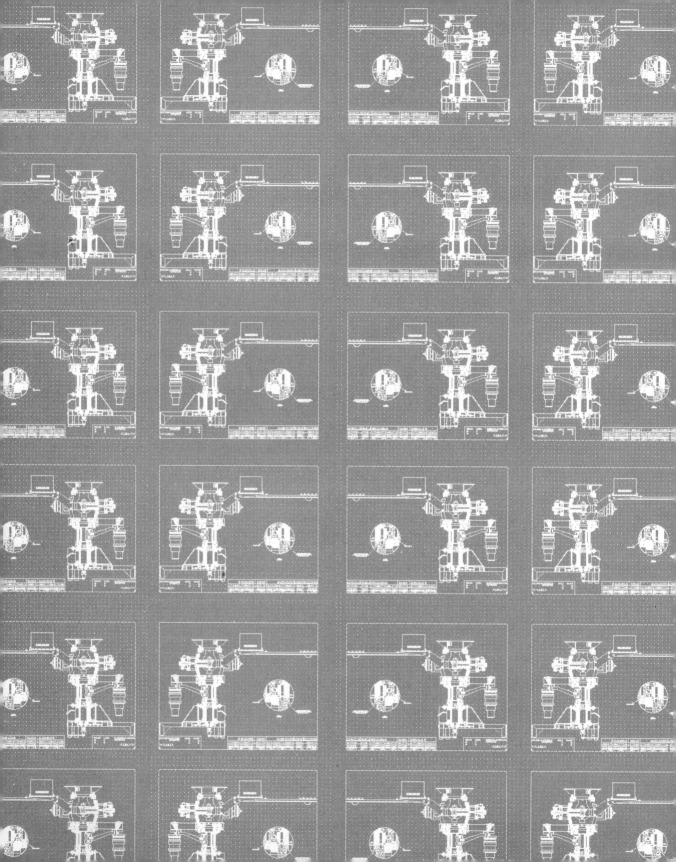

CHAPTER

SIX

Enhancing Your Drawing Skills

- Reducing Wait Times by Controlling Regeneration

- Improving Performance through Smart Use of the Display

- Improving Performance Using Layers

- Saving Views of Your Drawing

6

As your drawing becomes more complex, you'll find that you need to use the Zoom and Pan commands more often. Mechanical design requires accuracy whenever you're creating geometry, and it's often difficult to see the detail when you're standing back looking at the big picture. Larger drawings also require some special editing techniques. You'll learn how to assemble and view drawings in ways that will save you time and effort as your design progresses.

You can use the Grid and Snap functions for simple layout and sketches. If you continue to use Grid for larger, more complex drawings, you'll find that the grids take time to regenerate as you zoom in and out. You'll have to set the grid to be more coarse and thus less helpful—or suffer the cost in productivity. Snap is difficult to use in larger drawings because the cursor resolution is relative to the screen, and it can get very difficult to move the cursor one snap interval on large drawings. For these reasons, we recommend that you dispense with using Snap and Grid for the rest of these tutorials—and for much of your production work as well.

Start by opening a new file using the Versions template.

1. Choose File ➤ New. At the Create New Drawing dialog box, click the Use a Template button.

2. Find and select the template titled Versions.dwt and click OK.

3. Find the Grid toggle button on the status bar and click it to toggle it off.

4. Do the same operation for Snap.

> **NOTE**   You may toggle the Grid and Snap functions on and off by the click method or by typing **grid** or **snap** and following the prompts. Both Grid and Snap will continue to use the 0.125 grid functions that you originally set, until you change them.

5. Type **saveas↵**. In the dialog box, click the Save As Type drop-down list and select the Drawing File Template (.dwt) option. The list of templates appears.

6. Select the template named Versions.dwt. Click the Save button. You'll get the warning <path>\versions.dwt already exists. Do you want to replace it? and three buttons marked Yes, No, and Cancel. You do want to replace the existing file, which will redefine it.

7. Click the Yes button. The Template Description dialog box appears. You don't need to change the description. Click OK to continue.

From now on, when you start a new drawing with this template, both Grid and Snap will be off. If you want to use them for any reason, just toggle them on. When you open the new version of the bar drawing named 6_bar.dwg, you'll find that it uses the new template.

Assembling the Parts

Start by using a slightly modified version of the file that you have been working on. We've used the version of the bracket with two views, and we've changed the size of the tapped holes to 3/8" in the bracket and bar. You'll also find three views of a 3/8" hex-head screw.

1. Open the 6_bar.dwg file from the companion CD-ROM. Use the Save As option from the File menu to save the file as 6_bar to your hard disk. The choice of directories is yours, but remember that you may have to find this file later, so make a note of the file's name and location.

2. Choose View ➤ Zoom ➤ All, or type **z⌐a⌐** to get an overall view of the drawing area. Zoom in to the end of the bar. Include the bar end, the bracket, and the bolt details.

3. Make sure Grid and Snap are turned off.

4. In Chapter 4 you learned to group objects. Create a group from the front view bracket objects. Type **group⌐** to open the Object Grouping dialog box (see Figure 6.1). In the Group Name box, enter the first group's name **bkt_front**. Create another group of the objects of the side view of the bracket. Name these **bkt_side**. Continue with the bolt and call the hex view **bolt_plan**, the lower view **bolt_front**, and the last view **bolt_side**.

NOTE   You may use whichever style of data entry (upper- or lowercase) that you like or find convenient at the time. However, notice that the names are all changed to uppercase in the Group Name list box. AutoCAD is not sensitive to upper- or lowercase names.

---

5. Move the bkt_front group and the bkt_side group together into position on the bar. Use the Center Osnap to help you align the top-mounting hole in the bracket with the top leftmost mounting hole on the bar.

FIGURE 6.1:

The Object Grouping dialog box showing the group names

6. Zoom All to see the overall view. You may need to use the scroll bar to see all of the right side of the objects.

7. Be sure that Ortho mode is on and move the bkt_side group slightly beyond the right end of the bar.

NOTE   This is an example of the great power of the Ortho mode for mechanical design. All of the views of a multiple-view drawing are supposed to align with one another horizontally and vertically. You'll find that it is necessary to move entire views to better place them for dimensioning and positioning on title blocks. Notice how easy it was to move the side view group horizontally and maintain alignment with the front view.

8. Select all three groups of the bolt and use the Center Osnap to move them so that the bolt plan view is aligned with front view of the bracket and bar.

9. With Ortho still on, move the bolt_side group of the bolt until it is nearly in position next to the side view.

10. Move all of the bolt groups down to align with the bottom hole in the bracket. For clarity, move the bolt_front group again to clear the bottom of the bracket.

11. If you've had to zoom in for any of these moves you should now Zoom All to see the overall effect of your work. Your drawing should be similar to Figure 6.2.

FIGURE 6.2:

The bar, bracket, and bolt views

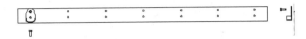

12. Array the brkt_front group along the bar. It is a rectangular pattern with one row and seven columns. At the Enter the distance between rows or specify unit cell (--): prompt of array, pick the center of the top-left hole; at the Other corner: prompt, pick the center of the top hole in the column to the right. This will give AutoCAD the incremental distance between the sets of objects.

13. Array the bolt_plan group using a rectangular pattern of two rows and seven columns. Use the existing geometry to provide the horizontal and vertical distances. At the Enter the distance between rows or specify unit cell (--): prompt, pick the center of the bottom-left hole; at the Specify opposite corner: prompt, pick the center of the top hole in the next column. This will give AutoCAD the incremental distance between the sets of objects, which will be applied to all of the arrayed patterns. See Figure 6.3 for the pick points.

FIGURE 6.3:

The array pick points for the unit cell distances between the bolt heads

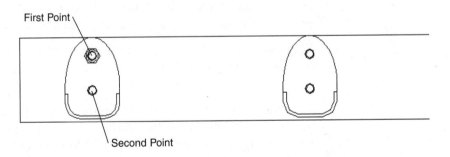

First Point

Second Point

The bar is a custom aluminum extrusion. You can find the cross section of the extrusion on the CD-ROM as Extrusion.dwg.

14. Use the Insert Block tool in the Draw toolbar to import the extrusion cross section and place it near the right end of the bar.

15. Click the Move tool in the Modify toolbar. Pick the extrusion, and then press ↵ to complete the selection set. Using the Running Osnaps (which have been

on since you started this exercise), pick the upper-left corner of the extrusion. At the `Specify second point...:` prompt, pick a point anywhere along the top of the bar.

Next insert an exploded block from the CD-ROM called `Washer.dwg`.

NOTE
The hardware is being provided as blocks here because most likely you will either create, purchase, or subscribe to a library of these parts. Very extensive libraries are becoming available on CD and via the Internet.

---

1. Select the Insert Block tool in the Draw toolbar.

2. Use the Files button to find the file named `Washer`.

3. Click the Explode check box. Then click OK.

4. Pick a point near the right end of the bar and insert the washers at full scale. Zoom in to get a close-up view of the washers and bolt.

Next create a stack of the washers and locate them on the bolt.

1. Highlight the grips of the edge view of the lock washer (the smaller washer) by picking any point on any object in the view. Note that the view is a block because only one grip appears. Pick the grip to make it hot.

2. Drag the cursor to the right, with the edge view in tow, until the midpoint triangle marker appears on the left side of the edge view of the flat washer, then click. The two side views are stacked.

3. Use either grips or the Move tool to move the stacked washers from the midpoint of the left edge of the lock washer to the midpoint of the vertical line of the right edge of the bolt. You may want to zoom in even closer to see that you have picked the right points. Your drawing should look like Figure 6.4.

4. Locate the bolt and washers on the bracket. You'll need two bolt-and-washer sets, so make a copy. You can use the distance between the tapped holes in the bar for the displacement of the copy.

5. Use Move and pick the bolt and both washers. At the `Specify base point or displacement:` prompt, pick the midpoint of the right side of the flat washer. At the `Specify second point of displacement or <use first point as displacement>:` prompt, Shift+right-click and select the Perpendicular Osnap, or type **per.**.

FIGURE 6.4:

The side view of the washers and bolt

6. Move the mouse until the Perpendicular Osnap marker appears, then click the mouse.

7. Click the Copy tool in the Modify toolbar and type **p**↵ to use the previous selection set of objects. Then press ↵ to complete the selection set and begin the next part of the command. At the Specify base point or displacement, or [Multiple]: prompt, use the Center Osnap to select any of the top row of circles that represent the tapped holes in the bar.

TIP   Look for the Center Osnap marker to appear as you hover the mouse over any of the circles in the bar, bracket, and bolts.

---

8. Next use the Center Osnap to pick the hole directly below the first one. Auto-CAD measures this distance and moves the bolt and washers accordingly.

9. Next, move the extrusion into place. Use the Move tool again and with the help of the Perpendicular Osnap, move the extrusion into position perpendicular to the horizontal line on the bottom of the bar. Your drawing should look like Figure 6.5.

FIGURE 6.5:

The side view of the bracket, bolt, washer, and bar assembly

Next you'll stack the front view of the washers, move them into place, and array them.

1. Select the front view of either washer and use the Running Center Osnap to identify and pick the center of the view as the location of the Specify base point or displacement: of the washer that you want to move. At the Specify second point of displacement: prompt, use the Running Center Osnap to identify and pick the new location of the front view of the washer.

2. Use the Move tool. Select both washers and use the center-to-center method again as you did in the previous step to position them at the lower-left bolt location on the bar.

3. Use the Array tool to create a rectangular array of the previous selection set. Create the array with two rows and seven columns, with a distance of 1.375 units between the rows and 9 units between the columns.

4. Zoom All to see the full effect of your handiwork. You should see something like Figure 6.6.

5. Save your work.

FIGURE 6.6:

The overall view of the bar, bracket, bolt, and washers

You could do much more to clean up these views. There are hidden lines and center lines to be added, and object lines to be removed before the views in this drawing could be considered finished. You've had to zoom in and out to see your work as you went along creating these views. Let's take a look at the tools Auto-CAD has provided to make this task easier.

TIP   If you happen to insert a block in the wrong coordinate location, you can use the Properties tool in the Standard toolbar to change the insertion point for the block.

Taking Control of the AutoCAD Display

By now you should be familiar with the Pan and Zoom functions in AutoCAD. There are many other tools at your disposal that can help you get around in your drawing. In this section, you'll get a closer look at the different ways you can view your drawing.

Understanding Regeneration and Redraw

AutoCAD uses two methods for refreshing your drawing display: drawing regeneration (or *Regen*) and *Redraw*. Each serves a particular purpose, although this may not be clear to a new user.

AutoCAD stores drawing data in two ways: one is like a database of highly accurate coordinate information and object properties. This is the core information you supply as you draw and edit your drawing. For the purposes of this discussion, we'll call this simplified database the *virtual display* because it's like a computer model of the overall display of your drawing. This virtual display is in turn used as the basis for what is shown in the drawing area. When you issue a Redraw, you're telling AutoCAD to reread this virtual display data and display that information in the drawing area. A Regen, on the other hand, causes AutoCAD to rebuild the virtual display based on information from the core drawing database.

As you edit drawings, you may find that some of the lines in the display disappear or otherwise appear corrupted. Redraw will usually restore such distortions in the display. In earlier versions of AutoCAD, the Blipmode system variable was turned on by default, causing markers called *blips* to appear wherever points were selected. Redraw was, and still is, useful in clearing the screen of these blips.

Regens are used when changes occur to settings and options that have a global effect on a drawing, such as a linetype scale change, a layer color change, or text style changes. (You'll learn more about text styles in Chapter 7, "Adding Text to Drawings.") In fact, Regens are normally issued automatically when such changes occur. You usually don't have to issue the Regen command on your own, except under certain situations.

TIP   If the screen doesn't look right, or if something is missing, try a **zoom⏎all⏎**. This will cause the entire drawing to be visible and force a Regen in the process.

Regens can also occur when you select a view of a drawing that is not currently included as part of the virtual display. The virtual display contains display data for a limited area of a drawing. If you zoom or pan to a view that is outside that virtual display area, a Regen occurs.

NOTE You may notice that Pan Realtime and Zoom Realtime won't work beyond a certain area in the display. When you've reached a point where these commands seem to stop working, you've come to the limits of the virtual display data. In order to go beyond these limits, AutoCAD must rebuild the virtual display data from the core data; in other words, it requires a drawing regeneration.

In past versions of AutoCAD, Regens were to be avoided at all cost, especially in large files. A Regen of a very large file could take several minutes to complete. Today, with faster processors, large amounts of RAM, and a retooled AutoCAD, Regens are not the problem they once were. Still, they can be annoying in multi-megabyte files, and if you are using an older Pentium-based computer, Regens can still be a major headache. For these reasons, it pays to understand the finer points of controlling Regens.

In this section, you'll discover how to manage Regens, thereby reducing their impact on a complex drawing. You can control how Regens impact your work in three ways:

- By taking advantage of AutoCAD's many display-related tools

- By setting up AutoCAD so that Regens don't occur automatically

- By freezing layers that don't need to be viewed or edited

We'll explore these methods in the upcoming sections.

Exploring Other Ways of Controlling AutoCAD's Display

Perhaps one of the easiest ways of avoiding Regens is by making sure you don't cross into an area of your drawing that falls outside of the virtual display's area. If you use Pan Realtime and Zoom Realtime, you're automatically kept safely within the bounds of the display list. In this section, you'll be introduced to other tools that will help keep you within those boundaries.

Controlling Display Smoothness

The Viewres command's sole purpose is to control the smoothness of linetypes, arcs, and circles when they appear in an enlarged view. With the display list active, linetypes sometimes appear as continuous even when they are supposed to be dotted or dashed. You may have noticed in previous chapters that on the screen, arcs appear to be segmented lines, though they are always plotted as smooth curves. You can adjust the Viewres value to control the number of segments an arc appears to have—the lower the value, the fewer the segments and the faster the redraw and regeneration. However, a low Viewres value will cause noncontinuous linetypes, such as dashes or center lines, to appear continuous.

TIP   A good value for the Viewres setting is 500. At this setting, linetypes display properly, and arcs and circles have a reasonably smooth appearance. At the same time, redraw speed is not noticeably degraded. However, you may want to keep Viewres lower still if you have a limited amount of RAM. High Viewres settings can adversely affect AutoCAD's overall use of memory.

In past versions of AutoCAD, when using Viewres, the user was prompted with a seemingly strange question, "Do you want fast zooms?" In other words, you were being asked to give AutoCAD permission to use memory resources to ensure that future zooms would be as fast as possible. Experienced AutoCAD users will be glad to know that this question is no longer a functioning option, and is retained only for script files.

TIP   You can set the Viewres value in the Display tab of the Options dialog box under the Arc and Circle Smoothness setting.

Another way to accelerate screen redraw is to keep your drawing limits to a minimum area. If the limits are set unnecessarily high, AutoCAD may slow down noticeably. Also, make sure that the drawing origin falls within the drawing limits.

Using the Aerial View Window

The Aerial View window is a useful tool that helps you navigate drawings that represent very large areas. Let's see how this tool works.

1. Click View ➤ Aerial View on the menu bar.

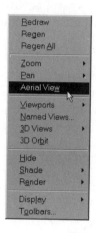

2. The Aerial View window appears, as shown in Figure 6.7. Click this window. Move your mouse and watch the drawing image on your screen move with it. You are panning. A bold rectangle, called the *view box*, represents your AutoCAD window view. The view box moves with your cursor.

3. Click the Aerial View window again. Now as you move your cursor from left to right, the view in the AutoCAD window zooms in and out. The view box in the Aerial View window shrinks and expands as you move the cursor from left to right, indicating the size of the area being displayed in the AutoCAD window.

4. Move the cursor to the left so that the view box is about half the size of the overall view of the bar, then right-click. Your AutoCAD view becomes fixed. Also notice that the magnification icon in the Aerial View toolbar becomes available.

With the Aerial View window, you can alternate between panning and zooming by clicking the mouse. The view box shows you exactly where you are in the overall drawing at any given time. When you want to fix your view in place, right-click.

FIGURE 6.7:

The Aerial View window

Zoom In
Increases the magnification of the drawing in Aerial View by zooming in a factor of 2, centered on the current view box.

Zoom Out
Decreases the magnification of the drawing in Aerial View by zooming out by a factor of 2, centered on the current view box.

Global
Displays the entire drawing and the current view in the Aerial View window.

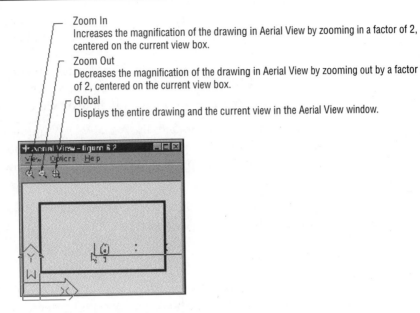

TIP   The View ➤ Zoom ➤ Dynamic option performs a similar function to the Aerial View window. Instead of opening a separate window, the Dynamic option temporarily displays the overall view in the drawing area.

The View menu in the Aerial View window offers several useful options for controlling the display:

Zoom In   Zooms in on the view defined by the current view box in the Aerial View window.

Zoom Out   Zooms out on the view defined by the current view box in the Aerial View window.

Global   Displays an overall view of your drawing in the Aerial View window.

The Options menu in the Aerial View window provides options for controlling the viewport display and dynamic updating of your drawing. (You can also right-click in the Aerial View window to pop up a menu with these options.)

Auto Viewport   Automatically displays the contents of a viewport in the Aerial View window when the viewport becomes active. (See Chapter 9, "Advanced Productivity Tools," for more on viewports.)

Dynamic Update   Updates the Aerial View window in real time as changes in the drawing occur. (When this option is not selected, changes in the drawing do not appear in the Aerial view until you click the Aerial View window.)

Realtime Zoom   Updates the AutoCAD display in real time as you zoom and pan in the Aerial View window.

Saving Views

Another way of controlling your views is by saving them. You might think of saving views as a way to create a bookmark or placeholder in your drawing. You'll see how to save views in the following set of exercises.

A few details in the side view of the drawing are not complete. You'll need to zoom in on the areas that need work to add the lines, but these areas are spread out over the drawing. You could use the Aerial View window to view each area. There is another way to edit widely separated areas. First, save views of the areas you want to work on, then jump from saved view to saved view. This technique is especially helpful when you know you will often want to return to a specific area of your drawing.

TIP
Stored views are especially useful for controlling what appears in your plot. The Plot command allows you to plot a saved view.

---

1. Close the Aerial View window by clicking the Close button in the upper-right corner of the window.

2. Choose View ➤ Zoom ➤ All, or type **z⏎a⏎**, to get an overall view of the plan.

3. Click View ➤ Named Views. The View dialog box appears.

NOTE   From this dialog box, you can call up an existing view (Set Current), create a new view (New), or get detailed information about a view (Details).

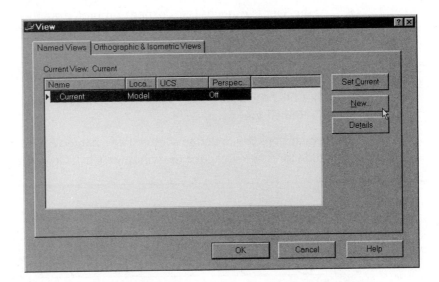

4.   Click the New button. The New View dialog box appears.

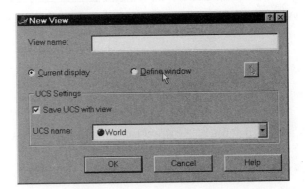

5.   Click the Define Window radio button. Notice that the Define View Window button became available.

6.   Click the Define View Window button. The dialog box momentarily disappears.

7. At the `Specify first corner:` prompt, click near one of the corners of the side view. (Don't pick a point where a marker has appeared; if you see one of the Osnap markers appear, press the F3 toggle.) You don't have to be exact because you're selecting view windows.

8. At the `Specify opposite corner:` prompt, drag the mouse to create a window around the view and click to pick the point. The dialog boxes will reappear.

9. Type **First** for the name of the view you just defined.

10. Click OK. The Define New View dialog box closes and you see First listed in the Named Views list.

11. Repeat steps 3–9 to define another view, named **Second**. Use Figure 6.8 as a guide for where to define the windows. Click OK when you're done.

FIGURE 6.8:

Save view windows in these locations for the plan drawing.

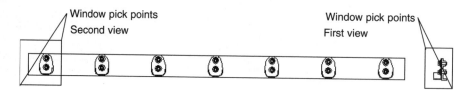

Window pick points
Second view

Window pick points
First view

TIP

A quick way to restore saved views is to type **-view↵r↵**, then enter the name of the view you wish to restore.

Now let's see how to recall the views that you've saved.

1. With the View dialog box open, click First in the list of views.

2. Click the Set Current button and then click OK. Your screen displays the first view you selected.

3. Use the View dialog box again to recall the view named Second.

Next, you'll zoom out to the extents of your drawing and then create a new view.

1. Choose View ➤ Zoom ➤ Extents or type z↵e↵.

2. Enter **view↵↵** at the command prompt and select New. In the New View dialog box select Current Display.

3. Type *Overall* for the name of the view. Click OK in the dialog boxes.

4. Save your work.

NOTE
Did you notice the New View dialog box has a UCS Name input box? For now, World is the correct setting, but when we work with 3D, this may change.

As you can see, this is a quick way to save a view. With the name Overall assigned to this view, you can easily recall the overall view at any time. (The View ➤ Zoom ➤ All option gives you an overall view, too, but it may zoom out too far for some purposes, or it may not show what you consider to be an overall view.)

TIP
Another useful tool for getting around in your drawing is the Zoom toolbar. It contains tools for Zoom Window, Dynamic, Scale, Center, In, Out, All, and Extents. To open the Zoom toolbar, right-click on any toolbar, then click the name *Zoom* in the dialog box that appears.

Opening a File to a Particular View

The Open Drawing dialog box contains a Select Initial View check box. If you open an existing drawing with this option checked, you're greeted with a Select Initial View dialog box just before the opened file appears on the screen. This dialog box lists any views saved in the file. You can then go directly to a view by double-clicking the view name. If you have saved views and you know the name of the view you want, using Select Initial View saves time when you're opening very large files.

Freezing Layers to Control Regeneration Time

You may wish to turn off certain layers altogether to plot a drawing containing only selected layers. But even when layers are turned off, AutoCAD still takes the time to redraw and regenerate them. The Layer Properties Manager dialog box offers the Freeze option, which acts like the Off option. However, Freeze causes AutoCAD to ignore frozen layers when redrawing or regenerating a drawing. By freezing layers that aren't needed for reference or editing, you can reduce the time AutoCAD takes to perform Regens.

TIP   You can freeze and thaw individual layers by clicking the sun icon in the layer drop-down list in the Object Properties toolbar.

Block Visibility with Freeze and Thaw

When the layer a block is inserted onto is frozen, the entire block is made invisible, regardless of the layer assignments of the objects contained in the block.

Keep in mind that when blocks are on layers that aren't frozen; the individual objects that are a part of a block are still affected by the status of the layer to which they are assigned.

You can take advantage of this feature by using layers to store parts of a drawing that you may want to plot separately. With respect to Freeze/Thaw visibility, external referenced files inserted using the external reference (Xref) command also act like blocks. For example, you can Xref several drawings on different layers. Then, when you want to view a particular Xref drawing, you can freeze all of the layers except the one containing that drawing.

Using layers and blocks requires careful planning and record keeping. If used successfully, this technique of managing block visibility with layers can save substantial time when you're working with drawings that use repetitive objects or that require similar information that can be overlaid.

Taking Control of Regens

Another way to control regeneration time is by setting the Regenmode system variable to 0 (zero). You can also use the Regenauto command to accomplish the same thing, by typing **regenauto.┘off.┘**.

If you then issue a command that normally triggers a Regen, AutoCAD will give the message About to Regen - Proceed? OK or Cancel. For example, when you globally edit attributes, redefine blocks, thaw frozen layers, change the Ltscale setting, or in some cases, change a text's style, you'll get the Regen queued message. You can *queue up* Regens at a time you choose, or you can issue a Regen to update all of the changes at once by choosing View ➤ Regen or by typing **re.┘**. This way, only one Regen occurs instead of several.

By taking control of when Regens occur, you can reduce the overall time you spend editing large files.

Creating Multiple Views

So far, you've looked at ways to help you get around in your drawing while using a single-view window. You also have the capability to set up multiple views of your drawing, called *viewports*. With viewports, you can display more than one view of at a time in the Auto-CAD drawing area. For example, you can have one viewport to show a close-up of the extrusion, another viewport to display the overall plan view, and yet another to display the bolt detail.

When viewports are combined with AutoCAD's Paper Space feature, you can plot multiple views of your drawing. Paper Space is a display mode that lets you "paste up" multiple views of a drawing, much like a page layout program. To find out more about viewports and Paper Space, see Chapter 9, "Advanced Productivity Tools" and Chapter 12, "Mastering 3D Solids."

Using the AutoTracking Feature

The AutoTracking feature is useful for controlling the placement of objects. AutoTracking displays alignment paths as dashed lines to help you draw objects at precise positions, such as at specific angles or in specific relationships to other objects. AutoTracking offers two options: Polar Tracking and Object Snap Tracking.

Polar Tracking

Polar Tracking helps you align your cursor to exact horizontal and vertical angles, much like a T-square and triangle. The Polar Tracking feature works in conjunction with the Polar Snap feature. When you turn on Polar Tracking and start to draw, alignment paths and tool tips appear when you move the cursor near polar angles. By default, Polar Tracking tracks along 90° polar angle increments, but you can change the angle it uses.

Let's try drawing with Polar Tracking.

1. Turn on Polar Tracking by pressing the F10 key or by clicking Polar in the status bar.

2. Click the Line tool in the Draw toolbar (or type l↵ or choose Draw ➤ Line).

3. At the Specify first point: prompt, select a point on the left side of the drawing area.

4. At the Specify next point or [Undo]: prompt, press the F8 key to turn off Ortho mode. Then move the crosshairs to the right until the coordinate readout is some distance, approximately 3, followed by < 0° (the distance isn't important, but the angle must be correct) and type **3**↵.

NOTE   Ortho mode restricts the cursor to a vertical or horizontal direction. You cannot have Ortho mode and Polar Tracking turned on at the same time. AutoCAD turns off Polar Tracking when you turn on Ortho mode.

Now you should have a line three units long. You'll continue placing line segments, but first you'll change the Polar Tracking angle to 45°. You change the angle through the Polar Tracking tab of the Drafting Settings dialog box. You don't need to exit the Line command; you can set angles "on the fly."

1. Choose Tools ➢ Drafting Settings and click the Polar Tracking tab (or right-click the Polar button in the status bar and select Settings from the pop-up menu).

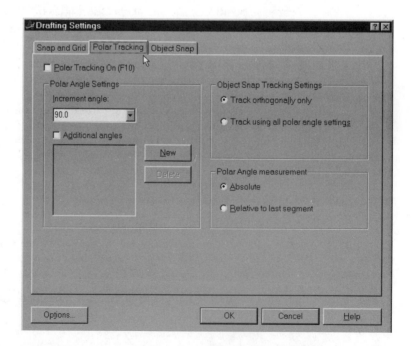

2. From the Increment Angle drop-down list, select 45.

3. For the Polar Angle Measurement setting, click the Relative to Last Segment radio button.

4. Click OK to close the Drafting Settings dialog box.

5. Draw lines with the following distances and angles, in the sequence shown, then close the drawing:

$$3 < 45°; 3 < 0°; 1 < 90°; 3 < 180°; 3 <135°; 3 < 180°$$

As this exercise demonstrated, you can change the Polar Tracking angle by selecting a predefined angle from the drop-down list in the Polar Tracking tab of the Drafting Settings dialog box. The predefined angle increments are 90, 60, 45, 30, 22.5, 18, 15, 10, and 5. You also can specify other angles and set the angle on which Polar Tracking bases its incremental angles.

To add a polar angle, display the Polar Angle tab (choose Tools ➤ Drafting Settings) and click the New button next to the Additional Angles list box. Then enter a new angle. The value you enter appears in the list box. To use your added angle, click the Additional Angles check box to put a check in it. To delete a value from the Additional Angles list box, highlight it and click the Delete button.

In the previous exercise, you checked the Relative to Last Segment option for the Polar Angle Measurement setting. This option uses the last-drawn object as the 0° angle. For example, if you draw a line at a 10° angle, with angle increments of 90°, Polar Tracking snaps to 10°, 100°, 190°, and 280°, relative to the actual 0° direction. The Absolute option uses the current AutoCAD setting for the 0° angle.

TIP You can enter a Polar Tracking angle that is valid for specifying one point. To enter a polar override angle, enter an angle preceded by a left angle bracket (<) whenever a command asks you to specify a point.

Object Snap Tracking

The Object Snap Tracking feature allows you to align a point to the geometry of an object instead of just selecting a point on an object. For example, with Object Snap Tracking, you can select a point that is based on an object midpoint or an intersection between objects.

The Object Snap Tracking feature works in conjunction with Osnaps. You must set an Osnap before you can track from an object's snap point. You can use the Endpoint, Midpoint, Center, Node, Quadrant, Intersection, Insertion, Parallel, Extension, Perpendicular, and Tangent Osnaps. If you use a Perpendicular or Tangent Osnap, Object Snap Tracking tracks to alignment paths perpendicular to or tangent to the selected object.

To turn on Object Snap Tracking, turn on an Osnap (for a single point or Running Osnap) and then press F11 or click Otrack in the status bar. Then, when a command prompts you to specify a point, move the cursor over the Osnap marker you want to use for tracking. Without clicking the mouse, hold the cursor there for a second until you see a small plus sign (+). Look carefully, because the plus sign is quite small. This is the Object Snap Tracking marker, indicating an "acquired" point. As you move the cursor, a dashed line appears, emanating from the Osnap marker. The cursor shows a small X following the dashed line as you move it. For example, to use Object Snap Tracking with an Endpoint Osnap, click the line's starting point, move the cursor over another line's endpoint to acquire it, move the cursor to endpoint you want for the new line, and click.

NOTE   To clear an acquired point, move the cursor back over the point's Object Snap Tracking marker or click Otrack in the status bar to turn off Object Snap Tracking.

The Object Snap tab of the Drafting Settings dialog box (select Tools ➤ Drafting Settings) has an Object Snap Tracking On check box. The Polar Tracking tab includes an Object Snap Tracking Settings section that lets you use Object Snap Tracking in conjunction with Polar Tracking. When the Track Using All Polar Angle Settings option is selected, Object Snap Tracking uses the angles set in the Polar Angle Settings section. If you select Track Orthogonally Only, Object Snap Tracking uses strictly orthogonal directions (0°, 90°, 180°, and 270°).

Changing AutoTracking Settings

You can change how AutoTracking displays the tracking vector used for Polar Tracking and Object Snap Tracking, and you can change how AutoCAD acquires Object Snap Tracking alignment points. These settings are in the Drafting tab of the Options dialog box (select Tools ➤ Options).

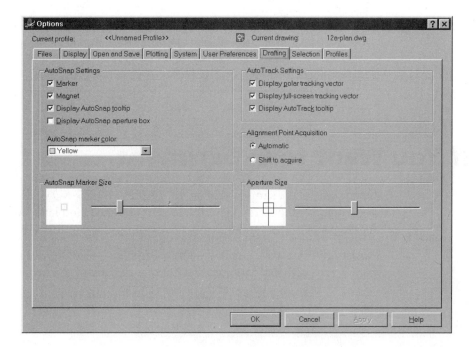

There are three options in the AutoTrack Settings section:

Display Polar Tracking Vector   Turns the Polar Tracking vector on or off (when cleared, no Polar Tracking path is displayed).

Display Full-Screen Tracking Vector   Controls whether the tracking vector appears across the full width of the drawing area or stops at the cursor location or the intersection of two tracking vectors. (When cleared, the Polar Tracking path is displayed only from the Osnap point to the cursor.)

Display AutoTracking Tool Tip   Turns the AutoTracking tool tips on or off.

There are two Alignment Point Acquisition options:

Automatic   Automatically acquires Object Snap Tracking alignment points.

Shift to Acquire   Acquires Object Snap Tracking alignment points only when you press Shift while the cursor is over an Osnap point.

NOTE   The Trackpath system variable stores AutoTrack alignment path display settings. The Polarmode system variable stores the Object Snap Tracking alignment point acquisition method.

If You Want to Experiment...

Open some of the sample files on your AutoCAD CD-ROM; Azimuth is a very good one. Try the Aerial view, the Regen command, and layer control with these files to better understand the options for improving productivity.

Also, give AutoTracking a try. Draw the exercises at the end of Chapter 5, "Editing for Productivity," using the techniques demonstrated here.

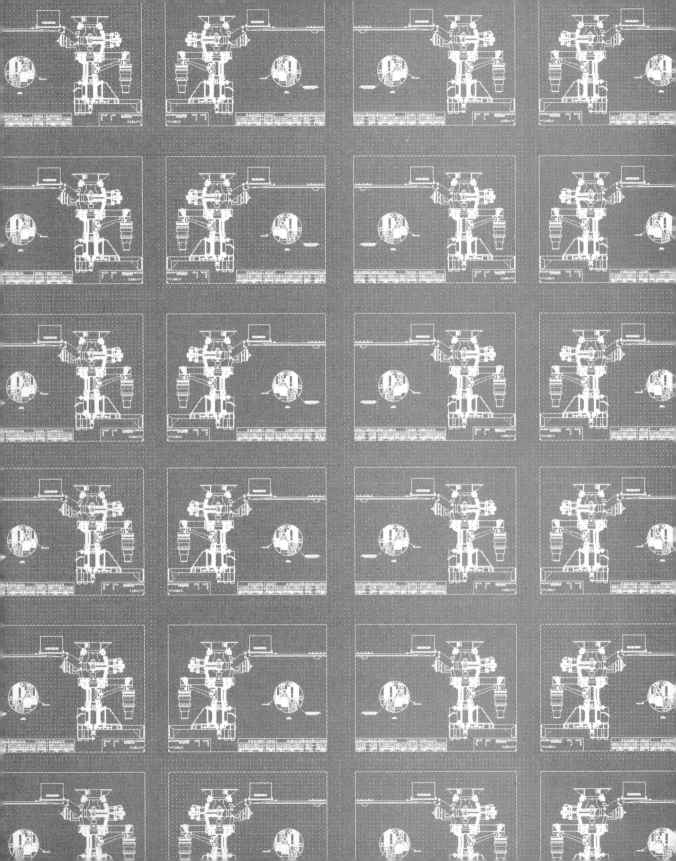

CHAPTER
SEVEN

Adding Text
to Drawings

- Adding Text

- Setting Justification of Text

- Setting Text Scale

- Selecting Fonts

- Creating a Text Style

- Editing Existing Text

- Importing Text from Outside AutoCAD

- Using the Spelling Checker

One of the more tedious drafting tasks is the addition of notes or annotations to your drawing. Anyone who has had to create a large drawing containing a lot of notes knows the true meaning of writer's cramp. AutoCAD not only makes this job go faster by allowing you to type your notes right into the same document as the corresponding drawing, it also helps you to create more professional-looking notes by using a variety of fonts, type sizes, and type styles. AutoCAD 2000 also provides an improved text tool that simplifies access to all of the text features.

In this chapter, you'll add notes to your bracket drawing. In the process, you'll explore some of AutoCAD's text features. You'll learn how to control the size, slant, type style, and orientation of text, and how to import text files. In Chapter 8, "Using Dimensions," we'll examine the effect of the current text settings on dimensions.

Adding Text to a Drawing

In this section, you'll type a set of notes that you might find on a working drawing.

1. Start AutoCAD and open the file named bkt_v1. You can find the file bkt_v1 .dwg on the CD-ROM. Use File ➢ Save As to save it as a file called notes.

2. Use the layer drop-down list to make the Dim layer current. Some users create a layer called Notes and make it the current layer. Dim is the layer on which you'll keep all of your text information in this exercise.

NOTE   You can place text on any layer. The text will inherit the color and linetype of the layer that you choose. If you placed text on the Hidden layer, it would be red and composed of dashed lines.

---

3. Click the Multiline Text tool in the Draw toolbar or type **mtext**⏎.

4. Click the first point indicated in the top image of Figure 7.1 to start the text boundary window. This boundary window indicates the area in which to place the text. Notice the arrow near the bottom of the window. It indicates the direction of the text flow.

NOTE   You don't have to be too precise about where you select the points for the boundary because you can make adjustments to the location and size later. The height of the window is unimportant. AutoCAD will use the width of the window to determine where the *word wrap* (the automatic width limit of the sentence or paragraph) will occur. AutoCAD will continue to write lines of text in the direction of the arrow as long as you do.

---

5. Click the second point indicated in the top image of Figure 7.1. The Multiline Text Editor appears.

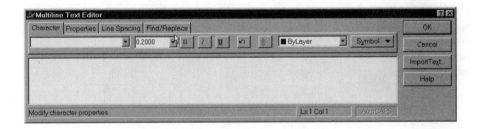

6. You could start typing the text, but first you need to select a size. Point to the font height drop-down list and click it. The default size highlights.

7. Enter **.12↵** to make the default height 0.12".

8. Click in the main text window and type **NOTES: UNLESS OTHERWISE SPECIFIED**. As you type, the words appear in the text window, just as they will appear in your drawing. As you'll see later when it's completed, the text also appears in the same font as the final text.

NOTE   The default font is a native AutoCAD font called Txt.shx. In this chapter, you'll see that you can also use TrueType fonts and PostScript fonts.

---

9. Press ↵ to advance one line, then enter **1. PERMANENT MARK PART NUMBER AND LATEST REVISION LETTER USING .10 MINIMUM HIGH CONTRASTING CHARACTERS APPROXIMATELY WHERE SHOWN.**

10. Press ↵ again to advance another line and enter **2. THIS DRAWING SHALL BE INTERPRETED PER ANSI Y14.5 1994**.

11. Click OK. The text appears in the drawing (see the second image in Figure 7.1).

FIGURE 7.1:

The top image shows the points to pick to place the text boundary window. The bottom image shows the completed text.

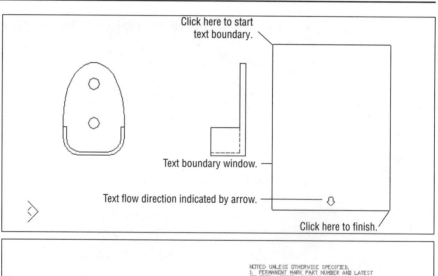

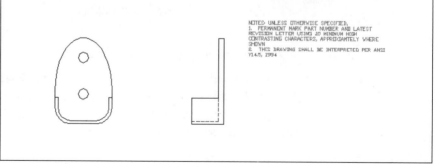

The Multiline Text Editor window works like any text editor, so if you make a typing error, you can highlight the error and then retype the letter or word. You can also perform many other functions such as searching and replacing, importing text, changing line spacing, or making font changes.

The following sections discuss some of the many options available for formatting text.

TIP   If text is selected in an area where a hatch pattern is to be placed, AutoCAD will automatically avoid hatching over the text. If you add text over a hatched area, you must re-hatch the area to include the text in the hatch boundary. (We'll discuss hatching in detail in Chapter 10, "Drawing Curves and Solid Fills.")

Understanding Text Formatting in AutoCAD

AutoCAD offers a wide range of text-formatting options. You can control fonts, text height, justification, and width. You can even include special characters such as degree symbols (°) or stacked fractions. In a departure from the somewhat clumsy text implementation of earlier AutoCAD versions, you now have a much wider range of control over your text.

Adjusting the Text Height and Font

Let's continue our look at AutoCAD text by making a proprietary note. You'll use the Multiline Text tool again, but this time you'll get to try out some of its other features. In this first exercise, you'll see how you can adjust the size of text in the editor.

1. Click the Multiline Text tool again. Then select a text boundary window, as shown in the top image of Figure 7.2.

FIGURE 7.2:

Placing the text boundary window for the proprietary note

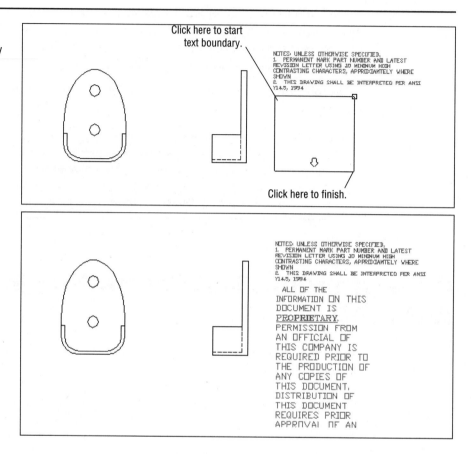

2. In the Multiline Text Editor, start typing the following text:

All of the information on this document is PROPRIETARY. Permission from an official of this company is required prior to the production of any copies of this document. Distribution of this document requires prior APPROVAL of an officer of this company.

As you type, you'll notice that the sentences break and appear on separate lines even though you did not press ↵ between them. AutoCAD uses word wrap to fit the text inside the text boundary area.

3. Highlight the text *All of the information* as you would in any word processor. For example, you can click the end of the line to place the cursor there; then Shift+click the beginning of the line to highlight the whole line.

4. Click the font height drop-down list and enter **.16**↵. The highlighted text changes to a smaller size.

5. Highlight the word *PROPRIETARY*.

6. Click the font drop-down list. A list of font options appears.

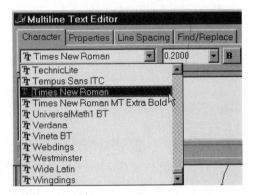

7. Scroll up the list until you find Complex. This font is available in all installations of AutoCAD. Notice that the text changes to reflect the new font.

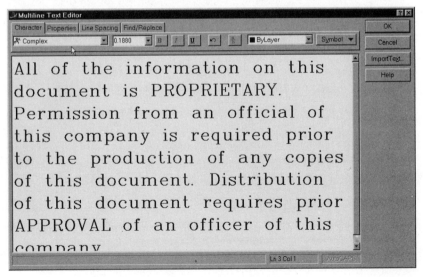

8. With *PROPRIETARY* still highlighted, click the Underline button.

9. Click OK. The note appears in the area you indicated in step 2 (see the bottom image of Figure 7.2).

Using PostScript Fonts

If you have PostScript fonts that you would like to use in AutoCAD, you'll need to compile them into AutoCAD's native font format. Here's how it's done.

1. Type **compile**↵. The Compile Shape or Font File dialog box appears.

2. Select PostScript Font (***.pfb**) from the File of Type drop-down list.

3. Double-click the PostScript font you want to convert into the AutoCAD format. AutoCAD will work for a moment, then you'll see the following message:

   ```
   Compiling shape/font description file
   Compilation successful. Output file E:\ACADR13\COMMMON\FONTS\
   fontname.shx contains 59578 bytes.
   ```

When AutoCAD is done, you'll have a file with the same name as the PostScript font file, but with the **.shx** filename extension. If you place your newly compiled font in AutoCAD's Fonts folder, it will be available in the font list.

When you work with AutoCAD's **.shx** font files, it is important to remember that AutoCAD cannot compile and load every Type 1 font. The PostScript font facilities in AutoCAD are intended to process a subset of Adobe fonts. If you receive an error while compiling a PostScript font, the resulting **.shx** file (if one is generated) may not load into AutoCAD. Also note the following:

- License restrictions still apply to the AutoCAD-compiled version of the PostScript font.

- Like other fonts, compiled PostScript fonts can use up substantial disk space, so compile only the fonts you need.

While using the Multiline Text tool, you may have noticed the [Height/Justify/ Rotation/Style/Width]: prompt immediately after you picked the first point of the text boundary. You can use any of these options to make on-the-fly modifications to the height, justification, rotation style, or width of the multiline text.

For example, after clicking the first point for the text boundary, you can type **r**↵ and then specify a rotation angle for the text windows, either graphically with a rubber-banding line or by entering an angle value. Once you've entered a rotation angle, you can resume selecting the text boundary.

Adding Color, Stacked Fractions, and Special Symbols

In the previous exercise, you were able to adjust the text height and font, just as you would in any word processor. You saw how you can easily underline portions of your text using the toolbar buttons in the editor. Other tools allow you to set the color for individual characters or words in the text, create stacked fractions, or insert special characters. Here's a brief description of how these tools work.

- To change the color of text, highlight the text and then select the color from the text color drop-down list.

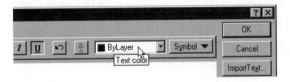

- To turn a fraction into a stacked fraction, highlight the fraction and then click the Stack/Unstack tool.

- To add a special character, place the cursor at the location of the character and then click the Symbol tool. A drop-down list appears, offering options for special characters.

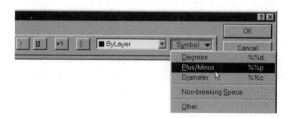

The Symbol tool offers three standard options that are typical for most technical drawings: the degree (°), plus (+)/minus (–), and diameter (n) signs. When you select these options, AutoCAD will insert the proper AutoCAD text code that corresponds to these symbols. Some symbols won't appear in the editor as symbols. Instead, they will appear as a special code. However, once you return to the drawing, you'll see the text with the proper symbol. You'll get a more detailed look at special symbols later in this chapter in the section "Adding Special Characters."

Adjusting the Width of the Text Boundary Window

Although your text font and height are formatted correctly, the text may appear stacked in a way that is too tall and narrow. The following exercise will show you how to change the boundary to fit the text.

1. Click any part of the *PROPRIETARY* text you just entered to highlight it.

2. Click the upper-right grip.

3. Drag the grip to the right to the location shown in Figure 7.3, then click that point.

4. Click any grip and then right-click the mouse and select Move.

5. Move the text down slightly.

AutoCAD's word-wrap feature automatically adjusts the text formatting to fit the text boundary window. This feature is especially useful to AutoCAD users because other drawing objects often affect the placement of text. As your drawing changes, you'll need to make adjustments to the location and boundary of your notes and labels.

FIGURE 7.3:

Adjusting the text boundary window

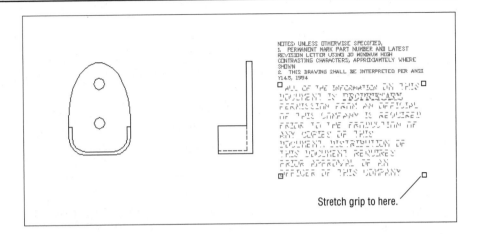

Adjusting the Text Alignment

The text is currently aligned on the left side of the text boundary. It would be more appropriate to align the text to the right. Here's how you can make changes to the text alignment.

1. If the text is not yet selected, click it.

2. Choose Modify ➤ Text from the pull-down menu. The Multiline Text Editor dialog box appears.

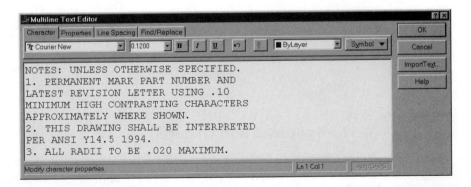

NOTE   If you click multiple text objects while using the Properties tool, you'll only get the abbreviated Change Properties dialog box.

---

3. Click the Properties tab. The editor changes to display a different set of options.

4. Click the Justification drop-down list. The alignment options appear.

5. Click Top Right. All of the text in the window is now aligned to the right.

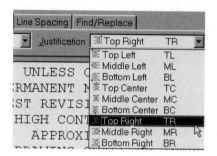

6. Click OK, then close the Properties dialog box. The text changes to align at the top right of the text, as shown in Figure 7.4.

FIGURE 7.4:

The text aligned using the Top Right alignment option

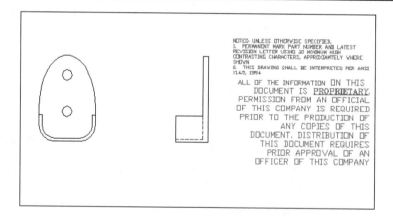

Text Alignment and Osnaps

While it's clear that the text is now aligned at its top right, one important, but less obvious, change also occurred. You may have noticed that the Justification list offered three centered options: Top Center, Middle Center, and Bottom Center. All three of these options will have the same effect on the text's appearance, but each has a different effect on how Osnaps act upon the text. Figure 7.5 shows where the Osnap point occurs on a text boundary, depending on which alignment option is selected. A Multiline text object will only have one insertion point on its boundary that you can access with the Insert Osnap.

FIGURE 7.5:

The location of the Insert Osnap points on a text boundary based on its alignment setting

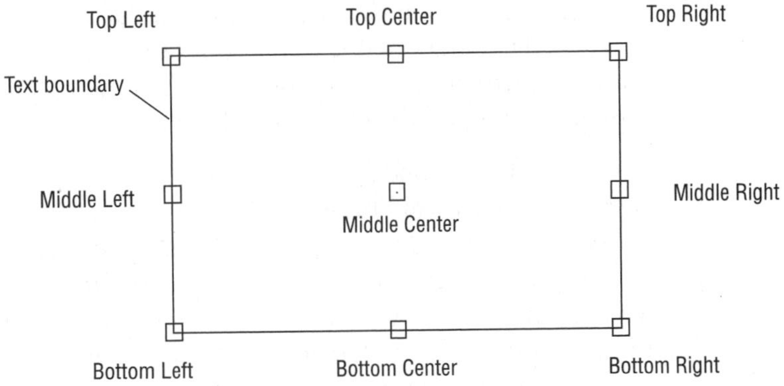

Top Left   Top Center   Top Right

Text boundary

Middle Left   Middle Right

Middle Center

Bottom Left   Bottom Center   Bottom Right

Knowing where the Osnap points occur can be helpful when you want to align the text with other objects in your drawing. In most cases, you can use the grips to align your text boundary, but the Center and Middle alignment options allow you to use the center and middle portions of your text to align the text with other objects.

Editing Existing Text

It is helpful to think of text in AutoCAD as a collection of text documents. Each text boundary window you place is like a separate document. To create and edit these documents, use the Multiline Text Editor. Let's add more text to the NOTES label.

1. Type **ddedit**↵ to issue the Ddedit command.

2. Click the words *NOTES: UNLESS*. The Multiline Text Editor dialog box appears.

3. Click the text cursor on the end of the line that reads *1994*, then press ↵ and type **3. ALL RADII TO BE .020 MAXIMUM.**

4. Click OK. The text appears in the drawing with the additional line.

5. The Ddedit command is still active, so press ↵ to exit Ddedit.

In step 5, the Ddedit command remained active, so you could continue to edit other text objects. Besides pressing ↵ to exit the command, you can select another text object.

As with the prior exercise, you can change the formatting of the existing or new text while in the Multiline Text dialog box. Notice that the formatting of the new text is the same as the text that preceded it. Similar to Microsoft Word, the formatting of text is dependent on the paragraph or word to which it is added. If you had added the text after the last line, it would have appeared in the AutoCAD Txt font and in the same 0.12-unit height.

You can also edit and format text using the Properties dialog box. Select the text, right-click, and select Properties. The Properties dialog box for the selected text appears. Make sure the Categorized tab is selected, and then scroll down the list until you can see all of the Text options. The text you selected appears in the Contents input box. If you want to make changes to the text, you can do it directly in the Contents box. To make formatting and other changes, click the Contents label in the left column. This exposes the button with three dots (the ellipsis button). Click the ellipsis button to bring up the Multiline Text Editor with the text. From here, you can edit and format the text as described in the previous sections.

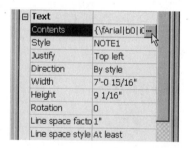

Text	
Contents	{\fArial|b0|iC ···
Style	NOTE1
Justify	Top left
Direction	By style
Width	7'-0 15/16"
Height	9 1/16"
Rotation	0
Line space facto	1"
Line space style	At least

NOTE   The code you see mixed in with the text in the Contents input box is normally hidden from you in the Multiline Text Editor, and you don't really need to concern yourself with it. However, if you edit the text in the Contents input box, make sure you don't change the coding unless you know what you're doing.

Finding and Replacing Text

AutoCAD's Find and Replace feature lets you search for and replace text any-where in your drawing. Click the Find and Replace tool in the Standard toolbar, type **find**⏎, or click the Find/Replace tab in the Multiline Text Editor.

The Find and Replace dialog box opens. Enter the text string you want to find in the Find Text String input box. If you want to replace the text with other text, enter the replacement text in the Replace With input box. The Search In drop-down list lets you choose between searching through the current selection or entire drawing. Click the Options button to select from options for refining your search. Click the Find Next button to start the search. The found text appears in the Search Results box.

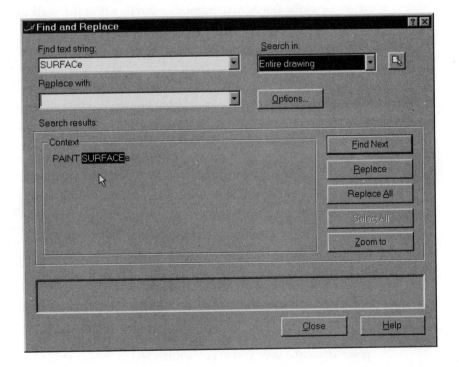

You can continue to find the next occurrences of the text string with the Find Next button. To replace the text, click the Replace button to replace occurrences one at a time or click the Replace All button to make all the replacements with one click (with the usual caveat of being careful when you make such global replacements). The Zoom To button zooms to the occurrence of the found text. The Help button takes you to help information about the Find and Replace feature. When you're finished, click the Close button.

Understanding Text and Scale

You'll sometimes need to make a scale conversion for your text size to make the text conform to the drawing's intended plot scale.

Text-scale conversion is a concept many people have difficulty grasping. As you may have discovered in previous chapters, AutoCAD allows you to draw at full scale—that is, to represent distances as values equivalent to the actual size of the object. When you plot the drawing later, you tell AutoCAD at which scale you wish to plot, and the program reduces or enlarges the drawing accordingly. This allows you the freedom to input measurements at full scale and not worry about converting them to various scales every time you enter a distance. Unfortunately, this feature can also create problems when you enter text and dimensions. Similar to converting the plotted sheet size to an enlarged size equivalent at full scale in the drawing editor, you must convert your text size to its equivalent at full scale.

To illustrate this point, imagine that you're drawing a large piping plan at full size on a very large sheet of paper. When you're done with this drawing, it will be reduced to a scale that will allow it to fit on an $8\frac{1}{2}$" × 11" sheet of paper. So you have to make your text quite large to keep it legible once it is reduced. This means that if you want text to appear $\frac{1}{8}$" high when the drawing is plotted, you must convert it to a considerably larger size when you draw it. To do this, multiply the desired height of the final plotted text by a scale conversion factor, which is the inverse of (1 divided by) the plot scale. For example, if you're going to plot the drawing at $\frac{1}{4}$ of the actual size of the object, use the formula text size=1 divided by the plot scale times the desired text size when plotted. To do the math, the text size=1 divided by $\frac{1}{4}$ times 0.125, giving you an actual size of 0.5.

If your drawing is at $\frac{1}{8}$"=1" scale, you multiply the desired text height, $\frac{1}{8}$", by the scale conversion factor of 1 divided by $\frac{1}{8}$" to get a height of 1 (Table 3.3 showed scale factors as they relate to standard drawing scales). This is the height you must make your text to get $\frac{1}{8}$"-high text in the final plot. Table 7.1 shows you some other examples of text height to scale.

TABLE 7.1: ⅛"-High Text Converted to Size for Various Drawing Scales

Drawing Scale	Scale Factor	AutoCAD Drawing Height for ⅛"-High Text
¹⁄₁₆"=1"	16.00	2.000"
⅛"=1"	8.00	1.000"
¼"=1"	4.00	0.500"
½"=1"	2.00	0.250"
¾"=1"	1.33	0.166"
1"=1"	11.00	0.125"
2"=1"	0.50	0.062"
4"=1"	0.25	0.031"

Organizing Text by Styles

As you expand your drawing skills and your drawings become larger, you'll want to start organizing your text into *styles*. AutoCAD used a Standard text style to create the text shape in the last exercise. You can think of text styles as a way to store your most common text formatting. Styles will store text height and font information, so you don't have to reset these options every time you enter text. But styles also include some settings not available in the Multiline Text Editor. You can also make global changes to all of the text created with a specific style. You could change the height, obliquing angle, justification, and font for all text of a certain style.

Creating a Style

In the previous examples, you entered text using the AutoCAD default settings for text. Whether you knew it or not, you were also using a text style: AutoCAD's default style called Standard. The Standard style uses the AutoCAD Txt font and numerous other settings that you'll learn about in this section. These other settings include width factor, obliquing, and default height.

TIP

If you don't like the way that the AutoCAD default style is set up, open the **Acad.dwt** file and change the Standard text style settings to your liking. Or add other styles that you use frequently.

The previous exercises in this chapter demonstrated that you could modify the formatting of a style as you entered the text. But for the most part, once you've set up a few styles, you won't need to adjust settings such as fonts and text height each time you enter text. You'll be able to select from a list of styles you've previously created, and just start typing.

To create a style, choose Format ➤ Text Style, and then select from the fonts available. The next exercise will show you how to create a style.

1. Choose Format ➤ Text Style or type **st↵**. The Text Style dialog box appears.

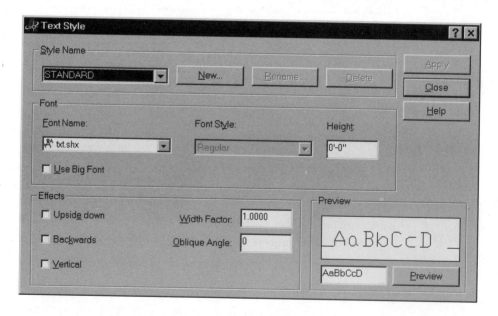

2. Click the New button in the Style Name group. The New Text Style dialog box appears.

3. Enter **note2** for the name of your new style, then click OK.

4. Now select a font for your style. Click the Font Name drop-down list in the Font group.

5. Locate the Courier New TrueType font and select it.

6. In the Height input box, enter **.12**.

7. Click Apply and then click Close.

Using a Type Style

A newly created style becomes the default style. All new text will be created with the Note2 style until you change it. You can also change existing text to a different style. Let's change some text to the new Note2 style.

1. Type **ed.↵** and select your notes text to open the Multiline Text Editor.

2. Click the Properties tab in the Multiline Text Editor.

3. Highlight the text you want to change.

4. Click the Style drop-down list and select Note2. Notice that all the text is converted to the new style.

5. Click OK. Your drawing looks like Figure 7.6.

FIGURE 7.6:

Modified notes text using the Note2 text style

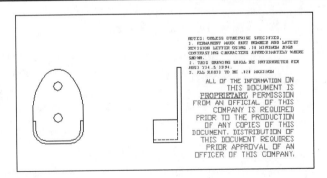

WARNING When you change the style of a text object, it loses any custom formatting it may have, such as font or height changes that are different from those of the text's default style settings.

Setting the Current Default Style

The last exercise showed you how to change the style of existing text. But suppose you want all of the new text you create to be of a different style than the current default style. You can change the current style by using the Text Style dialog box. Here's how it's done.

1. Click Format ➤ Text Style or type **st.**↵. The Text Style dialog box appears.

2. Select a style name from the Style Name drop-down list. It will be the default for all new text objects. For this exercise, select Standard to return to the Standard style.

3. Click Close.

AutoCAD will record the current default style with the drawing data when you issue a File ➤ Save command, so the next time you work on the file it will still have the same default style.

Understanding the Text Style Dialog Box Options

Now you know how to create a new style. This section describes the other settings in the Text Style dialog box. Some of them, such as the width factor, can be quite useful. Others, such as the Backward and Vertical options, are rarely used.

The Style Name Settings

The Text Style dialog box allows you to enter a custom name to identify the settings you'll use. This can be General for general text or Title for text in a title.

> **New**   Lets you create a new text style.
>
> **Rename**   Lets you rename an existing style. This option is not available for the Standard style.
>
> **Delete**   Deletes a style. This option is not available for the Standard style.

The Font Settings

The Font area is where you choose from a large selection of AutoCAD and True-Type fonts. This will allow you to make your drawing text match other documents you create.

Font Name   Lets you select a font from a list of available fonts. The list is derived from the font resources available to the Windows NT, 98, or 95 system fonts plus the standard AutoCAD fonts.

Font Style   Offers variations of a font such as italic or bold, when they are available.

Height   Lets you enter a font size. A 0 height has special meaning when you're entering text using the Dtext command described in the section "Adding Simple Text Objects," later in this chapter.

> **TIP**
> It is usually wise to leave the text height set to 0 for this reason. When we review dimensioning in Chapter 8, "Using Dimensions," you may wish to look at style and text height again.

The Effects Settings

Here, you may choose from the following options:

Upside down   Prints the text upside down.

Backwards   Prints the text backwards.

Width Factor   Adjusts the width and spacing of the characters in the text. A value of 1 keeps the text at its normal width. Values greater than 1 will expand the text; values less than 1 will compress the text.

```
This is the Simplex font expanded by 1.4
This is the simplex font using a width factor of 1
This is the simplex font compressed by .6
```

Oblique Angle   Skews the text at an angle. When this option is set to a value greater than 0, the text appears to be italicized. A value of less than 0 (–12, for example) will cause the text to "lean" to the left.

> *This is the simplex font*
> *using a 12-degree oblique angle*

Renaming a Text Style

You can use the Rename option in the Text Style dialog box to rename a style. An alternate method is to use the Rename command. This is a command that allows you to rename a variety of AutoCAD settings. Here's how to use it.

NOTE   This exercise is not part of the main tutorial. If you're working through the tutorial, make note of it and then try it out later.

---

1. Click Format ➤ Rename or enter **ren.**⏎ at the command prompt. The Rename dialog box appears.

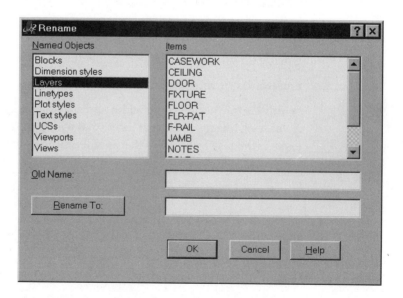

NOTE　The Rename command allows you to rename blocks, dimension styles, layers, linetypes, user coordinate systems, viewports, and views, as well as text styles.

---

2. In the Named Objects list box, click Style.

3. Click the name of the style you wish to change from the right-hand list; the name appears in the Old Name input box below the list.

4. In the input box next to the Rename To button, enter the new name, click the Rename To button, and then click OK.

What Do the Fonts Look Like?

You've already seen a few of the fonts available in AutoCAD. Chances are, you're familiar with the TrueType fonts available in Windows. You have some additional AutoCAD fonts from which to choose. In fact, you may want to stick with the AutoCAD fonts for all but your presentation drawings, as other fonts can consume more memory.

Figure 7.7 shows the basic AutoCAD text fonts. The Roman fonts are perhaps the most widely used, because they are easy to read and don't consume much memory.

FIGURE 7.7:

The Standard AutoCAD text fonts

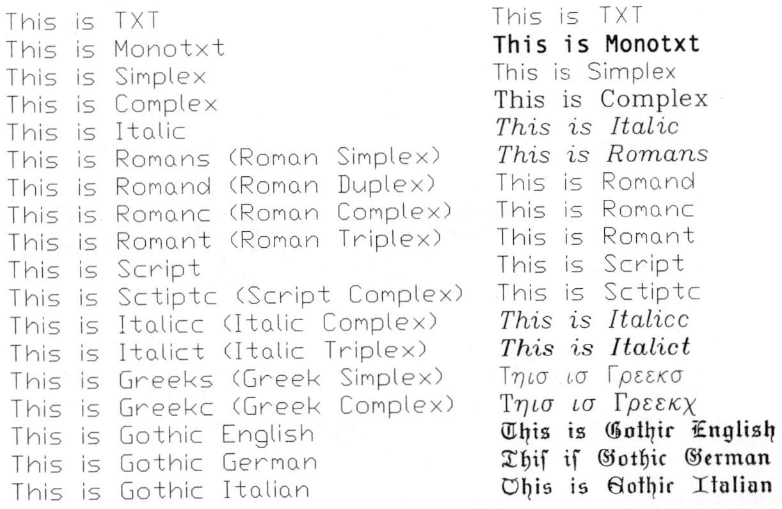

This is TXT	This is TXT
This is Monotxt	**This is Monotxt**
This is Simplex	This is Simplex
This is Complex	This is Complex
This is Italic	*This is Italic*
This is Romans (Roman Simplex)	*This is Romans*
This is Romand (Roman Duplex)	This is Romand
This is Romanc (Roman Complex)	This is Romanc
This is Romant (Roman Triplex)	This is Romant
This is Script	This is Script
This is Sctiptc (Script Complex)	This is Sctiptc
This is Italicc (Italic Complex)	*This is Italicc*
This is Italict (Italic Triplex)	*This is Italict*
This is Greeks (Greek Simplex)	Τηισ ισ Γρεεκσ
This is Greekc (Greek Complex)	Τηισ ισ Γρεεκχ
This is Gothic English	This is Gothic English
This is Gothic German	This is Gothic German
This is Gothic Italian	This is Gothic Italian

The Textfill System Variable

Unlike the standard stick-like AutoCAD fonts, TrueType and PostScript fonts have filled areas. These filled areas take more time to generate, so if you have a lot of text in these fonts, your Redraw and Regen times will increase. To help reduce Redraw and Regen times, you can set AutoCAD to display and plots these fonts as outline fonts, even though they are filled in their true appearance.

To change its setting, type **textfill⏎** and then type **0⏎**. This turns off text fill for PostScript and TrueType fonts. For plots, you can remove the check mark on the option labeled Text Fill (this is the same as setting the Textfill system variable to 0).

We've shown you samples of the AutoCAD fonts in this section. You can see samples of all the fonts, including TrueType fonts, in the Preview window of the Text Style dialog box. If you use a word processor, you're probably familiar with at least some of the TrueType fonts available in Windows and AutoCAD.

Adding Special Characters

Remember that you can add special characters using the Symbol button in the Multiline Text Editor. For example, you can use the degree symbol (°) to designate angles, the +/− symbol for showing tolerance information, and the diameter characters. AutoCAD also offers a nonbreaking space. You can use the nonbreaking space when there is a space between two words but you don't want the two words to be separated by a line break.

By clicking the Other option in the Symbols drop-down list, you can also add other special characters from the Windows Character Map dialog box.

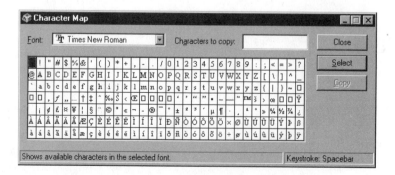

Characters such as the trademark (™) and copyright (©) symbols are often available. The contents of the list will vary depending on the current font. You can click and drag or just click your mouse over the character map to see an enlarged view of the character you are pointing to.

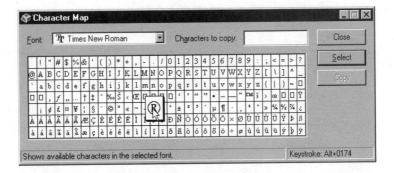

To use the characters from the Character Map dialog box, follow these steps. (This is not part of the regular tutorial in this chapter, although you can experiment with these steps on your own.)

1. Choose Other from the Symbols drop-down list in the Multiline Text Editor.

2. Highlight the character you want.

3. Either double-click the character or click the Select button. The character appears in the box at the upper-right corner of the dialog box.

4. Click Copy to copy the character to the Clipboard.

5. Close the dialog box.

6. In the editor, place the cursor where you want the special character to appear.

7. Press Ctrl+v to paste the character into your text. You can also right-click the mouse and choose Paste from the pop-up menu.

Importing Text Files

With Multiline text objects, AutoCAD allows you to import ASCII text or Rich Text Format (RTF) files. Here's how you go about importing text files.

1. From the Multiline Text Edit or dialog box, click Import Text.

Continued on next page

2. At the Open dialog box, locate a valid text file. It must be a file in a raw text (ASCII) format, such as a Notepad (`.txt`) file, or a Rich Text Format (`.rtf`) file. RTF files are capable of storing formatting information, such as bold and varying point sizes. They can be created in WordPad.

3. Once you've highlighted the file you want, double-click it or click OK. The text appears in the Edit Mtext window.

4. You can then click OK and the text will appear in your drawing.

In addition, you can use the Windows Clipboard and cut-and-paste features to add text to a drawing. To do this, take the following steps:

1. Use the Cut or Copy option in any other Windows program to place text onto the Windows Clipboard.

2. Go back to AutoCAD and choose Draw ➤ Text ➤ Multiline Text. At the `Specify first corner:` prompt, select a location for the new text. You'll then be prompted to `Specify second corner`. Try to allow enough room for the new text. (If you don't allow enough room, you can always use the Mtext objects Grips feature to make corrections after the text is placed in the drawing.)

3. The Mtext Editor appears. Use Window's Paste command (Ctrl+v) to import the text.

Text on the Clipboard may be pasted directly into AutoCAD by using the Edit ➤ Paste Special, and choosing Text as the format. This text is brought in as editable Mtext. Because AutoCAD is an OLE client, you can also attach other types of documents to an AutoCAD drawing file. See Chapter 18, "Getting and Exchanging Data from Drawings," for more on AutoCAD's OLE support.

Adding Simple Text Objects

You might find that you're entering a lot of single words or simple labels that don't require all the bells and whistles of the Multiline Text Editor. AutoCAD offers the *single-line text object*, which is simpler to use and can speed text entry if you're adding only small pieces of text.

Continue the tutorial by trying the following exercise:

1. Enter **dt.**⏎ or choose Draw ➤ Text ➤ Single Line Text. This issues the Dtext command.

2. At the DTEXT Specify start point of text or [Justify/Style]: prompt, pick the starting point for the text you're about to enter, just below the side view.

3. At the Specify height: prompt, enter **.12.**⏎ to indicate the text height.

4. At the Specify rotation angle of text <0>: prompt, press ⏎ to accept the default, 0°. You can specify any angle other than horizontal (for example, if you want your text to be aligned with a rotated object). You'll see a text I-beam cursor at the point you picked in step 2.

5. At the Enter Text: prompt, enter **VERSION -01 SHOWN.**⏎**SEE VERSIONS TABLE.**⏎**FOR OTHER OPTIONS**. As you type, the words appear in the drawing as well as in the Command window.

NOTE

If you make a typing error, use the left arrow (←) and right arrow (→) keys to move the text cursor in the Command window to the error; then use the Backspace key to correct the error. You can also paste text from the Clipboard into the cursor location by using Ctrl+v or by right-clicking in the Command window to access the pop-up menu.

6. Click above the front view to move the cursor there.

7. This time, type **PAINT SURFACE** ⏎. Figure 7.8 shows how your drawing should look.

8. Press ⏎ again to exit the Dtext command.

TIP

If for some reason you need to stop entering single-line text objects to do something else in AutoCAD, you can continue entering text where you left off by pressing ⏎ at the Start point: prompt of the Dtext command. The text will continue immediately below the last line of text entered.

Here you were able to add two single lines of text in different parts of your drawing fairly quickly. Dtext will use the current default text style settings (remember that earlier you set the style to Standard), so the VERSION and PAINT labels used the Standard style.

FIGURE 7.8:

Adding simple labels using
the Dtext command

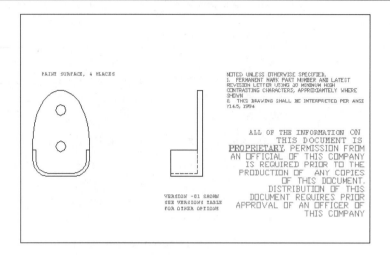

Editing Single-Line Text Objects

Editing single-line text uses the same tools as those for multiline text, although the dialog boxes that result are different. In this exercise, you'll change the PAINT label using the Ddedit command.

1. Type **ed**↵ or choose Modify ➤ Text.

2. Click the PAINT text. A small Edit Text dialog box appears.

3. Using the cursor, click the end of the line and type **4 PLACES**.

4. Click OK and then press ↵ to exit the Ddedit command.

As you can see, even the editing is simplified. You're limited to editing the text only. This can be an advantage, however, when you need to edit several pieces of text. You don't have other options to get in the way of your editing.

You can change additional properties of a single-line text object using the Properties dialog box. For example, suppose you want to change the PAINT label to a height of .18 inches.

1. Click the Properties tool in the Standard toolbar.

2. Click the PAINT text and press↵. The Properties dialog box appears.

3. Double-click the Height input box and enter **.18**.

4. Press ↵. The text increases in size to .18" high.

5. Click the Undo tool in the menu bar to undo the change in text height.

6. Choose File ➢ Save to save the changes you've made thus far.

The Properties dialog box lets you change the height, rotation, width factor, obliquing, justification, and style of a single-line text object. You can also modify the text content.

Justifying Single-Line Text Objects

Justifying single-line text objects works in a slightly different way from justifying multiline text. For example, if you change the justification setting to Center, the text will move so that the center of the text is placed at the text insertion point. In other words, the insertion point stays in place while the text location adjusts to the new justification setting.

To set the justification of text as you enter it, you must enter **j**↵ at the `Specify start point of text or [Justify/Style]:` prompt after issuing the Dtext command.

NOTE   You can also change the current default style by entering **s**↵ and then the name of the style at the `Specify start point of text or [Justify/Style]:` prompt.

Once you've issued the Dtext's Justify option, you'll then get the prompt:

```
Enter an option [Align/Fit/Center/
Middle/Right/TL/TC/TR/ML/MC/MR/BL/BC/BR]:
```

These options format text as follows:

Center   Causes the text to be centered on the start point, with the baseline of the text on the start point.

Middle   Causes the text to be centered on the start point, with the baseline slightly below the start point.

Right   Causes the text to be justified to the right of the start point, with the baseline on the start point.

TL, TC, and TR   Stand for top left, top center, and top right. Text using these justification styles appears entirely below the start point and justified left, center, or right, depending on which of the three options you choose.

ML, MC, and MR   Stand for middle left, middle center, and middle right. These styles are similar to TL, TC, and TR, except that the start point will determine a location midway between the baseline and the top of the lowercase letters of the text.

BL, BC, and BR   Stand for bottom left, bottom center, and bottom right. These styles, too, are similar to TL, TC, and TR, but here the start point determines the bottom-most location of the letters of the text (the bottom of letters that have descenders, such as p, q, and g). Figure 7.9 shows the relationship between the text start point and text justified with these options.

Align and Fit   Require you to specify a dimension within which the text is to fit. For example, suppose you want the text *SECTION A-A* to fit within the 1.25"-wide area at the bottom of a section view. You can use either the Fit or the Align option to accomplish this. With Fit, AutoCAD prompts you to select start and endpoints, and then stretches or compresses the letters to fit within the two points you specify. Use this option when the text must be a consistent height throughout the drawing and you don't care about distorting the font. Align works like Fit, but instead of maintaining the current text-style height, Align adjusts the text height to keep it proportional to the text width, without distorting the font. Use this option when it is important to maintain the font's shape and proportion. Figure 7.10 demonstrates how Fit and Align work.

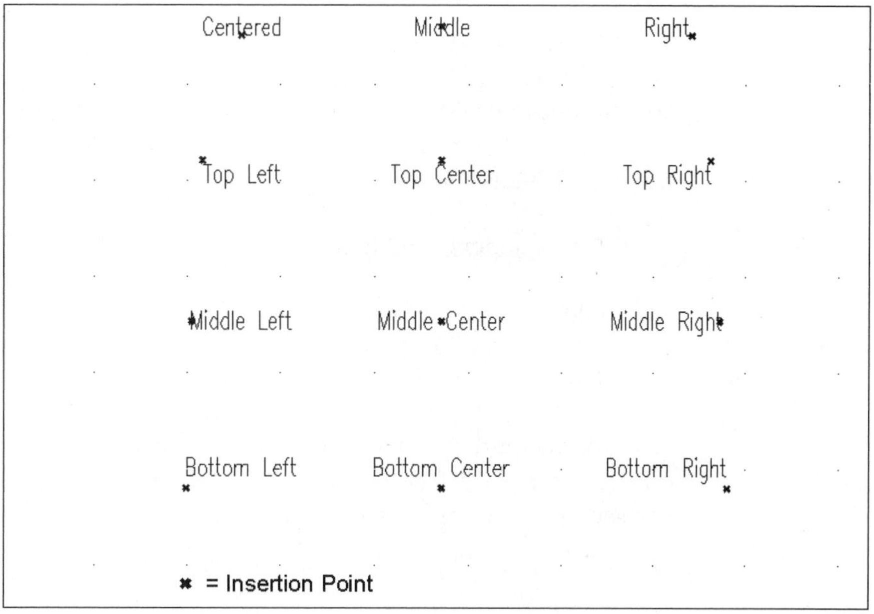

FIGURE 7.9:

Text inserted using the various Justify options. The X indicates the location of the start point in relation to the text. If you use the Properties tool to change the text justification, the text will move while the text insertion point remains in place.

Centered   Middle   Right

Top Left   Top Center   Top Right

Middle Left   Middle Center   Middle Right

Bottom Left   Bottom Center   Bottom Right

✻ = Insertion Point

FIGURE 7.10:

The phrase "SECTION A-A" as it appears normally and as it appears with the Fit and Align options selected

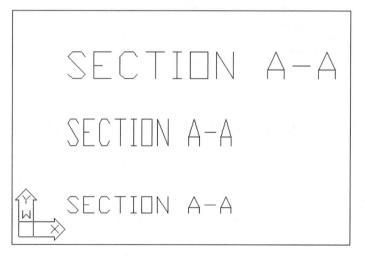

Using Special Characters with Single-Line Text Objects

You can add a limited set of special characters to single-line text objects. For example, you can place the degree symbol (°) after a number, or you can underscore (underline) text. To accomplish this, use double percent signs (%%) in conjunction with a special code. For example, to underscore text, enclose that text with the %% signs and follow it with the underscore code. So, to get this text:

This is <u>underscored</u> text.

you would enter this at the prompt:

This is %%uunderscored%%u text.

Overscoring (putting a line above the text) operates in the same manner. To insert codes for symbols, just place the codes in the correct positions for the symbols they represent. For example, to enter 100.5°, type **100.5%%d**.

Table 7.2 shows a list of the codes you can use.

TABLE 7.2:   Codes for Inserting Special Characters into Single-Line Text Objects

Code	Special Character
%%o	Toggles overscore on and off.
%%u	Toggles underscore on and off.
%%d	Places a degree symbol (°) where the code occurs.
%%p	Places a plus/minus sign where the code occurs.
%%%	Forces a single percent sign; useful when you want a double percent sign to appear, or when you want a percent sign in conjunction with another code.
%%nnn	Allows the use of extended Unicode characters when these characters are used in a text-definition file; nnn is the three-digit value representing the character.

Using the Character Map Dialog Box to Add Special Characters

You can add special characters to a single line of text in the same way you would with multiline text. You may recall that to access special characters, you use the Character Map dialog box. This dialog box can be opened directly from Windows Explorer.

Using Explorer, locate the file `Charmap.exe` in the Windows folder. Double-click it and the Character Map dialog box appears. You can then use the procedure spelled out in the "Adding Special Characters" section earlier in this chapter to cut and paste a character from the Character Map dialog box. If you find you use the Character Map dialog box often, create a shortcut for it.

Keeping Text from Mirroring

At times, you'll want to mirror a group of objects that contain some text. This operation will cause the mirrored text to appear backward. You can change a setting in AutoCAD to make the text read normally, even when it is mirrored.

1. Enter **mirrtext**↵.

2. At the `Enter new value for MIRRTEXT <1>:` prompt, enter **0**↵.

Now any mirrored text that is not in a block will read normally. The text's position, however, will still be mirrored, as shown here.

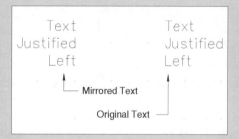

Mirrtext is set to 0 by default.

Checking Spelling

Although AutoCAD is primarily a drawing program, you'll find that some of your drawings contain more text than graphics. If you've ever used the spelling checker in a typical Windows word processor, such as Microsoft Word, the AutoCAD spelling checker's operation will be familiar to you. Here's how it works.

1. Choose Tools ➤ Spelling or type **spell**↵.

2. At the `Select objects:` prompt, select any text object you want to check. You can select a mixture of multiline and single-line text. When the spelling checker finds a word it doesn't recognize, the Check Spelling dialog box appears.

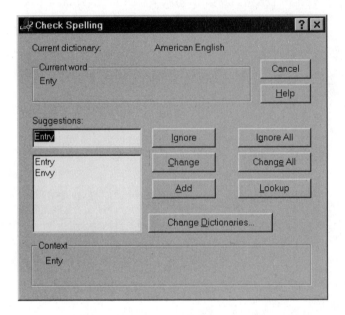

In the Check Spelling dialog box, you'll see the word in question, along with the spelling checker's suggested alternate word in the Suggestions input box. If the spelling checker finds more than one suggestion, a list of suggested replacement words appears below the input box. You can then highlight the desired replacement and click the Change button to fix the misspelled word, or click Change All to correct all occurrences of the word in the selected text. If the suggested word is inappropriate, choose another word from the replacement list (if any), or enter your own spelling in the Suggestions input box. Then choose Change or Change All.

Here is a list of the options available in the Check Spelling dialog box.

Ignore   Skips the word.

Ignore All   Skips all occurrences of the word in the selected text.

Change   Changes the word in question to the word you've selected (or entered) from the Suggestions input box.

Change All   Changes all occurrences of the current word when there are multiple instances of the misspelling.

Add   Adds the word in question to the current dictionary.

Lookup   Checks the spelling of the word in question. This option is for the times when you want to find another word that doesn't appear in the Suggestions input box.

Change Dictionaries   Lets you use a different dictionary to check spelling. This option opens the Change Dictionaries dialog box, described in the next section.

Choosing a Dictionary

The Change Dictionaries option opens the Change Dictionaries dialog box, where you can select a particular main dictionary for foreign languages, or create or choose a custom dictionary. Main dictionary files have the .dct extension. The Main dictionary for the U.S. version of AutoCAD is Enu.dct.

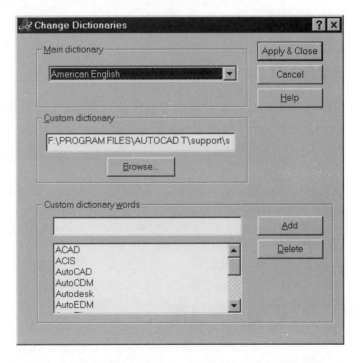

In this dialog box, you can also add or delete words from a custom dictionary. Custom dictionary files are ASCII files with the .cus extension. Because they are ASCII files, they can be edited outside of AutoCAD. The Browse button lets you view a list of existing custom dictionaries.

If you prefer, you can also select a main or custom dictionary using the Dctust and Dctmain system variables. See Appendix D, "System Variables," for more on these system variables.

A third place where you can select a dictionary is in the Files tab of the Options dialog box (Tools ➤ Options ➤ Files). You can find the Dictionary listing under Text Editor, Dictionary, and Font File Names. Click the plus sign (+) next to this listing and then click the plus sign next to the Main Dictionary listing to expose the dictionary options.

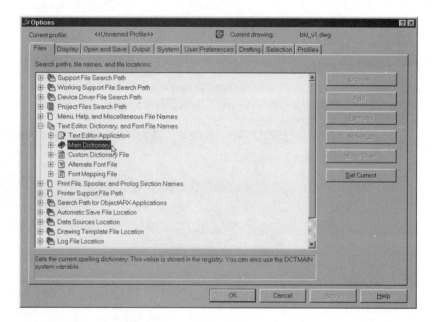

From here, you can double-click the dictionary you prefer. The pointing hand icon will move to the selected dictionary.

Substituting Fonts

There will be times when you'll want to change all of the fonts in a drawing quickly. For instance, you may want to convert PostScript fonts into a simple Txt.shx font to help shorten redraw times while you're editing. Or you may need to convert the font of a drawing received from another office to a font that conforms to your own office standards. In AutoCAD 2000, the Fontmap system variable works in conjunction with a font-mapping table, allowing you to substitute fonts easily in a drawing.

The font-mapping table is an ASCII file called Acad.fmp. You can also use a file you create yourself. You can give this file any name you choose, as long as it has the .fmp extension.

This font-mapping table contains one line for each font substitution you want AutoCAD to make. A typical line in this file would read as follows:

    romant; C:\acad2000\fonts\Txt.shx

In this example, AutoCAD is directed to use the Txt.shx font in place of the Roman font. To execute this substitution, you would type **fontmap**↵ *Fontmap_ filename*↵ where *Fontmap_filename* is the font-mapping table you've created. This tells Auto-CAD where to look for the font-mapping information. Then you would issue the Regen command to view the font changes. To disable the font-mapping table, type **fontmap**↵, a period (.), and then press ↵.

You can also specify a font-mapping file in the Files tab of the Options dialog box. Look for the Text Editor, Dictionary, and Font File listing. Click the plus sign (+) next to this listing and then click the plus sign next to the Font Mapping File listing to expose the current default font-mapping filename.

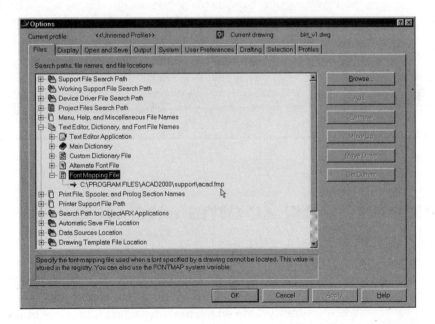

NOTE You can double-click this filename to open a Select a File dialog box. From there you can select a different font-mapping file.

Making Substitutions for Missing Fonts

When text styles are created, the associated fonts don't become part of the drawing file. Instead, AutoCAD loads the needed font file at the same time that the drawing is loaded. So if a text style in a drawing requires a particular font, AutoCAD looks for the font in the AutoCAD search path; if the font is there, it is loaded. Usually, this isn't a problem if the drawing file uses the standard fonts that come with AutoCAD or Windows. But occasionally you'll encounter a file that uses a custom font.

In earlier versions of AutoCAD, when you attempted to open such a file, you would see an error message. This missing-font message would often send the new AutoCAD user into a panic.

Now, AutoCAD offers a solution: it automatically substitutes an existing font for the missing font in a drawing. By default, AutoCAD substitutes the Txt.shx font, but you can specify another one using the Fontalt system variable. Type **fontalt**↵ at the command prompt and then enter the name of the font you want to use as the substitute. (Roman is usually an acceptable substitute, space permitting.) You can also select an alternate font through the Files tab of the Options dialog box. Locate the Text Editor, Dictionary, and Font File Names listing and then click the plus sign (+) at the left. Locate the Alternate Font File listing that appears and click the plus sign at the left. The current alternate is listed. You can double-click the font name to select a different font through a standard File dialog box.

Be aware that the text in your drawing will change in appearance, sometimes radically, when you use a substitute font. If the text in the drawing must retain its appearance, you'll want to substitute a font that is as similar in appearance to the original font as possible.

Accelerating Zooms and Regens with Qtext

If you need to edit a drawing that contains a lot of text, but you don't need to edit the text, you can use the Qtext command to help accelerate Redraws and Regens when you're working on the drawing. Qtext turns lines of text into rectangular boxes, saving AutoCAD from having to form every letter. This allows you to see the note locations so you don't accidentally draw over them.

Selecting a large set of text objects for editing can be annoyingly slow. To improve the speed of text selection (and object selection in general), turn off the Highlight and Dragmode system variables. This will disable certain convenience features but may improve overall performance, especially on large drawings. See Appendix D, "System and Dimension Variables," for more information.

To turn on Qtext, choose Tools ➤ Options ➤ Display, turn on the Show Text Boundary Frame Only check box, select Apply, and then click OK to exit Options. Another method is to enter **qtext**↵ at the command prompt. At the `Enter mode [ON/OFF] <OFF>:` prompt, enter **on**↵. Regardless of the method, to display the results of Qtext, issue the Regen command from the prompt. When Qtext is off, text is generated normally. When Qtext is on, rectangles show the approximate size and length of text, as shown in Figure 7.11.

FIGURE 7.11:

View of the bracket drawing with the Qtext system variable turned on

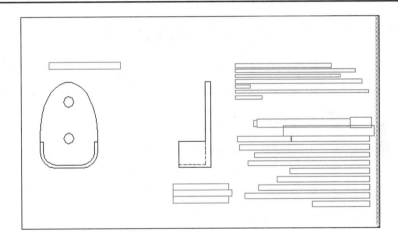

Manipulating Text beyond Labels

This chapter concentrates on methods for adding labels to your drawing, but you also use text in other ways with AutoCAD. Many of the inquiry tools in AutoCAD, such as Dist and List, produce text data. You can use the Windows Clipboard to manipulate such data to your benefit.

Continued on next page

For example, you can duplicate the exact length of a line by first using the List command to get a listing of its properties. Once you have the property list in the AutoCAD Text window, you can highlight its length listing and then press Ctrl+c to copy it to the Windows Clipboard. Next, you can start the Line command and then pick the start point for the new line. Press Ctrl+v to paste the line-length data into the Command window; then add the angle data or use the direct-distance method to draw the line.

Any text data from dialog box input boxes or the AutoCAD Text window can be copied to the Clipboard using the Ctrl+c keyboard shortcut. That data can likewise be imported into any part of AutoCAD that accepts text.

Consider using the Clipboard the next time you need to transfer data within AutoCAD, or even when you need to import text from some other application.

Bonus Text-Editing Utilities

Finally, before finishing this chapter, you'll want to know about a set of bonus utilities that give you the following capabilities:

- Draw text along an arc. If the arc changes, the text follows.

- Globally change the height, justification, location, rotation, style, text, and width factor of a set of text objects you select.

- Adjust the width of a single-line text object to fit within a specified area.

- Explode text into lines.

- Mask areas behind text so the text is readable when placed over hatch- or solid-filled patterns.

- Search and replace text for a set of single-line text objects.

These functions can save hours of your time when editing a complex drawing that is full of text. You can find out how to access these tools in Chapter 19, "Introduction to Customization."

If You Want to Experiment...

Try adding some notes to drawings you've created in other "If You Want to Experiment" sections of this book. Most companies don't like to have mechanical drawings created with multiple fonts or text sizes. Change all of the text to one size and one font.

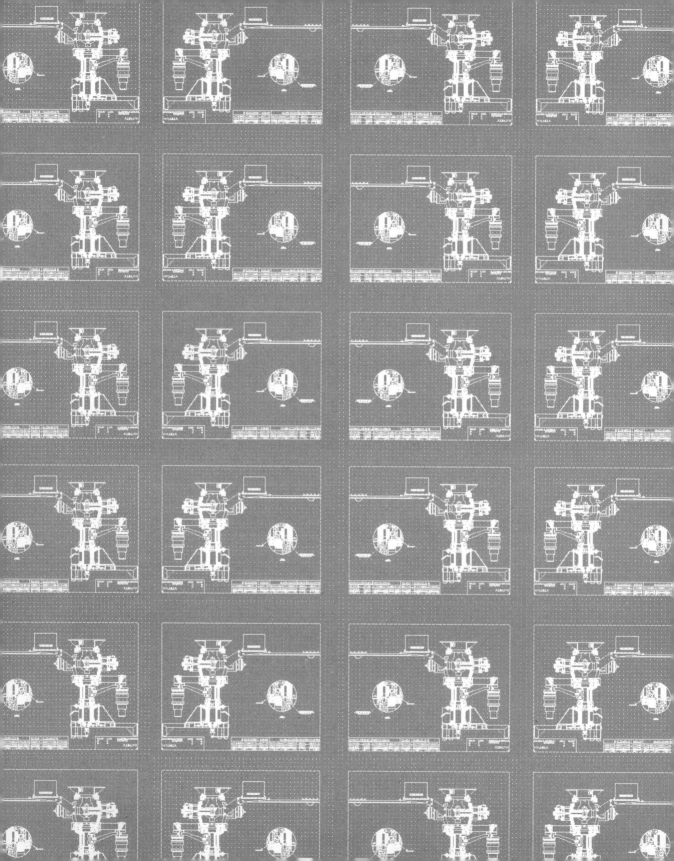

CHAPTER
EIGHT

Using Dimensions

- Setting Up a Dimension Style

- Drawing Linear Dimensions

- Continuing a Dimension String

- Drawing Dimensions from a Common Baseline

- Appending Data to a Dimension's Text

- Dimensioning Radii, Diameters, and Arcs

- Using Ordinate Dimensions

- Adding Tolerance Notation

In the exercises of previous chapters, you've diligently drawn your design to scale. In this chapter, you'll learn how easy it is to create dimensions in AutoCAD. The importance of dimensions cannot be overstated. The reason that you draw these models is to communicate concepts and specifications. Dimensioning can be crucial to how well a design works and how quickly it develops. The dimensions answer questions about tolerances, fit, and interference. When a design team is making decisions, communicating even tentative dimensions to others on the team can accelerate design development.

AutoCAD lets you add dimensions to any drawing easily. AutoCAD gives you an accurate dimension without your having to take measurements. You simply pick two points of an object to be dimensioned and the location of the dimension line, and AutoCAD does the rest. AutoCAD's *associative dimensioning* capability automatically updates dimensions whenever the size or shape of the dimensioned object is changed. These dimensioning features can save you valuable time and reduce the number of dimensional errors in your drawings.

AutoCAD's dimensioning feature has a substantial number of settings. Though they give you an enormous amount of flexibility in formatting your dimensions, all of these settings can be somewhat intimidating to the new user. We'll ease you into dimensioning by first showing you how to create a dimension style.

Creating a Dimension Style

Dimension styles are similar to text styles. They determine the look of your dimensions as well as the size of dimensioning features, such as the dimension text and arrows. You might set up a dimension style to have special types of arrows, for instance, or to position the dimension text above or in line with the dimension line. Dimension styles also make your work easier by allowing you to store and duplicate your most common dimension settings.

AutoCAD gives you a default dimension style called Standard, which is set up fairly close to a mechanical drafting standard. You'll doubtless add many other styles to suit the style of drawings you're creating. You can also create variations of a general style for those situations that call for only minor changes in the dimension's appearance.

In this section, you'll see how to set up a dimension style that is appropriate for ANSI drafting standards. Figure 8.1 shows AutoCAD's Standard dimension style compared to an ANSI-style dimension

FIGURE 8.1:

AutoCAD's standard
dimension style compared
to an ANSI-style dimension

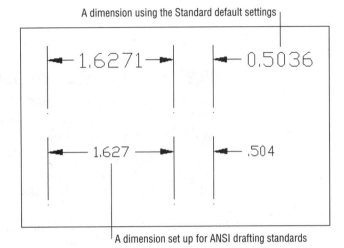

A dimension using the Standard default settings

A dimension set up for ANSI drafting standards

1. Open the `Plate.dwg` file from the companion CD-ROM. The plate drawing is a better choice to demonstrate dimensioning than the parts you've drawn so far.

2. Issue a Zoom All command to display the entire plate. Type **z↵ a↵** or click the Zoom tool in the Standard toolbar.

3. Choose Format ➢ Dimension Style, or type **dimstyle or ddim**↵ at the command prompt. The Dimension Style Manager dialog box appears.

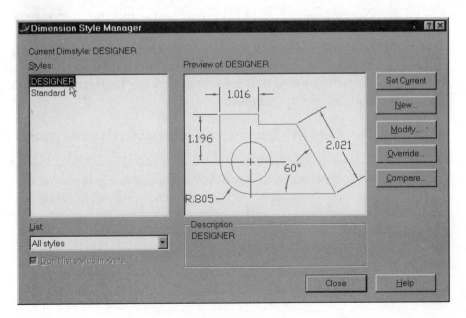

4. In the Styles list box, highlight the Standard style. We will use AutoCAD's Standard style as a basis for our new dimension style.

5. Click the New button. The Create New Dimension Style dialog box appears.

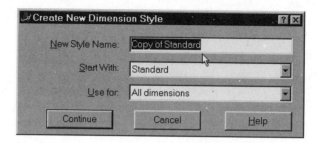

6. Notice the New Style Name input box shows the name Copy of Standard. Replace this name by typing **Layout**.

7. Click Continue. The New Dimension Style dialog box appears.

You've just created a dimension style called Layout, but at this point, it is identical to the Standard style on which it is based. (Nothing has happened to the Standard style; it is still available if you need to use it.)

Setting the Dimension Unit Style

Now you need to modify the new Layout dimension style so that it conforms to the ANSI dimension style. Let's start by changing the unit style for the dimension text. Just as you changed the overall unit style of AutoCAD to three-place decimal style for the bracket in Chapter 3, "Learning the Tools of the Trade," you must do the same for your dimension styles. Setting the overall unit style doesn't automatically set the dimension unit style.

1. In the New Dimension Style dialog box, click the Primary Units tab at the top.

TIP The Primary Units tab also offers Fraction Format options. These options display fractions three different ways: horizontal, diagonal, or not stacked.

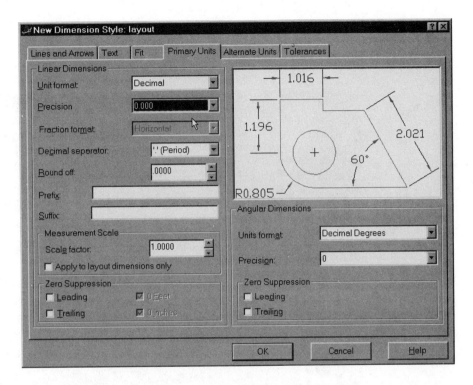

2. In the Linear Dimensions section, open the Unit Format drop-down list and choose Decimal. Notice that this drop-down list contains the same unit styles as the main Units dialog box.

3. Open the Precision drop-down box in the Linear Dimensions section and select the options for three decimal places.

4. In the Zero Suppression check box group at the bottom left of the dialog box, click the Leading check box to put a check mark in it. This turns on leading-zero suppression, so that dimensions with a value of less than one will not have a leading zero. (The ANSI standard is not to have leading zeros.)

5. Click the Tolerances tab at the top of the New Dimension Style dialog box.

6. In the Tolerance Format section, open the Method drop-down list and choose Symmetrical. The preview box to the right shows a sample of typical dimensions and how the tolerance method will affect them.

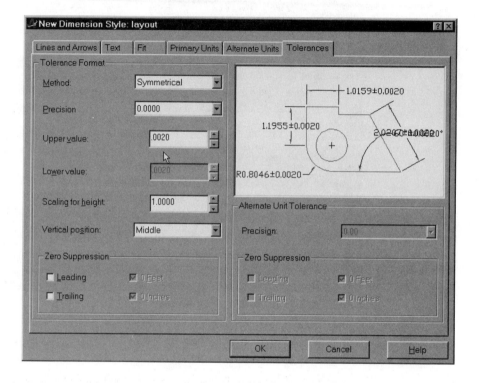

7. Open the Precision drop-down list in the Tolerance Format section and choose the option for three decimal places.

8. Below the Precision options are the Upper Value and Lower Value settings. Because we selected Symmetrical, only Upper Value is available. It will control both. Set Upper Value to 0.002.

Setting the Height for Dimension Text

Along with the unit style, you'll want to adjust the style that is used for the dimension text. The Text tab of the New Dimension Style dialog box lets you select a text style and set other text appearance and alignment options. For now, we'll just change the text height. You can experiment with the options to see how they affect your dimension style text.

1. In the New Dimension Style dialog box, click the Text tab.

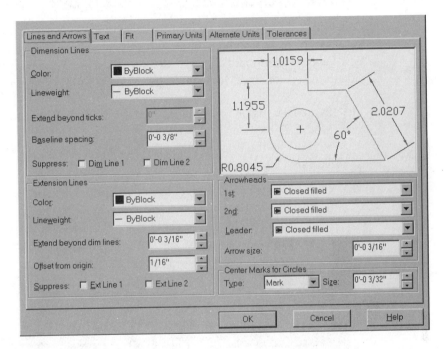

2. In the Text Height input box, type **.12** to make the text height 0.12-inch.

3. Click OK to close the New Dimension Style dialog box. The Dimension Style Manager dialog box reappears.

4. In the Dimension Style Manager dialog box, select the Layout style and click the Set Current button.

5. Click Close to close the Dimension Style Manager.

You've set up a new dimension style named Layout. This style shows dimensions and tolerances to three decimal places, rather than four, and has a text height of 0.12 inch. You've made the Layout style the current style, so you can use it immediately. You could have continued setting options on the tabs of the New

Dimension Style dialog box. The next sections show how to make changes to an existing style using the Modify Dimension Style dialog box.

Setting the Location of Dimension Text

AutoCAD's default setting for the placement of dimension text puts the text centered between the extension lines, as shown in the example at the top of Figure 8.1. The new Layout style should allow you to place the text anywhere between or even outside the extension lines. To do that, you'll use the Modify Dimension Style options.

1. In the Dimension Style Manager dialog box, click the Modify button. The Modify Dimension Style dialog box appears. (Here also, the name Layout is included in the title bar of the dialog box to indicate the name of the dimension style being edited.)

2. Click the Text tab. The Text Placement options will allow you to position the text. In the Vertical box open the drop-down list and choose Above. The sample graphic changes to show you what this format will look like in your dimensions.

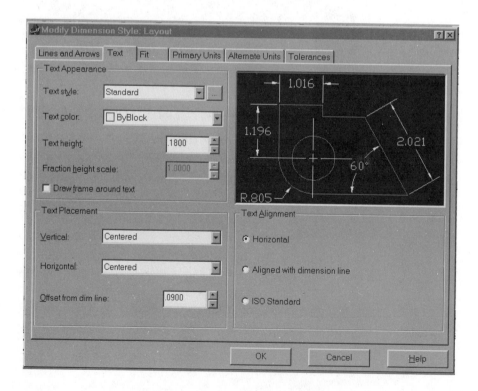

3. In the Text Alignment section (lower-right corner of the dialog box), verify that the Horizontal radio button is chosen. This forces the dimension text to be horizontal without regard to the orientation of the dimension.

4. Now click OK to return to the Dimension Style Manager dialog box.

Choosing an Arrow Style

Next, you'll want to verify that you're using the correct type of arrow for your new dimension style. For linear dimensions in mechanical drawings, a filled arrowhead is normally used; this is the AutoCAD default on start-up. If you're not the only one to use the computer or if someone else has configured AutoCAD, make sure that you're using the filled arrowhead. For some layout and architectural purposes a diagonal line or tick mark is used, rather than an arrow. We'll also take a look at this style of arrow.

In the Dimension Style Manager dialog box, confirm that the Layout dimension style is selected, click Modify, and then go to the Lines and Arrows tab. You'll see the Arrowheads group in the lower-right corner. You'll notice three drop-down lists labeled 1st, 2nd, and Leader.

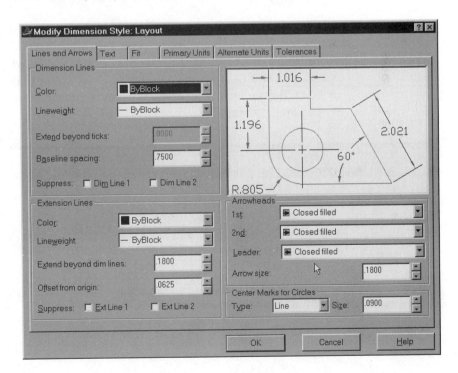

While you have the Lines and Arrows tab open, try various settings. Notice that the 1st arrowhead's setting controls both arrowheads, while changing the 2nd setting affects only the second one. Try different combinations until you feel comfortable, and then set 1st back to Closed Filled. Ensure that Leader is also set to Closed Filled, unless you wish to use a special arrowhead for notes with leaders.

Setting the Dimension Scale

Often you'll work on projects that are too large for the sheet size you wish to plot on. Calculating the plot scale will help in this situation. Usually, scales of 1/4, 1/2, full, or double (2x) will do the trick.

A drawing that will fit on a full size drawing sheet is no problem. But a drawing that must be scaled to 1/2 size (or some other fractional scale) will need its text, dimensions, and extension line offsets adjusted or these things will be too small. You can set the dimension scale as part of your dimension style.

1. If you closed the Dimension Style Manager dialog box after completing the last exercise, reopen it now. Ensure that the current style is Layout. Then click Modify.

2. Select the Fit tab. In the Scale for Dimension Features group, locate the Use Overall Scale Of input box, and change this value to 1.000. This is the scale factor for a full size, or 1"=1", scale drawing. If the drawing needed to be plotted half size (1/2), you would set this scale factor to the reciprocal 2/1, or 2.000. This will resize all of the features including dimension text to suit the required plot scale.

Calculating Scale Factors

Here's a simple way to figure out the required scale factors.

Architectural scales   Divide 12 by the decimal equivalent of the inch scale. So for 1/4" per foot, divide 12 by 0.25 to get 48. For 1/8", divide 12 by 0.125 to get 96. For 1 1/2", divide 12 by 1.5 to get 8, and so on.

Engineering scales   Multiply the scale of the drawing by 12. For example, for a scale of 1"=10' scale, multiply the 10 by 12 to get 120.

We'll call the result of the calculation (48, 96, 8, or 120) the magic number. These numbers will help us later when we work out our plot scale.

Setting the Center Mark Size

Often, you'll wish to add a center mark to the circles and arcs in your drawing. Or, you may wish to control the appearance of the center marks produced by the Dimension and Radius commands. You can specify center mark type and size as part of your dimension style.

1. If you closed the Dimension Style Manager dialog box after completing the last exercise, reopen it now. Ensure that Layout is selected and click Modify.

2. Select the Lines and Arrows tab. In the Center Marks for Circles group, you can tell AutoCAD to use two different types of center marks or none. The Size box defines the size of the mark. The Size box also controls how far beyond the circle the lines extend.

3. Make your selections and then click Close in the Dimension Style Manager dialog box.

In this section, we've introduced you to the various dialog boxes that let you set the appearance of a dimension style.

TIP If your application is strictly mechanical, you may want to make these same dimension style changes to the Acad.dwt template file, or create a set of template files specifically for mechanical drawings of differing scales.

Drawing Linear Dimensions

The most common type of dimension you'll be using is the *linear dimension*, which is an orthogonal dimension measuring the width and length of an object. AutoCAD offers three dimensioning tools for this purpose: Linear (Dimlinear), Continue (Dimcont), and Baseline (Dimbase). These options are readily accessible from the Dimension toolbar.

Finding the Dimension Toolbar

Before you apply any dimension, you'll want to open the Dimension toolbar. This toolbar contains nearly all of the commands necessary to draw and edit your dimensions.

Right-click any toolbar; then at the Toolbars list, click Dimension. The Dimension toolbar appears. Click OK to exit the Toolbars dialog box.

The Dimension commands are also available from the Dimension pull-down menu. To help keep your you screen organized, you may want to dock the Dimension toolbar to the right side of the AutoCAD window, but if you display this toolbar horizontally you'll receive a pleasant surprise. AutoCAD has added a drop-down list displaying all of the currently loaded dimension styles. See Chapter 1, "This Is AutoCAD," for more about docking toolbars.

Now you're ready to begin dimensioning.

Placing Horizontal and Vertical Dimensions

Let's start by looking at the basic dimensioning tool, Linear. The Linear Dimension button (the Dimlin command) on the Dimension toolbar accommodates both the horizontal and vertical dimensions.

In this exercise, you'll add a vertical dimension to the left side of the plate.

WARNING

Dimensions should be placed accurately. If you've disabled Running Osnaps, turn them on now.

1. To start either a vertical or horizontal dimension, click Linear Dimension from the Dimension toolbar, or enter **dli**↵ at the command prompt. You can also choose Dimension ➤ Linear from the menu bar.

2. The Specify first extension line origin or <select object>: prompt is asking you for the first point of the distance to be dimensioned.

An *extension line* is the line that connects the object being dimensioned by the dimension line. Pick the lower-left corner of the plate at the bottom of the fillet. Care is required here because you don't want to select the other end of the fillet.

3. At the `Specify second extension line origin:` prompt, pick the top-left corner of the plate.

NOTE Notice that the prompt or <select object>: in step 2 gave you the option of pressing ↵ to select an object. If you do this, you are prompted to pick the object you wish to dimension, rather than the actual distance to be dimensioned. If you select a line, AutoCAD will immediately add a dimension to it. Move the cursor about and notice how the dimension will change from horizontal to vertical. Selecting a place to locate the dimension text will complete the dimension. This is called *automatic dimensioning.*

4. In the next prompt, `Specify dimension line location or [Mtext/Text /Angle/Horizontal/Vertical/Rotated]:`, the dimension line is the line indicating the direction of the dimension and containing the arrows or tick marks. Move your cursor from left to right, and you'll see a temporary dimension appear. This allows you to visually select a dimension line location.

NOTE In step 4, you have the option to append information to the dimension's text or change the dimension text altogether. You'll see how to do this in "Editing Dimensions," later in this chapter.

5. Enter **@1.5<180**↵ to tell AutoCAD you want the dimension line to be 1.50" to the left of the last point you selected. (You could pick a point using your cursor, but this doesn't let you place the dimension line as accurately.) After you've done this, the dimension is placed in the drawing, as shown in Figure 8.2.

NOTE The prompt `Dimension text` = shows the horizontal or vertical distance between the points you've picked. If this value isn't what you think it should be, make sure you've picked the correct points. If you have selected the correct points, the problem may be in the drawing. You should try to find the error now to avoid bigger problems later.

FIGURE 8.2:

The dimension line added
to the plate drawing

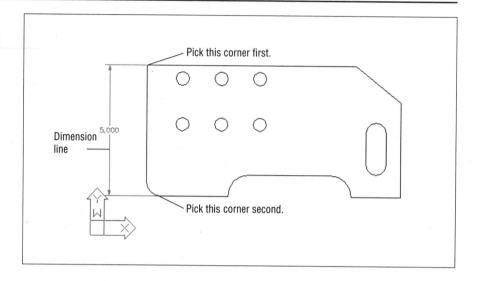

Continuing a Dimension

You'll occasionally want to input a group of dimensions strung together in a line. For example, you may want to continue dimensioning a feature because a special dimension tolerance is required. To do this, use the Continue option found in both the Dimension toolbar and the Dimension pull-down menu. We'll dimension the vertical location of the six-hole pattern.

Using Osnaps While Dimensioning

You may find that when you pick intersections and endpoints frequently during dimensioning, it is a bit inconvenient to use the Osnap pop-up menu. In situations where you know you'll be using certain Osnaps frequently, you can use Running Osnaps. Click Tools ➤ Drafting Settings, and then click the Object Snap tab. Select the Endpoint, Midpoint, and Center Osnap modes by clicking in the check box for each Osnap setting. With Running Osnaps set in this way, your cursor will automatically select the endpoint of a line or an arc, the midpoint of a line or an arc, or the center of a circle or an ellipse.

TIP

After setting Running Osnaps, if you have difficulty obtaining the one you want, you can use the Osnap Cycle mode. To do this, hover the cursor over the objects until the snap marker appears, then press the Tab key repeatedly until you see the marker representing the Osnap that you want.

WARNING

You can pick as many of the Osnaps as you like to be Running Osnaps; however, a few Osnaps can preclude the use of others. For example, suppose you have Center, Quadrant, and Tangent set. When you hover the cursor over a circle, only one marker will show (according to Murphy's law, the marker that you see will rarely be the right one). If you find that you must frequently use the Tab key to select another Osnap, are frequently using the Osnap cursor menu (Shift+right-click), or are frequently typing in Osnap overrides, you should take a look at changing the Running Osnap settings.

Once you've set your Running Osnaps, the next time you're prompted to select a point, the selected Osnap modes will be activated automatically. You can still override the default settings using the Osnap pop-up menu (Shift+right-click). However, you don't have to worry about accidentally using a Running Osnap during panning or zooming. Only explicitly issued Osnap overrides affect pans and zooms.

WARNING

There is a drawback to setting a Running Osnap mode: When your drawing gets crowded, you might end up picking the wrong point by accident. However, you can easily toggle the Running Osnap mode off by clicking Osnap in the status bar,or by pressing F3.

TIP

You may want to use Zoom to get a closer look when trying to select geometry. Dimensions are composed of lines, and you could as easily snap to one of these as to the intended geometry.

The Continue mode of linear dimensioning assists in making chain dimensions. When you use this mode, you'll be asked only for the second extension line because the chain continues from the second extension line origin of the last dimension that you created before beginning the Continue mode.

1. Click the Linear Dimension tool in the Dimension toolbar.

2. At the `Specify First extension line origin or <select object>:` prompt, pick the upper-left corner of the plate.

3. At the `Specify second extension line origin:` prompt, pick the top-left circle.

4. At the `Specify dimension line location or [Mtext/Text/Angle /Horizontal/Vertical/Rotated]:` prompt, type **from**↵ to use the From Osnap. At the `Base point:` prompt, pick the upper-left corner of the plate. At the `<offset>` prompt, enter **.75**↵ to tell AutoCAD you want the dimension line to be .75" to the left of the From point. Now, use direct-distance entry and with Ortho mode on, simply select a point to the left (of the upper-left corner of the plate) and down slightly. The dimension will be placed .75 units to the left of the plate.

5. Click the Continue Dimension tool in the Dimension toolbar, or enter **dco**↵. You can also choose Dimension ➢ Continue from the menu bar.

6. At the `Specify a second extension line origin or [Undo/Select] <Select>:` prompt, pick the circle just below the last one you picked. See Figure 8.3 for the results.

7. Press ↵ twice to exit the command.

TIP   If you find you've selected the wrong location for a continued dimension, you can click the Undo tool or press **u**↵ to back up your dimension.

The Continue Dimension option adds a dimension from where you left off. The last drawn extension line is used as the first extension line for the continued dimension. AutoCAD will keep adding dimensions as you continue to pick points, until you press ↵.

FIGURE 8.3:

The dimension chain using the Continue Dimension option

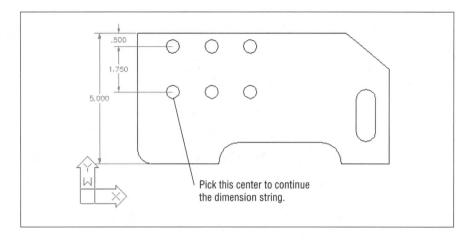

Pick this center to continue the dimension string.

Continuing a Dimension from a Previous One

If you need to continue a string of dimensions from an older linear dimension instead of the most recently added one, press ↵ at the Specify a second extension line origin or [Undo/Select] <Select>: prompt you saw in step 6 of the previous exercise. Then, at the Select continued dimension: prompt, click the extension line from which you wish to continue.

Using the Quick Dimension Feature

AutoCAD 2000 offers the Quick Dimension (Qdim) method for creating a series of continued dimensions with a single operation.

1. Click the Quick Dimension tool in the Dimension toolbar, type **qdim**, or select Dimension ➤ Qdim to start Quick Dimension.

2. At the Select geometry to dimension: prompt, select *all* of the objects you wish dimensioned. You can make your selections by clicking or by using a crossing window.

Continued on next page

3. Press ⏎ to finish your selection. The next prompt is `Specify dimension line position, or [Continuous/Staggered/Baseline/Ordinate/Radius/Diameter/datumPoint/Edit] <Continuous>:` prompt.

4. You can click a point to place a string of dimensions, or choose one of the options, such as Staggered, for some very interesting results.

Drawing Dimensions from a Common Base-Extension Line

Another method for dimensioning objects is to have several dimensions originate from the same extension line. To accommodate this, AutoCAD provides the Baseline option. To see how this works, you'll start another dimension—this time a horizontal one—across the top of the plate.

1. Click the Linear Dimension tool in the Dimension toolbar. Or, just as you did for the vertical dimension, you can type **dli**⏎ to start the horizontal dimension. This option is also on the Dimension pull-down menu.

2. At the `Specify first extension line origin...` prompt, use the Endpoint Osnap to pick the upper-left corner of the plate.

3. At the `Specify second extension line origin...` prompt, pick the upper-left circle again.

4. At the `Specify dimension line location...` prompt, use the From Osnap override. At the `Base point:` prompt, pick the upper-left corner of the plate. At the `<offset>` prompt, enter **@0.75<90**⏎ to tell AutoCAD you want the dimension line to be 0.75" above the From point. After placing the dimensions, Zoom All to see the dimensioned plate so far. It should look like the one in Figure 8.4.

Now you're all set to draw another dimension continuing from the first extension line of the dimension you just drew.

5. Click the Baseline Dimension tool in the Dimension toolbar. Or, you can type **dba**⏎ at the command prompt to start a baseline dimension.

FIGURE 8.4:

The plate with a horizontal dimension

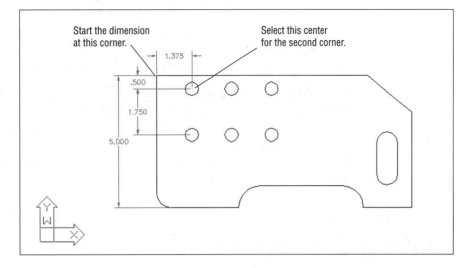

6. At the Second extension line... prompt, pick the upper-middle circle in the set of six circles.

7. Continue to pick the upper-right circle, the upper-left of the bevel, and the upper-right end of the plate. Your drawing will look like Figure 8.5.

FIGURE 8.5:

The plate with the baseline dimensions

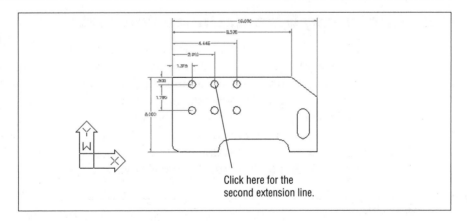

8. Press ↵ twice to exit the Baseline Dimension command.

9. Pan and zoom your view so that it looks similar to Figure 8.5.

In this example, you see that the Baseline Dimension option is similar to the Continue Dimension option, except that baseline allows you to use the first extension line of the previous dimension as the base for a second dimension. You can also use the Quick Dimension (Qdim) command from the Dimension toolbar or Dimension pull-down menu to create baseline dimensions.

Continuing from an Older Dimension   You may have noticed in step 8 that you had to press ↵ twice to exit the command. As with Continue Dimension, you can draw a baseline dimension from an older dimension by pressing ↵ at the Specify a second extension line origin or [Undo/Select] <Select>: prompt. You then get the Select base dimension: prompt, at which you can either select another dimension or press ↵ again to exit the command.

Editing Dimensions

As you begin to add more dimensions to your drawings, you'll find that AutoCAD will occasionally place a dimension line or text in an inappropriate location, or you may need to make a modification to the dimension text. In this section, you'll take an in-depth look at how dimensions can be modified to suit those special circumstances that always crop up.

Appending Data to Dimension Text

So far in this chapter, you've been accepting the default dimension text. You can append information to the default dimension value, or change it entirely if you need to. When you see the temporary dimension dragging with your cursor, enter **t**↵. The text inside the default <> brackets is the actual value of the dimension you're placing, modified by the current dimension style. You can add text either before or after the default dimension like so: **before<>after** (the brackets represent the value, you don't need to enter a number within them), or replace the brackets entirely to substitute for the default text.

> **WARNING**   Avoid changing the value of the dimension text, especially if it isn't correct. Please be aware that if you placed the dimension correctly, this is the actual object size as drawn. Look for the cause of the inaccuracy instead of just changing it. Zoom in close and examine the position of the dimension definition point. Select the dimension and turn on its grips. If they don't match the object, move the dimension grip until it does.

The Properties button in the Standard toolbar lets you modify existing dimension text in a similar way. Let's see how this works by changing an existing dimension's text in your drawing.

1. Click the Properties tool in the Standard toolbar.

2. Next, click the horizontal dimension to the bevel you added to the drawing in step 7 of the last exercise. The dimension is at the top of the screen.

TIP

With the Dimension Edit tool dialog box you can append text to several dimensions at once. You can also use the Multiline Text Editor (**ed**↵ or Modify ➤ Text) to edit the dimension text.

3. Press ↵. The Properties dialog box appears.

4. Click in the Text override input box and then type **<> PAINTED SURFACE**. Exit the Properties dialog box.

TIP

The Multiline Text Editor or DimEdit are simpler to operate if editing text is all you have in mind. We'll look at them in the sidebar at the end of this section.

5. The dimension has changed to read 8.235 PAINTED SURFACE. The text you entered is appended to the dimension text.

6. Because you don't really need the new appended text for the tutorial, click the Undo button in the Standard toolbar to remove the appended text.

TIP

Place your appended text in front of the <> symbols if you want to add text to the beginning of the dimension text. You can also replace the dimension text entirely with new text by replacing the <> sign in the Text Override input box. If you want to restore a dimension that has been modified to the default value, replace everything in the Text Override input box, including any spaces, with the default brackets<>. Or, include a space to leave the dimension text blank. Close the dialog box to exit and press Escape (Esc) twice to clear any grips.

NOTE   In this exercise, you were only able to edit a single dimension. To append text to several dimensions at once, you need to use the Dimension Edit tool. See the "Making Changes to Multiple Dimensions" sidebar for more on this command.

You can also have AutoCAD automatically add a dimension suffix or prefix to all dimensions, instead of just a chosen few, by using the Modify ➤ Primary Units option in the Dimension Style Manager dialog box.

Besides appending text to a dimension, the Properties dialog box lets you modify a dimension's other properties. The Properties dialog box offers access to all of the object properties and the dimension style settings of this specific dimension.

Making Changes to Multiple Dimensions

The Dimension Edit tool offers a quick way to edit existing dimensions. It adds the ability to edit the text of more than one dimension at one time. One common use would be to change a string of dimensions to read *equal*, instead of showing the actual dimensioned distance.

1. Click the Dimension Edit tool in the Dimension toolbar or type **ded**↵.

2. At the `Enter type of dimension editing [Home/New/Rotate/Oblique]` `<Home>:` prompt, type **n**↵ to use the New option. The Multiline Text Editor appears, showing the <> brackets in the text box.

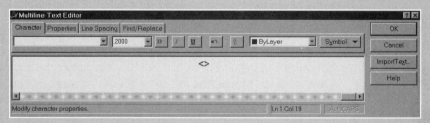

3. Click the space behind or in front of the <> brackets, and then enter the text you want to append to the dimension. Or, you can replace the brackets entirely to substitute the dimension with your text.

4. Click OK.

Continued on next page

5. At the `Select objects:` prompt, pick the dimensions you wish to edit. The `Select objects:` prompt remains, allowing you to select several dimensions. When you see `Select Objects:`, you may use any selection tool you choose. You may now select as many dimensions as required for the change.

6. Press ↵ to finish your selection. The dimension(s) change to include your new text.

The Dimension Edit tool is useful for editing dimension text, but you can also use this command to make graphical changes to the text. Here is a listing of the other options.

Home   Moves the dimension text to its standard default position and angle.

Rotate   Allows you to rotate the dimension text to a new angle.

Oblique   Skews the dimension extension lines to a new angle. See "Skewing Dimension Lines" later in this chapter.

Locating the Definition Points

AutoCAD provides the associative dimensioning capability to automatically update dimension text when a drawing is edited. Locations on the dimension called *definition points* are used to determine how edited dimensions are updated.

The definition points are located at the same points you pick when you determine the dimension location. For example, the definition points for linear dimensions are the extension-line origin and the intersection of the extension line/dimension line. The definition points for a circle diameter are the points used to pick the center of the circle and the points locating the diameter of the circle. The definition points for a radius are the points used to pick the circle's radius, plus the center of the circle.

Definition points are actually point objects. They are very difficult to see because they are usually covered by the feature they define. You can, however, see them indirectly, using grips. The definition points of a dimension are the same as the dimension's grip points. You can see them by simply clicking a dimension. Try the following:

1. Make sure the Grips feature is turned on (see Chapter 2 to refresh your memory on the Grips feature).

2. Click the dimension you drew in the earlier exercise that defines the vertical distance between the top of the plate and the nearer of the two rows of circles. You'll see the grips of the dimension, as shown in Figure 8.6.

FIGURE 8.6:

The grip points are the same as the definition points on a dimension.

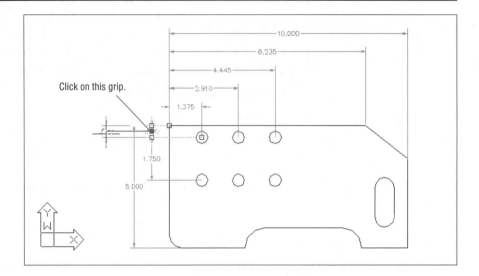

Making Minor Adjustments to Dimensions Using Grips

The definition points, whose locations you can see through their grips, are located on their own unique layer, called Defpoints. Definition points are displayed regardless of whether the Defpoints layer is on or off. To give you an idea of how these definition points work, try the following exercises, which show you how to directly manipulate the definition points:

1. With the grips still visible from the previous exercise, click the grip near the dimension text.

TIP Because the Defpoints layer has the unique feature of being visible even when turned off, you can use it as a layer for laying out your drawing. While Defpoints is turned off, you can still see objects assigned to it, but the objects won't plot.

| TIP | New in AutoCAD 2000 is the ability to make any layer a non-plotting layer. Check the Layer Properties Manager dialog box, and look for a column labeled Plot. Notice Defpoints is already labeled as non-plotting. |

2. Move the cursor around. Notice that when you move the cursor vertically, the text moves along the dimension line. When you move the cursor horizontally, the dimension line and text move together, keeping their parallel orientation to the dimensioned plate.

| NOTE | Here the entire dimension line moves, including the text. If you select the grip attached to the dimension text and move it to the side, you'll see how you can move the dimension text independently of the dimension line. |

3. Enter @.38<180↵. The dimension line, text, and the dimension extension lines move to the new location to the left of the previous location (see Figure 8.7).

FIGURE 8.7:

Moving the dimension line using its grip

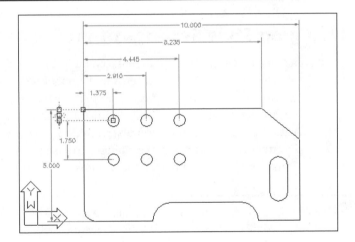

TIP
If you need to move several dimension lines at once, select them all at the same time; then Shift+click one set of dimension-line grips from each dimension. Once you've selected the grips, click one of the hot grips again, making it the base grip. You can then move all of the dimension lines at once as you move the base grip.

In step 3 of the last exercise, you saw that you could specify an exact distance for the dimension line's new location by entering a relative polar coordinate. Cartesian coordinates work just as well, and with Ortho on you can use dynamic-distance entry (DDE). You can even use object snaps to relocate dimension lines. Next, try moving the dimension lines back using the Perpendicular Osnap.

1. Click the grip at the bottom of the dimension line.

2. Shift+right-click and choose Perpendicular from the Osnap pop-up menu.

3. Place the cursor on the vertical dimension line that dimensions the distance between the two rows of circles.

4. The selected dimension line moves back to its original location to align with the other vertical dimension.

Changing Style Settings of Individual Dimensions

In some cases, you'll have to make changes to an individual dimension's style settings in order to edit that dimension. For example, if you try to move the text of a typical linear dimension, you'll find that the text and dimension lines are inseparable. You need to make a change to the dimension's style setting that controls how AutoCAD locates dimension text in relation to the dimension line. This section describes how you can make changes to the style settings of individual dimensions to facilitate changes in the dimension.

TIP
If you need to change the dimension style of a dimension to match that of another, you can use the Match Properties tool.

Moving Fixed-Dimension Text

In some instances, you'll want to manually move dimension text away from the dimension line, but as you saw in an earlier exercise, this cannot be done with the current settings.

In the next exercise, you'll make a change to a single dimension's style settings. Then you'll use grips to move the dimension text away from the dimension line.

1. Press Escape (Esc) twice to cancel the grip selection from the previous exercise.

2. Click Properties in the Standard toolbar and then click the vertical dimension that measures the overall plate height—the 5.00 dimension—and press ↵ to finish your selection. The Properties dialog box appears. This dialog box contains all of the properties and dimension-style settings.

3. Scroll down the list until you see Fit.

4. In the Fit section, choose Dim Line Forced; this will open a drop-down list. Choose On from the list. Also in the Fit section, choose Text Movement, and choose Move Text, Add Leader from the list.

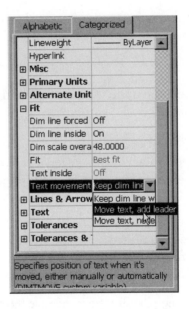

5. Next, continue searching the list until you see Text. Now look for Text Pos Vert and verify that the Vertical Justification group setting is centered. We'll explain why in the following paragraphs.

6. Make sure Ortho mode is off. Click the grip at the center of the dimension text and move the dimension text above and to the left of its current location, as

shown in Figure 8.8. The dimension text is now horizontal and shows a leader from the text to the dimension line.

FIGURE 8.8:

Moving the dimension text using grips

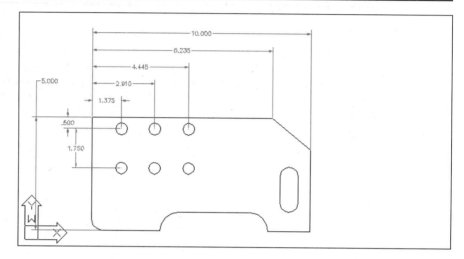

7. Type **u↵** to undo the last command or use the Undo tool in the Standard toolbar.

In the Properties dialog box, the Text Movement option you chose in the Fit section lets you move the dimension text independently of the dimension line. It also causes a leader to be drawn from the dimension line to the text. We asked you to change the Vertical Justification option to Centered; otherwise, the leader line would be drawn as an underline beneath the dimension text.

Both the Fit and Vertical Justification settings can be made using system variables. See Appendix D, "System Variables," for more on these settings.

TIP   To change an existing dimension to the current dimension style, use the Dimension Update tool. Click Dimension Update in the Dimension toolbar, or choose Dimension ➤ Update from the menu bar. Then select the dimensions you want to change. Press ↵ when you're done selecting dimensions. The selected dimensions will be converted to the current style.

Rotating Dimension Text

Once in a while, dimension text works better if it is in a vertical orientation, even if the dimension itself is not vertical. If you find that you need to rotate dimension text, here's how to do it.

1. Click the Dimension Text Edit tool in the Dimension toolbar.

NOTE You can also use the Dimension Edit tool (Dimedit command) in the Dimension toolbar to rotate the dimension text. Click Dimension Edit or type **ded↵r↵**, enter the rotation angle, and then select the dimension text.

2. At the Select Dimension: prompt, click the 5.00" dimension text again.

3. At Specify new location for dimension text or [Left/Right/Center/ Home/Angle]:, type **a↵**.

4. At the Specify angle for dimension text: prompt, type **45↵** to rotate the text to a 45° angle.

TIP You can also choose Dimension ➤ Align Text ➤ Angle, select the dimension text, and then enter an angle. A 0 angle will cause the dimension text to return to its default angle.

The Dimension Text Edit tool (Dimtedit command) also allows you to align the dimension text to either the left or right side of the dimension line. This is similar to the Alignment option in the Multiline Text Editor.

NOTE You can use the Home option of the Dimension Text Edit tool or Dimension ➤ Align Text to move dimension text back to its original location.

You may want to make other adjustments to the dimension text, such as its location along the dimension line or its rotation angle.

As you've seen in this section, the Grips feature is especially well suited to editing dimensions. With grips, you can stretch, move, copy, rotate, mirror, and scale dimensions.

Modifying the Dimension-Style Settings for Groups of Dimensions

In "Moving Fixed-Dimension Text" earlier in this chapter, you used the Properties button to facilitate the moving of the dimension text. You can also use the Dimension ➤ Override option (Dimoverride command) to accomplish the same thing. The Override option allows you to make changes to an individual dimension's style settings. The advantage to Override is that it allows you to effect changes to groups of dimensions, not just one dimension. Here's an example showing how Override can be used in place of the Properties button in the first exercise of the "Moving a Fixed Dimension Text" section.

1. Press Escape (Esc) twice to make sure you're not in the middle of a command. Then choose Dimension ➤ Override from the menu bar.

2. At the next prompt

   Dimension variable name to override (or Clear overrides):

   type **dimfit**↵.

3. At the Enter new value for dimension variable <3>: prompt, enter **4**↵. This has the same effect as selecting options from the Dimension Style Manager, such as Modify, Fit, Text Placement, Over the Dimension Line, With a Leader.

4. The Dimension variable name to override...: prompt appears again, allowing you to enter another dimension variable. Press ↵ to move to the next step.

5. At the Select objects: prompt, select the dimension you want to change. You can select a group of dimensions if you want to change several dimensions at once. Press ↵ when you're done with your selection. The dimension settings will be changed for the selected dimensions.

Continued on next page

As you can see from this example, Dimoverride requires that you know exactly which dimension variable to edit in order to make the desired modification. In this case, setting the Dimfit variable to 4 will let you move the dimension text independently of the dimension line. If you find the Dimoverride command useful, consult Appendix D, "System Variables," to find which system variable corresponds to the Dimension Styles dialog box settings. If you've already applied a Dimoverride to a dimension, you can also use the Match Properties icon or type **matchprop**↵ to allow other dimensions to inherit the properties. Of course, they will also inherit the layer and linetype of the source object if those properties are set for matching as well.

Editing Dimensions and Other Objects Together

Certainly, it's helpful to be able to edit a dimension directly using its grips. But the key feature of AutoCAD's associative dimensions is their ability to automatically adjust themselves to changes in the drawing. As long as you include the dimension's definition points when you select objects to edit, the dimensions themselves will automatically update to reflect the change in your drawing. To see how this works, try moving the six-hole pattern away from left edge of the plate. You can move a group of objects using the Stretch command with the crossing window option.

1. Click the Stretch tool in the Modify toolbar, or type **s**↵.

2. At the `Select objects to stretch by crossing-window or crossing-polygon... Select objects:` prompt, use your mouse to create a crossing window by picking one of the right corners first, then dragging the cursor diagonally to the left, as illustrated in Figure 8.9. Next press ↵ to confirm your selection. You must use either a crossing window or a crossing polygon to select objects (the default for the prompt is a crossing window). Remember that a crossing window is created from right to left and a standard window is created from left to right.

3. At the `Specify base pointor displacement:` prompt, enter **.5,0** and then press ↵↵ to move the circles 0.5" in the positive X direction. The circles move, and the dimension text changes to reflect the new dimension, as shown in Figure 8.10.

FIGURE 8.9:

The Stretch crossing window

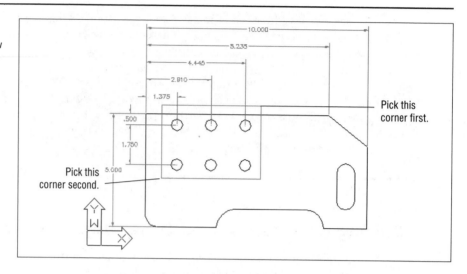

FIGURE 8.10:

The moved circles, with the updated dimensions

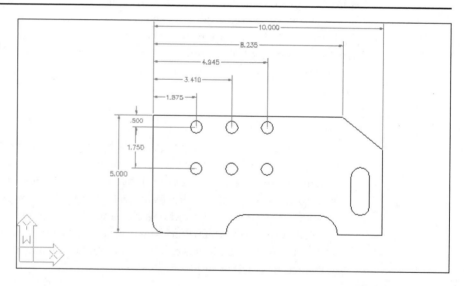

4. When you're done reviewing the results of this exercise, close the file without saving it.

When you selected the crossing window corners, you included the definition points of the horizontal dimensions. This allowed you to move the dimension-extension lines along with the circles, thereby updating the dimensions automatically.

Understanding the Stretch Command

The tool you used for moving the line and the dimension-line extensions is the Stretch command. We introduced the Stretch command in Chapter 5, "Editing for Productivity." Now let's take another look at it. This is one of the most useful yet least understood commands offered by AutoCAD. Think of Stretch as a vertex mover: Its sole purpose is to move the vertices (or endpoints) of objects.

Stretch actually requires you to do two things: select the objects you want to edit, and then select the vertices you wish to move. The crossing window and the crossing polygon window offer a convenient way to do two things at once, because they select objects and vertices in one operation. But when you want to be more selective, you can click objects and window vertices instead. For example, consider the exercise in this chapter that you just completed where you moved the circles with the Stretch command. If you wanted to move the left edge and lower fillet but not the dimension-line extensions, you would do the following:

1. Click Stretch in the Modify toolbar or the Modify pull-down menu. You may also type **s↵**.

2. At the `Select objects:` prompt, enter **w↵** (Window) or **wp↵** (Window Polygon) or pick an implied automatic window starting from the left.

3. Window the vertices you wish to move. Because the Window and Window Polygon selection options select objects completely enclosed within the window, enclose the vertical line and the fillet.

4. Click the top and bottom adjacent horizontal lines to include them in the set of objects to be edited.

5. Press ↵ to finish your selection.

6. Indicate the base point and second point for the stretch.

You could also use the Remove selection option and click the dimensions to deselect them in the previous exercise. Then, when you enter the base and second points, the line and fillet would move but the dimensions would stay in place.

Stretch will stretch only the vertices included in the last window, crossing window, crossing polygon, or window polygon (see Chapter 2, "Creating Your First Drawing," for more on these selection options). Thus, if you had attempted to window another part of your drawing in the circle-moving exercise that we just completed, nothing would have moved. Before Stretch will do anything, objects need to be highlighted (selected) and their endpoints windowed.

Continued on next page

The Stretch command is especially well suited to editing dimensioned objects, and when you use it with the Crossing Polygon (CP) or Window Polygon (WP) selection options, you have substantial control over what gets edited.

You can also use the Mirror, Rotate, and Stretch commands with dimensions. The polar arrays will work as well, and Extend and Trim can be used with linear dimensions.

When editing dimensioned objects, be sure you select the dimension associated with the object being edited. As you select objects, using the Crossing (C) or Crossing Polygon (CP) selection options will help you include the dimensions. For more on these selection options, see the "Other Selection Options" sidebar in Chapter 2, "Creating Your First Drawing."

WARNING Although it is possible to turn off the associative feature, it is not recommended. Associativity is controlled by the system variable dimaso (enter **dimaso** ↵ **off** ↵) in your drawing or in any template file (`acad.dwt`), and you can explode an existing dimension. If you create a dimension with associativity turned off, AutoCAD draws the components of the dimension as individual objects (lines, arrowheads, and text). If you explode an associative dimension, AutoCAD turns the dimension into individual objects. You'll undoubtedly find drawings in industries where this has happened. You cannot restore associativity to dimensions by turning this feature back on in a drawing. Exploding dimensions was a common industry practice a few years ago, but today it is mostly unnecessary because of the wide range of control over dimensions now in place. You cannot restore associativity to dimensions, but keep in mind that dimensions you create with this feature on will be associative—you can replace or add new dimensions to a drawing that has exploded dimensions.

Dimensioning Nonorthogonal Objects

So far, you've been reading about how to work with linear dimensions. You can also dimension nonorthogonal objects, such as circles, arcs, triangles, and trapezoids. Dimaligned, Dimdiameter, Dimradius, and Dimangular are powerful automatic tools that will assist you in dimensioning the myriad of shapes not describable orthogonally. AutoCAD can read the size of circles, arcs, and angles, allowing you to dimension these objects, just as it could read the distance between objects when you were placing linear dimensions.

Dimensioning Nonorthogonal Linear Distances

Now you'll dimension the true length of the beveled corner. The unusual shape of the bevel prevents you from using the horizontal or vertical dimensions you've used already. However, the Aligned Dimension option will allow you to dimension at an angle.

1. Click the Aligned Dimension tool in the Dimension toolbar or enter **dal**⏎ to start Aligned Dimension. You can also select Dimension ➤ Aligned from the menu bar.

2. At the Specify first extension line origin or <select object>: prompt, press ⏎. You could have picked extension-line origins as you did in earlier examples, but using the ⏎ will show you firsthand how the Select option works.

3. At the Select object to dimension: prompt, pick the line that represents the bevel. As the prompt indicates, you can also pick an arc or circle for this type of dimension.

4. At the Specify dimension line location or [Mtext/Text/Angle]: prompt, pick a point clear of the overall horizontal length dimension. You'll now see Dimension text = 2.240 and the dimension appear in the drawing, as shown in Figure 8.11.

TIP   Similar to linear dimensions, you can enter **t**⏎ at step 4 to input alternate text for the dimension.

FIGURE 8.11:

The aligned dimension of a nonorthogonal line

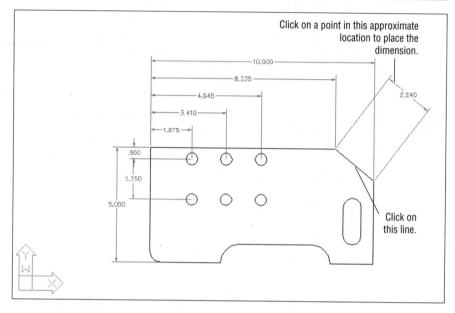

Next, you'll dimension a face of a hexagon. Instead of its actual length, however, you'll dimension a distance at a specified angle—the distance from the center of the face. First, we'll draw a polygon in the space to the right of the plate.

1. Click the Polygon tool in the Draw toolbar or type **pol**↵.

2. At the Number of sides: prompt, enter **6**↵.

3. At the Specify center of polygon or [Edge]: prompt, pick a point for the center of the polygon approximately three units to the right of the plate. Use the coordinate display to locate this point. You may want to scroll to the right to see the area.

4. Enter **c**↵ at the Enter an option [Inscribed in circle/Circumscribed about circle] <I>: prompt to select the Circumscribe option. This tells AutoCAD to place the polygon outside the temporary circle used to define the polygon.

5. At the Specify radius of circle: prompt, you'll see the hexagon drag along with the cursor.

NOTE   You could pick a point with your mouse to determine its size.

6.   Enter **1**↵ to get an exact-size hexagon. Your drawing will look like Figure 8.12.

FIGURE 8.12:

The completed polygon

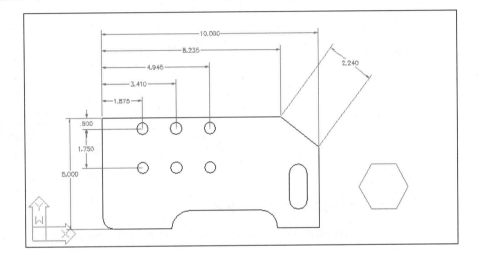

Now let's locate the dimension.

1.   Click the Linear Dimension tool in the Dimension toolbar, or enter **dimlin**↵.

2.   At the `Specify first extension line origin or <select object>:` prompt, press ↵.

3.   At the `Select object to dimension…:` prompt, pick the lower-right face of the hexagon near the coordinate.

4.   At the `Specify dimension line location or [Mtext/Text/Angle/Hor-izontal/Vertical/Rotated]:` prompt, type **r**↵ to select the rotated option.

5.   At the `Specify angle of dimension line <0>:` prompt, enter **30**↵.

6.   At the `Specify dimension line location or [Mtext/Text/Angle /Horizontal/Vertical/Rotated]:` prompt, pick a point a little below and to the right of the hexagon. You'll see `Dimension text = 1.000`. Your drawing should look like Figure 8.13.

FIGURE 8.13:

A linear dimension using the Rotated option

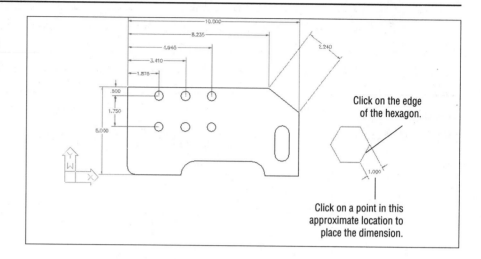

Dimensioning Angles

To dimension angles, you use the Angular Dimension option.

1. Click the Angular Dimension tool in the Dimension toolbar. Alternatively, you can enter **dan**↵ or choose Dimension ➢ Angular from the menu bar to start the Angular Dimension tool.

2. At the Select arc, circle, line, or <specify vertex>: prompt, pick the top line of the plate.

3. At the Select second line: prompt, pick the bevel line.

4. At the Specify dimension arc line location or [Mtext/Text/Angle]: prompt, notice that as you move the cursor around the apex of the angle formed by the two edges, the dimension changes to measure either the acute angle or the obtuse angle in four quadrants.

5. Pick a point to the right of the apex, clear of the object and measuring the acute angle. You'll see, `Dimension text = 38.` The dimension is fixed in the drawing (see Figure 8.14).

FIGURE 8.14:

The angular dimension added to the plate

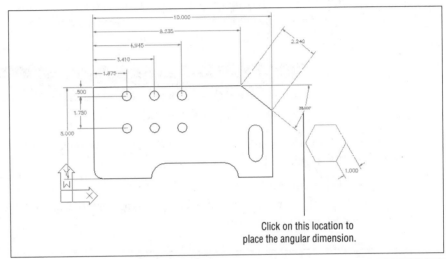

Click on this location to place the angular dimension.

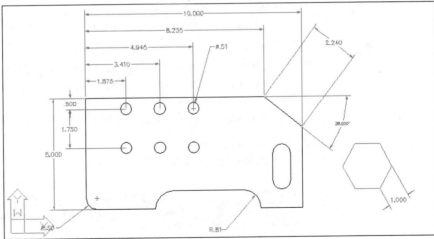

TIP   If you need to make subtle adjustments to the dimension line or text location, you can do so using grips, after you've placed the angular dimension.

Dimensioning Radii, Diameters, and Arcs

To dimension circular objects, you use another set of tools in the Dimension toolbar. Now try the Diameter tool, which shows the diameter of a circle.

1. Click the Diameter Dimension tool in the Dimension toolbar. Alternatively, you can select Dimension ➤ Diameter from the menu bar or type **ddi**↲ to start the Diameter Dimension tool.

2. At the Select arc or circle: prompt, pick the top-right circle in the set of six circles.

3. At the Specify dimension line location or [Mtext/Text/Angle]: prompt, you'll see the diameter dimension drag along the circle as you move the cursor.

NOTE   With Dimfit set to 3, AutoCAD gives you the option to place dimension text outside the circle as you drag the temporary dimension to a horizontal position.

4. Pick a point parallel with the third horizontal baseline dimension at about a 60° angle from the circle. AutoCAD has drawn a center mark at the center of the circle and a leader pointing at the circle toward the center of the circle; it has also included the diameter symbol in the text, as shown in Figure 8.15.

The Radius Dimension tool in the Dimension toolbar gives you a radius dimension just as Diameter provides a circle's diameter.

Figure 8.15 shows a radius dimension on the outside of the fillet at the bottom left of the plate. The leader points at the center of the arc, and AutoCAD has added the *R* prefix to the text. AutoCAD will also dimension the inside of an arc, as shown by

the arc dimension to the inside of the recess at the bottom of the plate. The Center Mark tool on the Dimension toolbar just places a cross mark in the center of the selected arc or circle.

FIGURE 8.15:

Dimension showing the diameter of a circle and the radius of two arcs

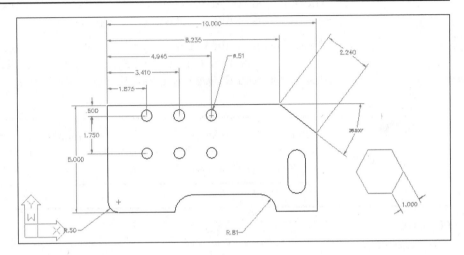

| TIP | You can alter the format of diameter dimensions by changing the Dimtix and Dimtofl dimension variable settings. For example, to have two arrows appear across the diameter of the circle, turn on both Dimtix and Dimtofl. |

Adding a Note with an Arrow

Another option for dimensions is to add a note with an arrow pointing to the object the note describes. Experienced AutoCAD users will look for the Leader command, but it has been replaced by the Quick Leader (Qleader) tool.

1. Click the Quick Leader tool in the Dimension toolbar, enter **Qleader**↵, or select Dimension ➣ Leader from the menu bar.

2. At the Specify first leader point, or [Settings]<Settings>: prompt, pick a point near the middle of the right side of the slot.

3. At the Specify next point: prompt, pick a point below the hexagon and away from the plate.

4. Press ↵ and you'll see Specify text width <0.0000>:. This is similar to Mtext, in that you may specify the width the note will occupy. If you don't specify a width, AutoCAD will not wrap the text.

5. Press ↵ again. At the Enter first line of annotation text <Mtext>: prompt, type **NO PAINT PERMISSIBLE IN THE SLOT**↵ ↵ as the label for this leader. Your drawing will look like Figure 8.16.

NOTE   If you press ↵ at the Enter first line of annotation text <Mtext>: prompt, you'll see the Multiline Text Editor dialog box. From there, you can enter and format text as required.

FIGURE 8.16:

The leader with a note added

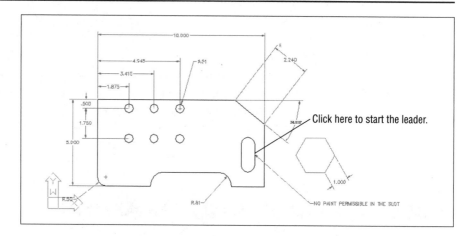

In this exercise, you used the annotation option to add the text for the note. However, the Quick Leader tool also lets you pick from annotation and leader options, as explained in the next section.

The Leader Options

When you use the Quick Leader tool, you are prompted to Specify first leader point, or [Settings] <Settings>:. If you press [↵][↵] or type

S here, AutoCAD displays the Leader Settings dialog box. This dialog box has three tabs: Annotation, Leader Line & Arrow, and Attachment.

The Annotation Tab

The options in the Annotation tab let you control the type of annotation that is attached to the leader.

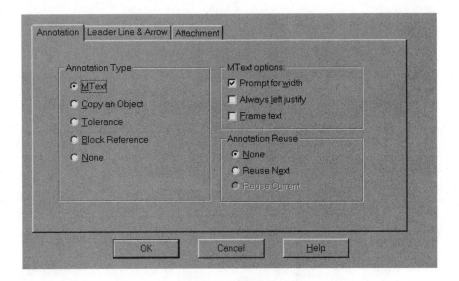

The Annotation Type section has five options:

Mtext   Allows you to edit text through the Multiline Text Editor dialog box.

Copy an Object   Prompts you to select an object to copy. You can select an existing text note, block reference, or tolerance object (feature control frame).

Tolerance   Opens the Geometric Tolerance dialog box. Use the dialog box to create a feature control frame, which AutoCAD attaches to the leader. (See "Adding Tolerance Notation" later in this chapter.)

Block Reference   Allows you to insert a block at the end of the leader.

None   Ends the leader without a note.

The MText Options section offers three choices for multiline text:

Prompt for Width   Asks you to select a width for multiline text.

Always Left Justify   Left justifies multiline text.

Frame text   Draws a frame around the text.

The Annotation Reuse section of the Annotation tab has three options:

None   Always prompts you for annotation text.

Reuse Next   Reuses the annotation text you enter for the next leader.

Reuse Current   Reuses the current annotation text.

The Leader Line & Arrow Tab

The options in the Leader Line & Arrow tab give you control over the leader line and arrow format.

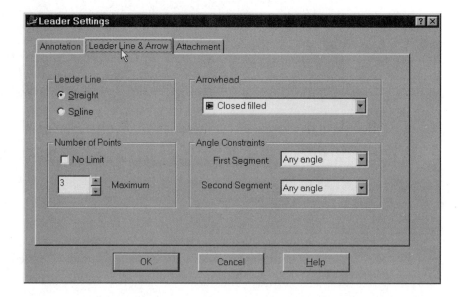

Leader Line   Lets you select either a straight line or spline for the leader line. Figure 8.17 shows the straight and spline choices, as well as the leader line without an arrowhead (the None option for Arrowhead).

FIGURE 8.17:

The Leader format options

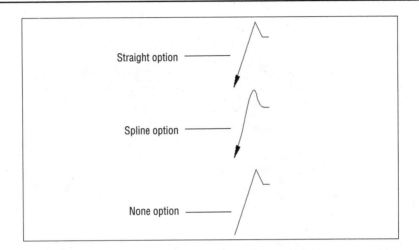

Number of Points   Lets you set the number of leader points you select before the command prompts you for the annotation. For example, if you set the points to 3, Quick Leader automatically prompts you to specify the annotation after you specify two leader points.

Arrowhead   Lets you select an arrowhead from a drop-down list. The arrowheads are the same ones that are available for dimension lines. If you choose User Defined, a list of blocks in the drawing is displayed. Select one of the blocks to use it as a leader arrowhead.

Angle Constraints   Lets you constrain the angle at which the leader line extends from the arrow.

The Attachment Tab

The options in the Attachment tab let you control how the leader connects to the annotation text, depending on which side of the leader the annotation appears. This allows you to set the location of the leader endpoint in relation to the note.

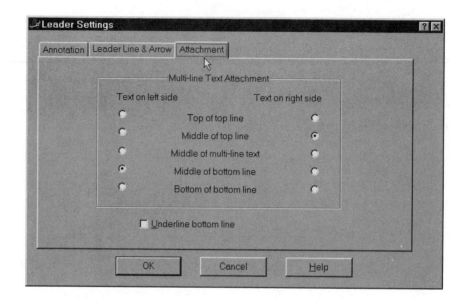

TIP　　As with other dimensions and objects in general, you can modify properties of a leader, including its text, using the Properties dialog box. For example, you can change a straight leader to a spline leader.

Skewing Dimension Lines

At times, you may find it necessary to force the extension lines to take on an angle other than 90° to the dimension line. This is a common requirement of isometric drawings, where most lines are at 30° or 60° angles instead of 90°. To facilitate nonorthogonal dimensions like these, AutoCAD offers the Oblique option.

1. Use the Linear Dimension tool in the Dimension toolbar to create a linear dimension between the lower-right two circles of the set of six circles. The dimension will look like the one in the top image of Figure 8.18.

FIGURE 8.18:

A dimension using the Oblique option

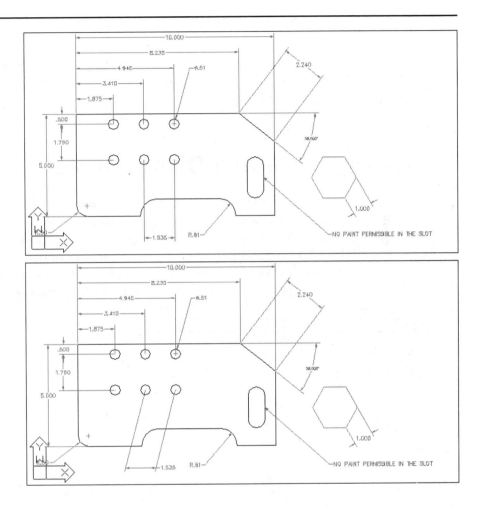

2. Choose Dimension ➢ Oblique. You can also click the Dimension Edit tool in the Dimension toolbar, and then type **o**↵.

3. At the `Select objects:` prompt, pick the dimension that you just created and press ↵ to confirm your selection.

4. At the `Enter obliquing angle (press ENTER for none):` prompt, enter **80↵** for 80°. The dimension will skew so that the extension lines are at 80°, as shown in Figure 8.18.

Applying Ordinate Dimensions

In mechanical drafting, *ordinate dimensions* are used to maintain the accuracy of machined parts by establishing an origin on the part. All major dimensions are described as X or Y coordinates from that origin. The origin is usually an easily locatable feature of the part, such as a machined bore or two machined surfaces. Figure 8.19 shows a typical application of ordinate dimensions. In the lower-left corner of the figure, notice the two dimensions whose leaders are jogged. Also notice the origin location in the upper-right corner.

FIGURE 8.19:

A drawing using ordinate dimensions

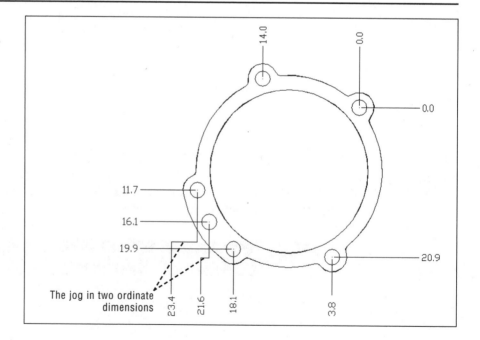

To use AutoCAD's Ordinate Dimension command, follow these steps:

1. Choose Tools ➣ UCS ➣ Origin or type **ucs.⏎or⏎**.

2. At the `Current ucs name: *WORLD* Enter an option [New/Move/ orthoGraphic/Prev/Restore/Save/Del/Apply/?/World] <World>: _move Specify new origin point or [Zdepth]<0,0,0>:` prompt, click the exact location of the origin of your part. Use an appropriate Osnap.

It is important for you to realize that you're moving the UCS origin temporarily, and only for your convenience. AutoCAD will still record all object locations by their World coordinates, even though the current display reflects the current UCS coordinates.

---

3. Toggle on the Ortho mode.

4. Click the Ordinate Dimension tool in the Dimension toolbar.

5. At the `Specify feature location:` prompt, click the item you want to dimension.

6. At the `Specify leader endpoint or [Xdatum/Ydatum/Mtext/Text/ Angle]:` prompt, indicate the length and direction of the leader. Do this by positioning the rubber-banding leader perpendicular to the coordinate direction you want to dimension, and then clicking on that point.

In steps 1 and 2, you used the UCS feature to establish a new origin in the drawing. The Ordinate Dimension tool then used that origin to determine the ordinate dimensions. You'll get a chance to work with the UCS feature in Chapter 11, "Introducing 3D."

You may have noticed options in the Command window for the Ordinate Dimension tool. The Xdatum and Ydatum options force the dimension to be of the X or Y coordinate no matter what direction the leader takes. The Mtext option opens the

Multiline Text Editor, allowing you to append or replace the ordinate dimension text. The Text option lets you enter replacement text directly through the Command window.

Similar to other dimensions, you can use grips to make adjustments to the location of ordinate dimensions.

With Ortho mode on, the dimension leader is drawn straight. If you turn off Ortho mode, the dimension leader will be drawn with a jog to whichever side of horizontal or vertical that you've picked for the text point. The first choice for an ordinate dimension should be straight from the object, and you should only use the jog to avoid writing over or through another dimension (see Figure 8.19).

Adding Tolerance Notation

In mechanical drafting, *tolerances* are a key part of a drawing's notation. They specify the allowable variation in size and shape that a mechanical part can have. To help facilitate tolerance notation, AutoCAD provides the Tolerance command, which offers common ISO tolerance symbols together with a quick way to build a standard *feature control symbol*. Feature control symbols are industry-standard symbols used to specify tolerances. If you're a mechanical engineer or drafter, AutoCAD's tolerance notation options will be a valuable tool.

To use the Tolerance command, choose the Tolerance tool from the Dimension toolbar, type **tol**↙ at the command prompt, or select Dimension ➢ Tolerance from the menu bar.

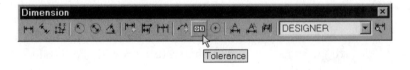

The Geometric Tolerance dialog box appears.

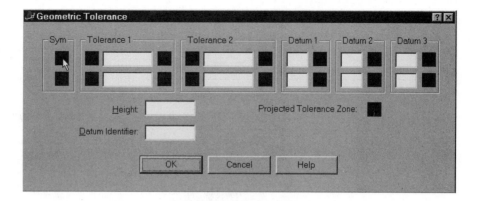

Selecting the Sym box opens the Symbol dialog box.

The top image of Figure 8.20 shows what each symbol in the Symbol dialog box represents. The bottom image shows a sample drawing with a feature symbol used on a cylindrical object.

Selecting a symbol will advance the cursor to the Tolerance 1 section.

Enter a value in the Tolerance 1 box and then select the black box on the right. This will open the Material Condition dialog box.

You may continue the process by selecting the left box of the Tolerance 2 section.

You can enter two tolerance values and three datum values. In addition, you can stack values in a two-tiered fashion

FIGURE 8.20:

The tolerance symbols

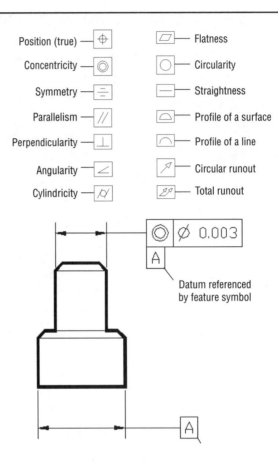

Position (true) — ⊕
Concentricity — ◎
Symmetry — ≡
Parallelism — //
Perpendicularity — ⊥
Angularity — ∠
Cylindricity — ⌀

Flatness — ▱
Circularity — ○
Straightness — —
Profile of a surface — ⌒
Profile of a line — ⌒
Circular runout — ↗
Total runout — ↗↗

◎ ⌀ 0,003
A

Datum referenced
by feature symbol

A

Understanding the Power of the Properties Tool

In this and the previous chapter, you've made frequent use of the Properties tool. By now, you may have recognized that the Properties tool is like a gateway to editing virtually any object. It allows you to edit the general properties of layer, color, and linetype assignments. When used with individual objects, it allows you to edit properties that are unique to the selected object. For example, using this tool, you can change a spline leader with an arrow into one with straight-line segments and no arrow.

Continued on next page

AutoCAD 2000 is Autodesk's clear effort to make AutoCAD's interface more consistent. The text-editing tools now edit text of all types—single-line, multiline, and dimension text—so you don't have to remember which command or tool you need for a particular object. Likewise, the Properties tool offers a powerful means of editing all types of objects in your drawing.

You may want to experiment with the Properties tool on new objects you learn about. In addition to allowing you to edit properties, the Properties tool can show you the status of an object, much like the List tool.

If You Want to Experiment...

At this point, you may want to experiment with the settings described in this chapter to identify the ones that are most useful for your work. You can then establish these settings as defaults in a template file or the Acad.dwt file.

It's a good idea to experiment with the settings you don't think you'll need often—chances are you'll have to alter them from time to time.

As an added exercise, try using the Ordinate Dimension tool to define the plate drawing. Start a new file using the plate drawing from the companion CD-ROM.

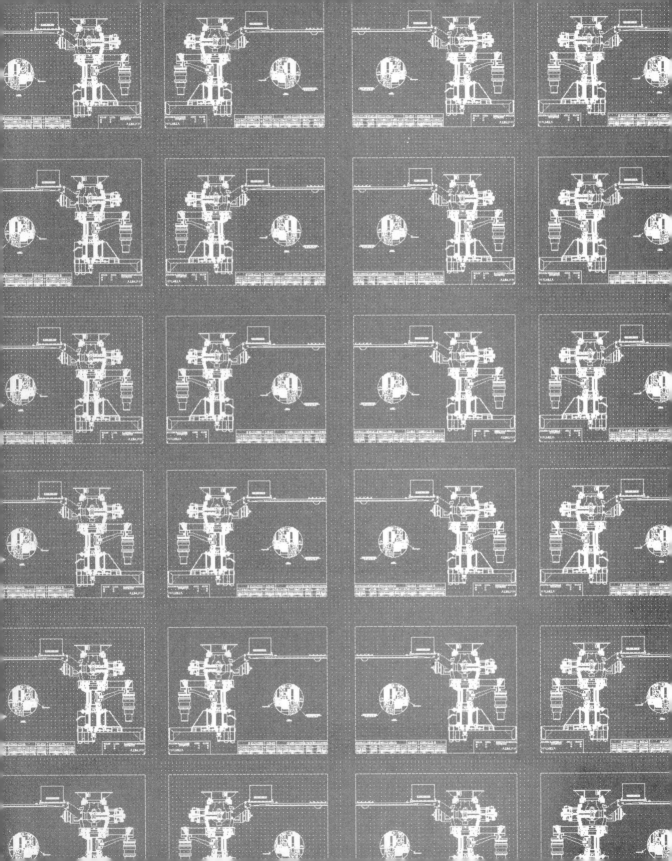

CHAPTER
NINE

Advanced Productivity Tools

- Editing More Efficiently

- Using Grips to Simplify Editing

- Using External References (Xrefs)

- Switching to Paper Space

- Advanced Tools—Selection Filter and Geometry Calculator

9

As you've seen so far, drawing in AutoCAD 2000 is more than just drawing a line, then drawing another, then drawing a circle. You can save a great deal of time by reusing existing geometry already created in this or another drawing. If there is symmetry in your design, you need only draw half of an object and use the mirror tool to make the other half. Editing is not used just for making changes—it is also used to create drawings. In this chapter, you'll learn some refinements of the editing tools you've already encountered.

You've worked with blocks in some of the previous chapters. Drawings imported as blocks become part of your drawing and can swell the size of a drawing file enormously. If you don't need to explode and use the geometry in a block, you can use an external reference instead. You'll have the opportunity to experiment with external references later in this chapter.

In preparation for 3D modeling and as an alternative to using views to move around your work, you'll experiment with Paper Space, Model Space, and the Tilemode system variable. You'll see that you can set up your work in viewports, which allow you to see multiple details of your work simultaneously. You'll also see how Paper Space can be used to help you format your drawing with a title block and standard notes.

Additionally, you'll take a tour of the geometry calculator and selection filters. The Calculator gives the AutoCAD 2000 user access to higher math functions. You can also use hooks into AutoCAD 2000 commands that enable you to use the position and size of existing geometry in your calculations. The selection filter allows you to choose specific objects by name from a selection set for editing.

Editing More Efficiently

The bar and bracket assembly that you have been working on is incomplete. The side view contains a number of hidden lines that are shown as solid. In the following exercise you'll change the layer so that the lines will be on the Hidden layer and will inherit the linetype of the Hidden layer. You'll be working in some tight areas full of detail, so you'll also take a look at how to be more precise when making your selections.

Quick Access to Your Favorite Commands

As you continue to work with AutoCAD 2000, you'll find that you use a handful of commands 90 percent of the time. You can collect your favorite commands into a single toolbar using AutoCAD 2000's toolbar customization feature. This way, you can have ready access to your most frequently used commands. Chapter 21, "Integrating AutoCAD into Your Projects and Organization," gives you all of the information that you need to create your own custom toolbars.

Editing an Existing Drawing

First, let's look at how you can modify the bolts in the bar plan. You'll begin by trimming the existing bolt lines.

1. Open the Bar file.

2. Make Layer 0 the current layer by clicking the Make Object's Layer Current button in the Object Properties toolbar, and then clicking an object line.

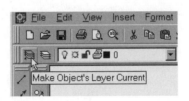

TIP   If you didn't create a bar, you can use `Bar_9.dwg` from the companion CD-ROM.

---

3. If they are not already on, turn on Noun/Verb Selection (Options ➤ Selection ➤ Selection Modes ➤ Noun / Verb or type **ddselect**⏎) and the Grips feature (Options ➤ Selection ➤ Grips ➤ Enable Grips or type **ddgrips**⏎).

4. Zoom in to the right side view and center the view in the viewport. Save this view as Right Side View.

The bolts and interior of the washer should not show through the extrusion or each other. You'll need to trim away the offending geometry and re-create it on the Hidden layer.

Using the Fence Object Selection in the Trim Command

Using the Fence object selection tool is similar to drawing a multi-segment line. Each point that you pick becomes a vertex, and the fence line continues until you press ⏎. At that time, any object crossed by any of the fence line segments will be selected. You don't "fence in" or surround the objects you wish to select.

1. Enter **trim** ⏎ to start the Trim command or click the Trim tool in the Modify toolbar.

2. Pick the right side of each boss on the extrusion shown in Figure 9.1. Complete the selection set.

FIGURE 9.1:

Lines on the extrusion to be used as cutting edges during the trim

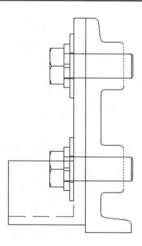

3. Type **f↵** to invoke the Fence option for selecting objects. Press ↵ to complete the selection set.

4. Pick the points for the fence vertices in the order shown in Figure 9.2.

FIGURE 9.2:

The points to pick to create the fence

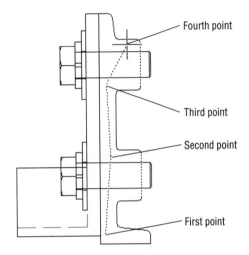

Fourth point

Third point

Second point

First point

5. Press ↵ to complete the selection set. All four lines representing the bolt have been trimmed.

NOTE   The Fence is the only tool that you can use to trim multiple objects simultaneously. This handy little tool works with any of the other commands that ask you to select objects. Remember that Trim and Extend are very similar commands. The Fence tool will also select multiple objects for the Extend command.

Changing the Layer of an Object

When you created the Versions template file, you created a layer named Hidden, giving it a hidden linetype property and the color red. You used the Versions template file when you started the Bar drawing file. The AutoCAD rule is that any object (that is not a block or Xref) whose linetype property and/or layer property is ByLayer (the default linetype and layer property) will inherit the linetype and/or color of the layer it is drawn or placed on.

The following exercise demonstrates how you can use this rule to change lines to be hidden lines and to change their color to make them easier to see on the screen. Let's change a few lines to the Hidden layer, then draw the lines for the top bolt back in and change them to the Hidden layer. You would normally erase the lines in the washers because they would cause confusion when the drawing was printed. For this exercise, we'll change them to the Hidden layer.

1. Zoom into the top bolt head and washers. You'll see six horizontal lines. Four of these represent the holes through the washers, and the other two (close together near the center line of the washer) represent the split in the lock washer. Pick the four lines that represent the holes in the washers, turning on their grips.

2. Look at each of the three property indicators on the Object Properties toolbar. The color and linetype should indicate ByLayer. If they don't, change them so they do. Open the layer drop-down list from the Object Properties toolbar and select the Hidden layer. The lines are changed to the Hidden layer, and they will turn red. Press Escape (Esc) twice to exit Grips.

3. Pick one of the new hidden lines and select the Make Object's Layer Current tool from the Object Properties toolbar. Notice that the Object Properties toolbar layer listing now shows the Hidden layer as current.

NOTE   Just a reminder: When using grips as you have been in the previous steps, you must press the Escape (Esc) key twice to complete the command. If an object's grips are still highlighted when you enter the next command, AutoCAD 2000 will assume that you want to apply the next command to the highlighted objects.

4. Use the Line tool to draw a line to replace the portion of the trimmed line that was the top of the bolt through the extrusion. You should already have a running Osnap endpoint setting. If you toggled off Osnaps by pressing the F3 key, toggle them on now. The line that you're drawing should terminate at the vertical line representing the base of the bolt head. To properly connect to it, use a temporary Perpendicular Osnap override.

5. Pick the line and highlight its right-end grip. Right-click to pop the Grips menu and select Move, right-click again to select Copy, then pick the endpoint of the remaining bottom of the bolt. Press ↵ once to complete the command, then press Esc twice to exit Grips. Your drawing will look like Figure 9.3.

FIGURE 9.3:

The hidden lines for the top bolt

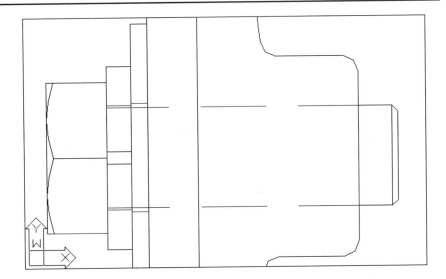

Using the Copy Command to Copy Objects on Any Layer

You've created the objects for the top bolt that can now be used for the bottom bolt. Next you'll copy the top bolt's hidden lines to the bottom bolt.

1. Zoom out until you see all of the side view again.

2. Use the Copy command, grips, or the Copy tool from the Modify toolbar and select the two new hidden lines for the bolt.

3. At the `Base point or displacement:` prompt, pick the endpoint of the line of the top bolt that was part of the bolt after it was trimmed. At the `Second point of displacement:` prompt, choose the corresponding point on the lower bolt. A copy appears like the one in Figure 9.4.

FIGURE 9.4:

The right side view of both bolts, with the hidden lines and Ltscale set to .25

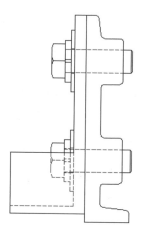

You can use the Match Properties (Matchprop) tool to make a set of objects match the layer of another object. Click Match Properties in the Standard toolbar, select the objects whose layer you want to match, and then select the objects you want to assign to the object's layer. Use this method to correct the objects around the lower bolt that should also be on the Hidden layer. There are several objects inside the U-shaped section of the ledge, which should be placed on the Hidden layer. After you've accomplished this, you should see a color change, but no change to the linetype. This is usually due to either the line being too short to show the hidden-line dashes, or duplicate geometry. Use Ltscale to correct the first condition. By setting it to a value of .25, the lines that are set to hidden should be displayed correctly. Now look for extra lines that don't respond. If there are any, try either erasing or trimming them.

Upon closer examination, you may find still other lines that are not shown correctly. They may be displayed incorrectly as hidden lines this time. If you do, simply repeat the process we just completed. Trim off the excess, make the object layer current, and draw new lines. Having the Hidden and object line layers set to different colors will be a great aid in this type of clean-up operation.

In this exercise, you used the Fence option to select the objects you wanted to trim. You could have selected each line individually by clicking it, but the Fence option offered you a quick way to select a set of objects without having to be too precise about where they are selected.

This exercise also shows that it's easier to trim back lines and then draw them back in than to try to break them precisely at each location. At first this may seem counterproductive, but trimming and then redrawing can be much faster because doing so requires fewer steps.

Singling Out Overlapping Objects

In Chapter 3, "Learning the Tools of the Trade," we mentioned that you'll encounter situations where you need to select an object that is overlapping or very close to another object. Often in such a situation, you end up selecting the wrong object. In some situations you may find two or more objects stacked up on top of each other. Often, erasing by selecting will only affect the top object. Upon completion of the erase operation it will appear that nothing has happened. You may even feel there is something wrong with the system. To help in the situations described, or to select the exact object you want, AutoCAD 2000 offers selection cycling and object sorting.

The Selection Cycling Tool

Selection cycling lets you cycle through objects that overlap until you select the one you want. To use this feature, hold down the Ctrl key and click the object you want to select. If the first object highlighted is not the one you want, click again, but this time don't hold down the Ctrl key. When several objects are overlapping, just keep clicking until the right object is highlighted and selected. When the object you want is highlighted, press ⏎, and then go on to select other objects or press ⏎ to finish the selection process. You may want to practice using selection cycling a few times to get the hang of it.

Object Sorting

If you're a veteran AutoCAD user, you may have grown accustomed to selecting the most recently created object of two overlapping objects by simply clicking it. In Release 12 and later versions of AutoCAD, by default, you don't always get to the most recently drawn object when you click overlapping objects. However, you can adjust settings that control the method of selecting overlapping objects.

If you prefer, you can set up AutoCAD to offer the most recently drawn object when you click. This setting is in the User Preferences tab of the Options dialog box. Select Tools ➤ Options, then click the User Preferences tab. The Object Sorting Methods group lets you set the sorting method for a variety of operations.

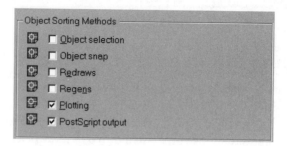

If you enable any of the operations listed, AutoCAD sorts objects based on the order in which they were added to the drawing. You probably won't want to change the sort method for Osnaps or Regens. But by checking the Object Selection check box, you can control which of two overlapping lines are selected when you click them.

These settings also can be controlled through system variables. See Appendix D, "System Variables," for details.

NOTE   When you use the Draworder command (Tools ➤ Display Order), all of the options in the Object Sorting Methods group in the User Preferences tab of the Options dialog box are turned on.

Using External References (Xrefs)

Concurrent engineering has been the goal of a lot of companies over the last decade. AutoCAD devised external references to facilitate design teams working in this manner. Think of a file with Xrefs as a room with many windows into other rooms where work is being done. The Xref is the window. As the work in another room progresses, each team can look through the window to see how the changes are affecting their work. Each time that you reload the Xref, it is refreshed and represents the current version of the file. For example, imagine that you're working on a hub and someone else is working on the shaft. The horsepower requirements of the shaft have been reduced, and the other designer has redrawn it with the new diameters and the new-sized bearings. You have the shaft in your layout as an Xref and you get e-mail informing you of the change. If you're working in the file, you must reload the Xref; if not, the latest version (the last-saved version) of the Xref will appear when you load your drawing.

Any file can be referenced into any other file. We'll call the file that is being referenced the *original* and the file that is using the Xref the *working drawing.* You can only change the Xref by opening the original file. If you have an Xref inserted into a working drawing and you want to change some aspect of the Xref, you must open the original and make the change. To see the change, you must reload the Xref in the file that is using it.

There are some problems with trying to use Xrefs in production working drawings, such as part and assembly drawings. ISO 9000 and nearly all company standards require that drawing changes be controlled so that when a drawing is distributed to the enterprise, the drawing cannot change unless the distribution channel is notified. If a drawing contains an Xref, and even if the Xref is changed correctly (with a change order and distribution), there is no process within AutoCAD 2000 to warn of the change's effect on other files that use it as an Xref. Each drawing contains a list of the Xrefs that it is using, but there is no AutoCAD command that can tell if other drawings are using a file as an Xref.

In this section, you'll see firsthand how to create and use Xrefs to help reduce design errors, share information in workgroups, and reduce the size of assembly and layout drawings.

Attaching a Drawing as an External Reference

The next exercise shows how you can use an external reference in place of an inserted block to construct a gearbox and motor assembly. You'll start by opening the gearbox file, then importing a motor.

1. Open the Gearbox.dwg file from the companion CD-ROM.

2. Choose Save As and save the file to your hard disk under the name Gearbox in your favorite directory. (My Documents will serve if you have trouble deciding which directory to use.) The current file is now Gearbox. Use Windows Explorer to locate the Motor.dwg file on the companion CD-ROM, then copy the file to your hard disk in Acad2000/Sample or any other directory that you like.

3.   Type **xr**↵ to open the Xref Manager dialog box.

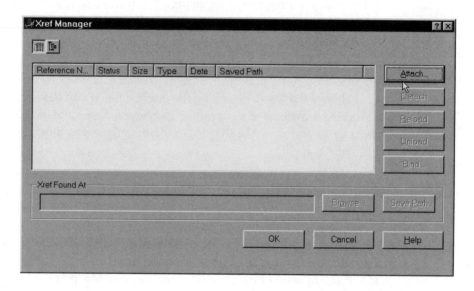

4.   Click the Attach button in the dialog box. The Select Reference File dialog box appears. This is a typical AutoCAD File dialog box, complete with a Preview window.

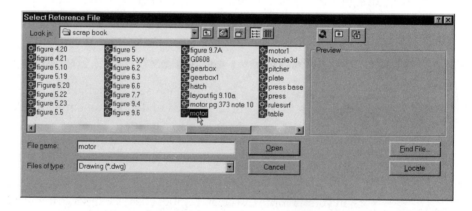

5.   Locate and select the Motor file from your hard disk, then click Open. The External Reference dialog box appears.

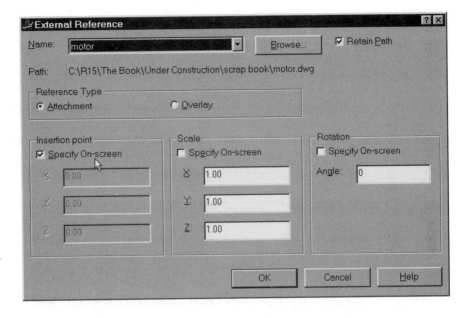

6. Click OK. You'll get a description of the other options in the Xref Manager and External Reference dialog boxes later in this chapter.

7. Pick a point near the gearbox for the insertion point.

8. Align the motor with the gearbox. Use grips to rotate and move the motor Xref, and Osnaps to align the top view of the motor and gearbox. Your drawing should look like Figure 9.5.

FIGURE 9.5:

The motor aligned with the gearbox

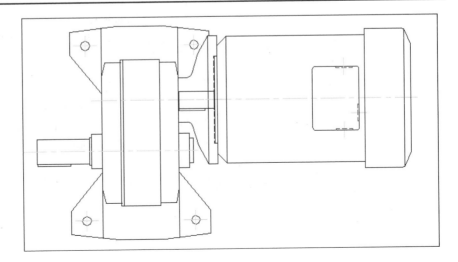

9. Save the `Gearbox` file and close it. Then open `Motor.dwg`.

10. Move the label onto the motor slightly above the electrical box. Save, then close `Motor`.

11. Click File in the menu bar. Read down the choices until you see a list of recently opened files. Pick the `Gearbox` file, then if prompted to save this file, click Yes. After the `Gearbox.dwg` file opens, notice that the Motor Xref has been updated to reflect the change that you made to the `Motor.dwg` file. Notice the message `Resolve Xref "motor": C:\..path.. \Motor.dwg" "motor"` loaded in the text screen.

> **NOTE**
>
> *Resolve Xref* means that AutoCAD has located and revised the Xref attachment named Motor.

Importing Named Elements from External References

In Chapter 5, "Editing for Productivity," we discussed how layers, blocks, linetypes, and text styles—called named elements—are imported along with a file that is inserted into another file. Xref files, on the other hand, don't import named elements. However, you can review their names and use a special command to import the ones you want to use in the current file.

> **NOTE**
>
> You can set the Visretain system variable to 1 to force AutoCAD to remember layer settings of Xref files. The layer settings affected are On, Off, Freeze, Thaw, Color, Ltype, LWeight, and PStyle (if PStylemode is set to 0). The system variables are discussed in Appendix D. You can also use the Layer Manager Express Tools utility to save layer settings for later recall. The Layer Manager is described in Chapter 19, "Introduction to Customization."

When the Xref is attached to the drawing, AutoCAD renames named elements to eliminate conflict between the Xref's named elements and the named elements of the current drawing. AutoCAD renames the elements by prefixing them with the name of the file they've come from. In our example, the Xref filename is `Motor` and there is a layer in the `Motor` file named Hidden. When the Xref is attached, the layer is added and renamed Motor | Hidden. You cannot make these layers current, so you're unable to draw on them. However, you can view and manage objects on Xref layers in the Layer Properties Manager dialog box.

Let's look at how AutoCAD identifies layers and blocks from Xref files and import a layer from an Xref.

1. With the Gearbox file open, display the Layer Properties Manager dialog box. Notice that the names of the layers from the Xref files are all prefixed with the filename and the vertical bar (|) character. Exit the Layer Properties Manager dialog box.

NOTE   You can also open the layer drop-down list in the Object Properties toolbar to view the layer names.

---

2. Click the External Reference Bind tool in the Reference toolbar or enter **xbind[cr]**.

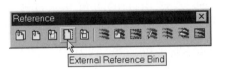

The Xbind dialog box appears. It contains a listing of the current Xrefs. Each listing shows a plus sign (+) to its left. This list box follows the Microsoft Windows 95/98/NT format for expandable lists, much like the directory listing in Windows Explorer.

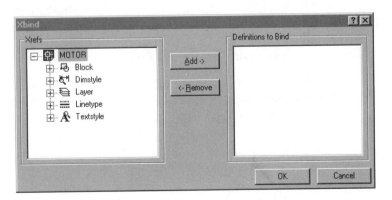

3. Click the plus sign next to MOTOR. The list expands to show the types of elements available to bind.

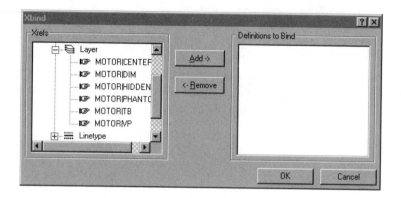

4. Click the plus sign next to Layer. The list expands further to show the layers in the available for binding.

5. Click MOTOR | HIDDEN, then click the Add button. MOTOR | HIDDEN is added to the list to the right, Definitions to Bind.

6. Click OK to bind the MOTOR | HIDDEN layer.

7. Open the Layer Properties Manager dialog box and scroll down the list and look for the MOTOR | HIDDEN layer. You won't find it. In its place is a layer called MOTOR0HIDDEN.

As you can see, when you use Xbind to import a named item, such as the MOTOR | HIDDEN layer, the vertical bar (|) is replaced by two dollar signs ($$) surrounding a number, which is usually zero. (If for some reason the imported layer name MOTOR0HIDDEN already exists, then the 0 in that name is changed to 1, as in MOTOR1HIDDEN.) Other named items are also renamed in the same way, using the 0 replacement for the vertical bar.

Although you used the Xbind dialog box to bind a single layer, you can also use it to bind multiple layers, as well as other items from Xrefs attached to the current drawing.

NOTE	You can bind an entire Xref to a drawing, converting it into a simple block. By doing so, you have the opportunity to maintain unique layer names of the Xref being bound or to merge the Xref's similarly named layers with those of the current file.

Another method for binding elements from Xref files is to start from the Xref Manager dialog box. After you've attached the Xref, the Bind button is available. Click Bind to open the Bind Xrefs dialog box, with the choices Bind and Insert. The Bind option binds the selected Xref definition to the current drawing in the same manner as the External Reference Bind tool. The Insert option binds the Xref to the current drawing in a way similar to detaching and inserting the reference drawing. Rather than being renamed using the $#$ syntax, these elements are stripped of the Xref name. For example, after an insert-like bind, the Xref-dependent layer MOTOR| HIDDEN becomes the locally defined layer HIDDEN.

After you bind an element, you can use Detach in the Xref Manager dialog box to remove the Xref file. The imported element will remain as part of the current file.

As another option, the AutoCAD DesignCenter lets you import settings and other drawing components from any drawing, not just Xref files. You'll learn more about the DesignCenter in Chapter 21, "Integrating AutoCAD into Your Projects and Organization."

Nesting External References and Using Overlays

External references can be nested. For example, the `Gearbox.dwg` file uses the `Motor.dwg` file as an Xref. In this situation, Motor.dwg is nested in the `Gearbox.dwg` file, which could in turn be an Xref in another drawing file.

This combination of nested external references could be used numerous times in another file being drawn to lay out the entire machine. AutoCAD would dutifully load each instance of the motor gearbox and thus use substantial memory, slowing down your computer.

If you don't need to see the motor information in your layout, you can use the Overlay option when you attach the Xref. The External Reference dialog box contains a Reference Type group with the two check boxes Attachment and Overlay. If you select the Overlay option when you attach the Gearbox.dwg file, the nested `Motor.dwg` Xref will be ignored.

Other Differences between External References and Blocks

You've seen that unlike inserted blocks, Xrefs don't need to be updated to reflect changes. You've also seen that to segregate layers on an Xref file from the ones in the current drawing, the Xref file's layers are prefixed with their file's name. A vertical bar separates the filename prefix and the layer name, as in motor | 0. Here are a couple other differences between Xrefs and inserted blocks that you'll want to keep in mind:

- Xrefs cannot be exploded. However, you can convert an Xref into a block, then explode the block. To do this, use the Bind button in the Xref Manager dialog box, as described in the previous section.

- If an Xref is renamed or moved to another location on your hard disk, Auto-CAD won't be able to find the moved or renamed file when it opens other files to which the Xref is attached. If this happens, use the Browse option in the Xref Manager dialog box to tell AutoCAD where to find the Xref file.

> **WARNING** Take care when relocating an Xref file with the Browse button. The Browse button can assign a file of a different name to an existing Xref as a substitution.

Other External Reference Options

There are many other features unique to Xref files. Let's look briefly at some of the options in the Xref Manager and External Reference dialog boxes that we haven't discussed yet.

Options in the Xref Manager Dialog Box

You'll find the following options in the Xref Manager dialog box. Click the External Reference tool in the Reference toolbar or type **xr[cr]** to open this dialog box. All but the Attach option are available only when an Xref is present in the current drawing and its name is selected from the list of Xrefs.

TIP   The two buttons in the upper-left corner of the dialog box let you switch between a list view of your Xrefs and a hierarchical tree view. The tree view can be helpful in determining how Xrefs are nested.

Attach   Allows you to select a file to attach and to set the parameters for the attachment.

Detach   Detaches an Xref from the current file. The file and all of its components (such as layers, dimension style, and so on) will then be completely disassociated from the current file.

Reload   Restores an unloaded Xref.

Unload   Similar to Detach, but maintains a link to the Xref file so that it can be quickly reloaded. This can reduce Redraw, Regen, and file-loading times.

Bind   Allows you to bind named elements from the Xref file to the current file, as described in the previous section.

Browse   Allows you to select a new file or location for a selected Xref.

Save Path   Saves the file path displayed in the Xref Found At input box.

TIP   The Xref Manager's list works like other Windows lists, offering the ability to sort by name, status, size, type, date, or path. To sort by name, for example, click the Reference Name button at the top of the list.

The External Reference Dialog Box

When you pick an Xref file to attach, you're presented with the External Reference dialog box, which offers the following options:

Retain Path   Checking this box has AutoCAD stores the path to the Xref's location information in the current drawing database. If this check box is not checked, AutoCAD will use the default file-search path stored in

Options ➤ File ➤ Support Files Search Path to locate the Xref the next time the current file is open. If you plan to send your files to someone else, you may want to turn off this option. Otherwise, the recipient of the file will need to duplicate the exact file-path structure of your computer before the Xref will load properly.

Reference Type   Selecting the Attachment option causes AutoCAD to include other Xref attachments that are nested in the selected file. This means that if your drawing contains other Xref attachments and you attach it to another drawing, they'll go with it. The Overlay option is similar to attaching, except that any other overlays nested in the drawing are ignored and dropped. It is recommended that you use overlaying when you're referencing geometry that is not helpful for other users to see when they reference your drawing. For example, you may have created a wiring plan for a house and need to reference the floor plan of the house. If you have chosen to overlay (rather than attach) the floor plan, another user could Xref your wiring plan without the floor plan attached.

Insertion Point   Selecting the Specify On-screen option causes AutoCAD to prompt you to select a location to anchor the incoming reference.

Scale   Selecting the Specify On-screen option allows you to enter a scale factor to size the incoming reference, relative to its original size. You can change its size relative to the original object in all three directions. If you change the X scale and press [cr], the Y and Z scale factors will be set equal to X.

Rotation   Selecting the Specify On-screen option allows you to rotate the incoming object around the insertion point. A rotation angle of 0~o will place it as drawn in the original drawing.

Clipping Xref Views and Improving Performance

Xrefs are frequently used to import large drawings for reference or backgrounds. Multiple Xrefs, such as weldments, purchased detail parts, and fabricated parts, might be combined into one file. One drawback to multiple Xrefs in earlier versions of AutoCAD was that the entire Xref was loaded into memory, even if only a small portion of the Xref was used for the final plotted output. In computers with limited resources, multiple Xrefs could slow the system to a crawl.

AutoCAD 2000 offers two tools that will help make display and memory use more efficient when using Xrefs: the Xclip command and the Demand Load option in the Options dialog box.

Clipping Views

Xclip is the name of a command accessed by choosing Modify ➤ Clip ➤ Xref. This command allows you to clip the display of an Xref or block to any shape you desire, as shown in Figure 9.6. For example, you may want to display only an L-shaped portion of a layout to be part of your current drawing. Xclip lets you define such a view.

FIGURE 9.6:

The first panel shows a polyline outline of the area to be isolated with Xclip. The second panel shows how the Xref appears after Xclip is applied. The third panel shows a view of the plan with the polyline's layer turned off.

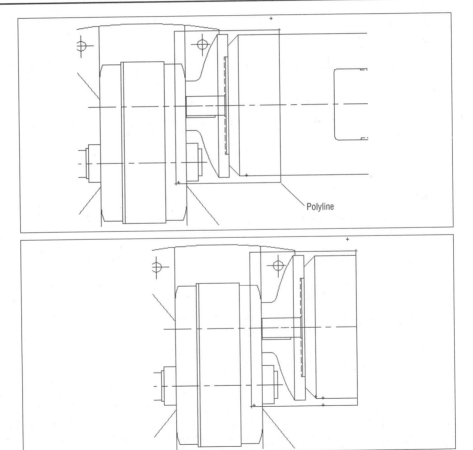

**FIGURE 9.6
CONTINUED:**

The first panel shows a polyline outline of the area to be isolated with Xclip. The second panel shows how the Xref appears after Xclip is applied. The third panel shows a view of the plan with the polyline's layer turned off.

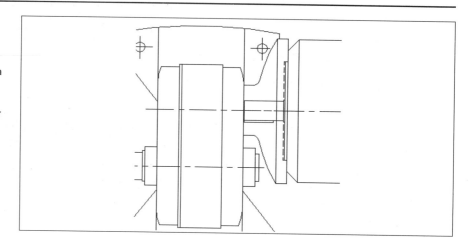

Blocks and multiple Xrefs can be clipped as well. You can also specify a front- and back-clipping distance so that visibility of objects in 3D space can be controlled. You can define a clip area using polylines or spline curves (see Chapter 10, "Drawing Curves and Solid Fills," for details).

Controlling Xref Memory Use

The Demand Load option in the Performance tab of the Options dialog box limits how much of an Xref is loaded into memory. Only the portion that is displayed gets loaded into memory.

Demand Load has a drop-down list with three settings: Disabled, Enabled, and Enabled with Copy. Demand Load is enabled by default in the standard AutoCAD drawing setup. Besides reducing the amount of memory an Xref consumes, Demand Load also prevents other users from editing the Xref while it is being viewed as part of your current drawing. This is done to aid drawing-version control and drawing management. The third Demand Load option, Enabled with Copy, creates a copy of the source Xref file, then uses the copy, thereby allowing other AutoCAD users to edit the source Xref file.

Demand loading improves performance by loading only the parts of the referenced drawing that are needed to regenerate the current drawing. You can set the location for the Xref copy in the Files tab of the Options dialog box under Temporary External Reference Location.

Special Save As Options That Affect Demand Loading

AutoCAD offers a few additional settings that will boost the performance of the Demand Load feature mentioned in the previous section. When you choose File ➢ Save As to save a file in the standard .dwg format, you'll see a button labeled Options. If you click the Options button, the Save As Options dialog box appears. This dialog box offers the Index Type drop-down list. The index referred to can help improve the speed of demand loading at the cost of a six percent or so increase in file size per index. These are the index options:

None　Creates no index.

Layer　Loads only layers that are both turned on and thawed.

Spatial　Loads only portions of an Xref or raster image within a clipped boundary.

Layer & Spatial　Turns on both the Layer and Spatial options.

WARNING　Remember, if you move an external reference file to another directory after you've inserted it into a drawing, AutoCAD may not be able to find it later when you attempt to open the drawing. If this happens, you can use the Browse option in the External Reference dialog box to tell AutoCAD the new location of the Xref file.

External references don't have to be permanent. You can attach and detach them easily at any time. This means that if you need to get information from another file—to see how well an object aligns, for example—you can temporarily Xref the other file to quickly check alignments, and then detach it when you're done.

Think of these composite files as final plot files that are only used for plotting and reviewing. Editing can then be performed on the smaller, more manageable external-referenced files. Figure 9.7 diagrams the relationship of these files.

The combinations of external references are limited only by your imagination, but avoid multiple external references of the same file.

TIP　Because Xref files do not become part of the file they are referenced into, you must take care to keep Xref files in a location where AutoCAD can find them when the referencing file is opened. This can be a minor annoyance when you need to send files to others outside your office. To help you keep track of external references, AutoCAD offers the Pack 'n Go tool in the Bonus Standard toolbar. See Chapter 19, "Introduction to Customization," for details.

Diagram of external referenced file relationships

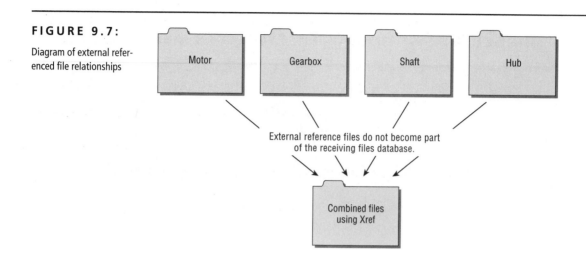

In-Place Editing of Blocks and Xrefs

You've seen different methods for editing inserted blocks and Xrefs as external files. For example, you can edit an Xref in its source file or explode a block to make corrections to the individual objects. An easier way to make changes to blocks and Xrefs is through in-place editing; that is, you can modify the Xref or block in the current drawing. In-place editing is available through the Refedit command, which is the In-Place Xref and Block Edit option on the Modify pull-down menu.

Refedit lets you select an Xref or inserted block in the current drawing. Auto-CAD temporarily extracts the objects you choose. This set of extracted objects is called the *working set* (the other objects in the current drawing are grayed). You can make changes to the working set and then save your changes back to update the Xref or block definition.

WARNING   You can't save your drawing while you're using the Refedit command. Once you start using Refedit, if you try to issue a Save command, you'll see the message SAVE command not allowed during reference editing. Be sure to save your drawing prior to starting Refedit.

To see how this works, we'll make a correction to the motor Xref, which we attached to the gearbox. We'll then update the source drawing. (If you didn't

attach the motor to the gearbox earlier, open the Gearbox1 drawing on the CD.)
Our goal is to rotate the junction box 90° in a clockwise direction.

1. Open the `Gearbox.dwg` file. Then select Modify ➤ In-Place Xref or Block
   Edit ➤ Edit Reference from the menu bar (or type **refedit[cr]**).

2. When you're prompted to select the reference, click anywhere on the motor.
   The Reference Edit dialog box appears.

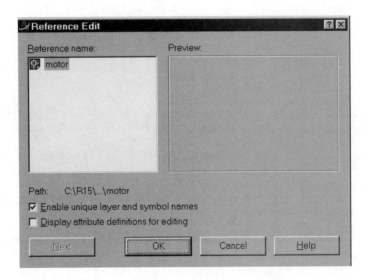

3. Click Motor in the Reference Name list box, then click OK to close the Reference Edit dialog box.

4. When the `Select nested objects:` prompt appears, select the junction
   box on the motor (see the top portion of Figure 9.6, earlier in the chapter).

5. The Refedit toolbar appears. You can now edit the junction box. Click the
   Rotate tool, select the objects that make up the junction box, and rotate them
   90° (clockwise). Try to use the center of the junction box as the base point for
   rotation.

TIP One method you might try to find the center is to use point filters (see Chapter 5).
Use the .X of the midpoint of the bottom of the junction box for the X coordinate
and the midpoint of either side of the junction box for the .YZ coordinates.

---

6. Once the junction box is rotated, it's time to save the changes to the Xref and send the changes to the drawing containing the original information. Click the Save Back Changes to Reference button in the Refedit toolbar.

The five buttons on the Refedit toolbar work as follows (from left to right):

Edit Block or Xref   Lets you select the items in the Xref or block to be edited.

Add Objects to Working Set   Lets you select objects from the Xref. The objects will be sent back to the source after editing.

Remove Objects from Working Set   Lets you delete objects from the selection set.

Discard Changes to Reference   Lets you undo any changes you've made during in-place editing.

Save Back Changes to Reference   Sends your edits back to the source drawing.

Here are some additional notes about in-place editing:

- If you choose an object that is part of a nested Xref, nested Xrefs will be listed in the Reference Edit dialog box.

- You can edit only one Xref or block at a time.

- You can edit the attributes of blocks. However, these changes will apply only to subsequent insertions of the block; the attributes in existing blocks are not affected.

- You can't edit a block inserted using the Minsert command.

Using Tiled Viewports

To gain a clear understanding of tiled viewports, imagine that the object you've been drawing is actually the full-sized object (it is, you know, because you've been drawing the object at full scale). Your computer screen is your window into a "room" where this object is being constructed, and the keyboard and mouse are your means of access to this room. There is one window in the beginning of your drawing session (your computer monitor). You can control your window's position in relation to the object through the use of Pan, Zoom, View, and other display-related commands. You can also construct or modify the model by using drawing and editing commands. Think of this room as your *Model Space*.

Now suppose that you have the ability to step back and add windows with different views looking into your room. The effect is the same as having several video cameras each connected to a different monitor. You can divide your monitor into several sections, and view all of your windows at once. Or, you may enlarge a single window to fill the entire screen. These sections of your monitor (or windows) are called *tiled viewports.* The overall environment in which you see tiled viewports is called *tiled Model Space*.

In Chapter 6, "Enhancing Your Drawing Skills," you learned how to save views so that you could more easily move around a large drawing. Tiled viewports offer a similar advantage. In addition, each tiled viewport can have different Zoom and Pan settings, which means that each tiled viewport can offer a different view of your drawing. We'll take a look at viewports now, and we'll return to them again in Chapter 12, "Mastering 3D Solids."

1. Open the file named `Azimuth.dwg` on the CD-ROM.

NOTE   Notice that the UCS icon looks quite different from what you've seen before. Instead of the familiar icon with two arrows pointing to X and Y, the icon is now a triangle with a W, a square, and an X inside. This new icon is found only in Paper Space. We'll discuss Paper Space later in this chapter.

---

2. Click the word *Model* in the tab at the bottom of the drawing. When you opened the drawing you were in Paper Space, on Layout1.

3. Now we'll divide the screen into several viewports. Go to View ➤ Viewports ➤ New Viewports. The Viewports dialog box appears.

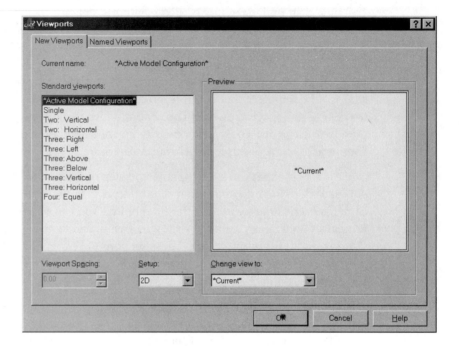

4. Click the Three: Right option. Notice that the three-box version with the large box on the right appears in the Preview box.

5. Click OK. The screen takes some time to regenerate. When the computer is finished, your drawing will look like the top image of Figure 9.8.

6. Click in each of the three viewports, and notice the outline change. As each is highlighted, it becomes the current viewport.

7. Click in the right tile then type **z↵2x↵** to zoom the viewport twice its previous zoom scale.

8. Click in the lower-left tile and type **zoom↵c↵52,38↵** to center the zoom on the portion of the drawing at that absolute coordinate. At the Enter magnification or height: prompt, type **8x↵**. (The image will appear to be eight times larger.) Your drawing should now look like the bottom image of Figure 9.8.

FIGURE 9.8:

The viewports before (top) and after (bottom) changing the zoom factor

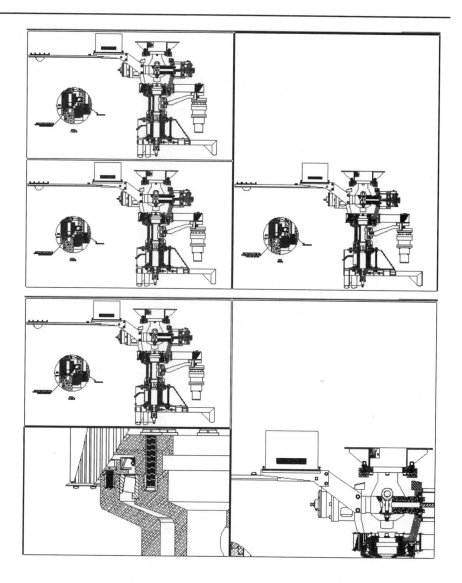

9. Next you'll draw a line from the middle of the roller in the bearing to the midpoint of the top flange. See the top of Figure 9.9 for these points. You'll need to use Osnaps for this part of the exercise. Set the Endpoint and Midpoint Running Osnaps.

FIGURE 9.9:

The Osnap points and the line in the Azimuth drawing

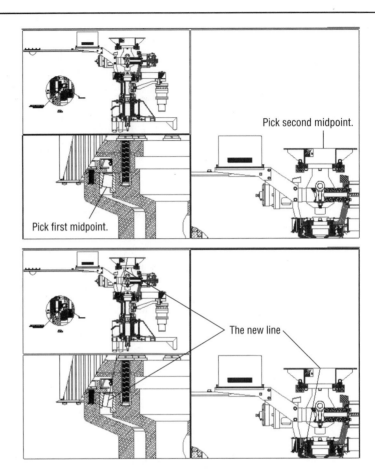

10. Click in the lower-left viewport to make it active. Click the Line tool in the Draw toolbar and find the midpoint of the lower side of the roller in the bearing. At the `Specify first point:` prompt, click in the large window to make it active, then click the midpoint of the top flange. Your drawing should look like the bottom of Figure 9.9.

NOTE   You've just seen that the tiled viewports can be used in real time. Notice that the new line appeared in all three tiled viewports.

You can change the number of viewports to a maximum of 48. You may also change the viewpoint or zoom scale in any of the tiled viewports whenever you wish. There are a few restrictions:

- Whatever combination of viewports you develop, they must fill the entire screen.

- If you print or plot your drawing, only the current active tiled viewport will plot. You cannot plot all of the viewports that you see.

You've just experienced tiled viewports in Model Space. There is another viewport option. You can create multiple viewports in Paper Space mode.

Understanding Model Space and Paper Space

You've probably noticed the tabs at the bottom of the drawing area labeled Model, Layout 1, and Layout 2. So far, you've done most of your work in the Model tab, also known as Model Space. The other two Layout tabs open views to your drawing that are specifically geared toward printing and plotting, called *Paper Space*. The Layout views allow you to control drawing scale, add title blocks, and even set up different layer settings from those in the Model tab. You can think of the Layout tab as page layout spaces that act like a desktop publishing program. You can have as many Layout tabs as you like, each set up for a different type of output.

NOTE   Model Space and Paper Space are controlled by the Tilemode system variable. When you select the Model tab, Tilemode is set to 1 (on), the default setting, and you're in Model Space, where your model is located. When you select a Layout tab, Tilemode is set to 0 (off), and you're in Paper Space.

Paper Space is similar to tiled Model Space but it is more versatile. In tiled Model Space, you can print only one viewport at a time. In Paper Space, you can position, size, and print as many viewports as you like. Figure 9.10 shows the Azimuth drawing set up in Paper Space to display different views.

FIGURE 9.10:

Different views of the same drawing in Paper Space

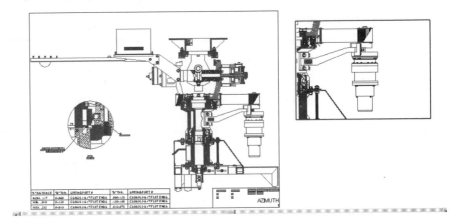

Paper Space is most valuable to mechanical engineers for plotting multiple views in different scales. You can create a viewport to show an enlarged detail of an area of the drawing, with the objects at a different scale from the rest of the views.

> **NOTE**   You may need to create a special dimension style for your scaled views, because the dimension text and arrows will be a different size from the other views. See Chapter 8 for more information about dimension styles.

Taking a Tour of Paper Space

Let's start with the basics of entering Paper Space.

1. Open the 6_Bar file, click the tab labeled Layout 1, or click on the Model indicator in the status bar.

2. The Page Setup Layout1 dialog box appears. Click OK.

Several things happen on your screen. The UCS icon changes to a triangular shape, and your screen now shows a blank sheet of paper, sized to match the sheet size for your current printer. A dashed line inside the sheet indicates the allowable printing area. Also notice the indicator on the status bar changes from Model to Paper. You've entered Paper Space.

NOTE If you click Paper in the status bar, you won't go back to the tiled Model Space view of your drawing. Instead, you'll toggle on viewports within Paper Space. To return to Model Space, click the Model tab.

Creating a Paper Space Floating Viewport

Before you can view your drawing, you must create the windows, or viewports, that let you see from Paper Space into Model Space.

1. Use the Layer Properties Manager to create a new layer named **Vports**. Change the layer's color to red and make it the current layer.

2. Select View ➤ Viewports ➤ 1 Viewport. You see these options:

   ```
   Specify corner of viewport or
   [ON/OFF/Fit/Hideplot/Lock/Object/Polygonal/Restore/2/3/4] <Fit>:
   ```

3. Type **f**[cr] to select the Fit option. Your drawing appears within a rectangle on your screen. The rectangle is the same size as the dashed line representing your printer's maximum print area. You've just created a Paper Space floating viewport.

This rectangular marker can have layer properties. If you set the current layer (Vports) to off, the border will disappear but not the viewport, and your drawing view will appear more normal.

TIP You can also create nonrectangular viewports consisting of polylines with or without arc segments, polygons, ellipses, or circles.

Because you used the Fit option, the viewport fills most of the sheet. However, the objects are not in the correct scale. Let's fix the relative scale.

1. Click the Paper indicator in the status bar. You'll notice that the triangular icon that represents Paper Space is replaced by a standard Model Space icon inside the viewport. Also, notice that the crosshairs no longer work outside the viewport. You've returned to Model Space.

2. Select the Zoom tool and use its scale option. At the Zoom prompt, type **.125xp** [cr]. The view of your model will shrink.

You've created a viewport in a layout and scaled it to 1/8 scale. You could now copy this layout and add more settings to it, as described in the next sections.

Using Layout Tools

Right-clicking the Layout tab pops up a menu with the following choices:

New Layout   Creates a new layout tab, where you can store new layout information, including a title block and sheet size.

From template   Brings an existing template into your drawing.

Delete   Removes a layout from your drawing.

Rename   Lets you change the name of a layout.

Move or Copy   Lets you move the selected layout tab to a different location or copy a layout (including its settings) to a new tab.

Select All Layouts   Selects all layout tabs (useful before deleting or copying layouts)

Page Setup   Opens the Page Setup dialog box.

Adding Floating Viewports

Now let's add more viewports for some extra views. Using Paper Space layouts, you can create these floating *viewports*, which can be placed in various configurations for plotting.

1. Create a new layout by right-clicking Layout1 and selecting New Layout. Click the new layout tab to make it current.

2. Select View ➢ Viewports ➢ New Viewports. The Viewports dialog box opens.

3. Select Three: Above, then click OK. At the Specify first corner or [Fit] <Fit>: prompt, enter **f**↵. Three new rectangles appear on the layout, as shown in Figure 9.11. The viewport at the top fills the whole width of the drawing area; the bottom half of the screen is divided into two viewports.

FIGURE 9.11:

FIGURE 9.11:

The newly created viewports

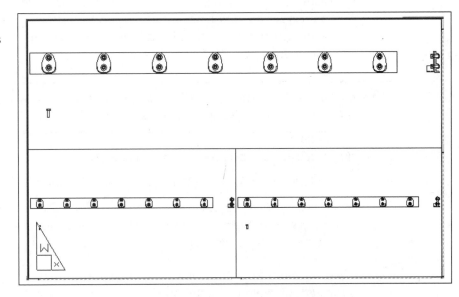

4. Click Paper in the status bar. This gives you control over Model Space even though you're in Paper Space. (You can also enter **ms**[↵] as a keyboard shortcut.) When you click Paper in the status bar, the UCS icon again changes shape; instead of one triangular-shaped icon, you have three arrow-shaped ones, one for each viewport on the screen. Figure 9.12 shows the three viewports.

FIGURE 9.12:

The three viewports, each with a different view of the bar assembly

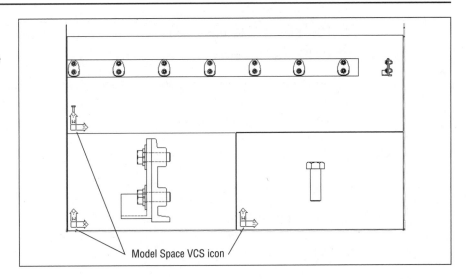

Model Space VCS icon

5. Move your cursor over each viewport. In the active viewport, the cursor appears as the crosshairs. In the other viewports, the cursor looks like an arrow pointer. You can pan and zoom, as well as edit objects, in the active viewport.

TIP   If your drawing disappears from a viewport, you can usually retrieve it by selecting View ➤ Zoom ➤ Extents (or typing **zoom↵e↵**).

6. Click the lower-left viewport to activate it.

7. Select View ➤ Zoom ➤ Window and window the side view of the bar.

8. Click the lower-right viewport and use View ➤ Zoom ➤ Window to enlarge your view of the front view of the bolt. You can also use the Pan Realtime and Zoom Realtime tools.

You can move from one viewport to another while you're in the middle of most commands (exceptions are the Snap, Zoom, Vpoint, Grid, Pan, Dview, and Vplayer commands). For example, you can issue the Line command, pick the start point in one viewport, then go to a different viewport to pick the next point. To activate a different viewport, simply click it. You can also switch viewports by pressing Ctrl+r.

Let's return to Paper Space and adjust the view of the viewports.

1. Click the Model indicator in the status bar or type **ps[↵]** to switch to Paper Space.

2. Use the Zoom Realtime tool to zoom the Paper Space view out so that it looks like Figure 9.13.

This brief exercise shows that you can use the Zoom Realtime tool in Paper Space just as you would in Model Space. All the display-related commands are available, including Pan Realtime.

FIGURE 9.13:

The view of Paper Space after zooming out

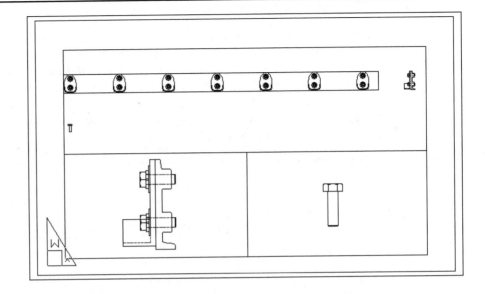

Getting Back to Full-Screen Model Space

Once you've created viewports, you can then reenter Model Space through the viewport using View ➤ Floating Model Space. But what if you want to quickly get back into the old, familiar, full-screen Model Space you were in before you entered Paper Space? The following exercise demonstrates how to do this.

1. Click the Model tab at the bottom of your screen. Your drawing returns to the original full-screen Model Space view and everything is back to normal.

2. Click the Layout tab to go back to your plot layout. You're back in Paper Space. Notice that all of the viewports are still there when you return to Paper Space. Once you've set up Paper Space, it remains part of the drawing until you delete all of the viewports.

You may prefer doing most of your drawing in Model Space and using Paper Space for setting up views for plotting. Because viewports are retained, you won't lose anything when you go back to Model Space to edit your drawing.

Working with Paper Space Viewports

Paper Space is intended as a page-layout or composition tool. You can manipulate viewports' sizes, scale their views independently of one another, and even set layering and linetype scale independently. Let's try manipulating the shape and location of viewports, using the Modify command options.

1.   Click the bottom edge of the lower-left viewport to expose its grips (see the top image of Figure 9.14).

FIGURE 9.14:

Stretching, erasing, and moving viewports

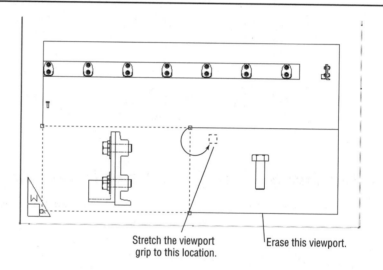

Stretch the viewport grip to this location.   Erase this viewport.

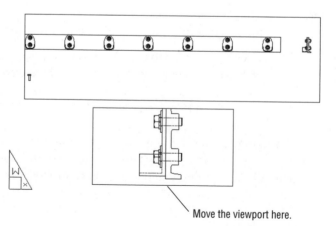

Move the viewport here.

2. Click the upper-right grip and then drag it down and to the left @–1,–1.

3. Press the Escape (Esc) key twice and then erase the lower-right viewport by clicking Erase in the Modify toolbar, then clicking the bottom edge of the viewport.

4. Move the lower-left viewport so it is centered in the bottom half of the window, as shown in the bottom image of Figure 9.14.

In this exercise, you clicked on the viewport edge to select it for editing. While in Paper Space, if you attempt to click the image within the viewport, you won't select anything. However, you can use the Osnap modes to snap to parts of the drawing image within a viewport.

Viewports are recognized as AutoCAD objects, so they can be manipulated by all of the editing commands similar to any other object. In the previous exercise, you moved, stretched, and erased viewports. Next, you'll see how layers affect viewports.

1. Remember the Vp (for Viewports) layer you created in Chapter 4? Turn it off; we don't want to make it current.

2. Use the Properties button in the Standard toolbar to change the viewport borders to the Vp layer. The Vp layer color is magenta; notice that the outlines of the viewports become magenta.

3. Turn off the Vp layer. The viewport borders will disappear.

A viewport's border can be assigned a layer, color, or linetype. If you put the viewport's border on a layer that has been turned off or frozen, that border will become invisible, just like any other object on such a layer. Making the borders invisible is helpful when you want to compose a final sheet for plotting. Even when the borders are turned off, all of the viewports will still display their views.

Disappearing Viewports

As you add more viewports to a drawing, you may discover that some of them blank out, even though you know you haven't turned them off. Don't panic. AutoCAD limits the number of active viewports at any given time to 64, with a default of 48. (An *active* viewport is one that displays.) This limit is provided because too many active viewports can bog down a system.

Continued on next page

If you're using a slow computer with limited resources, you can lower this limit to two or three to gain some performance. Then only two or three viewports will display their contents. (All viewports that are turned on will still plot, however, regardless of whether or not their contents are visible.) Zooming into a blank viewport will restore its visibility, thereby allowing you to continue to work with enlarged Paper Space views containing only a few viewports.

The Maxactvp system variable controls this value. Type **maxactvp**↵ and then enter the number of viewports you want to have active at any given time.

Scaling Views in Paper Space

Paper Space has its own unit of measure. You can set the limits of Paper Space independently of Model Space. When you first enter Paper Space, regardless of the area your drawing occupies in Model Space, you're given limits that are 12 units wide by 9 units high. This may seem incongruous at first, but if you keep in mind that Paper Space is like a paste-up area, then this difference of scale becomes easier to comprehend. Just as you might paste up photographs and maps representing several square feet onto an 11" × 17" board, you also use Paper Space to paste up views of scale drawings representing huge, complete assembly lines or micromechanisms measuring only a few thousandths of an inch. In AutoCAD, however, you have the freedom to change the scale and size of the objects that you are pasting up.

NOTE While in Paper Space, you can edit objects in a Model Space viewport, but to do so, you must enter Model Space. You can then click a viewport and edit within that viewport. While in this mode, objects that were created in Paper Space cannot be edited. Selecting Model in the status bar brings you back to the Paper Space environment.

If you want to be able to print your drawing at a specific scale, you must carefully consider scale factors when composing your Paper Space pasteup. Let's see how to put together a sheet in Paper Space and still maintain accuracy of scale.

1. If Model is visible on the status bar, you may enter the viewport. If Paper is visible, click it and toggle over to Model, or enter **ms**↵ to return to Model Space. This will allow you to manipulate the views of each viewport.

2. Click the top view to activate it.

3. Click View ➤ Zoom ➤ Scale or enter **z↵s↵**.

4. At the `Enter a scale factor (nX or nXP):` prompt, enter **1/8xp↵**. The *xp* suffix appended to the 1/8 tells AutoCAD that the current view should be scaled to 1/8 of the Paper Space scale. The 1/8 scale factor reduces the size of the image so that it will fit on the sheet.

5. Click in the lower viewport.

6. Choose View ➤ Zoom ➤ Scale again and enter **1xp↵** at the prompt. Your view of the unit will be scaled to full size in relation to Paper Space (see Figure 9.15).

FIGURE 9.15:

Paper Space viewport views scaled to 1/8 =1 and 1/2 =1

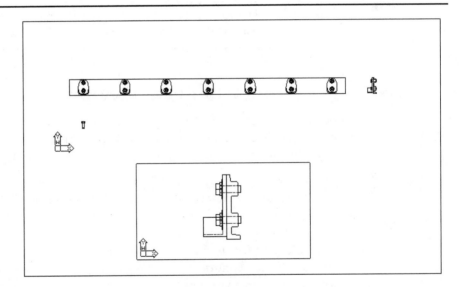

It's easy to adjust the width, height, and location of the viewports so that they display only the parts of the unit you want to see. Just go back to the Paper Space mode and use the Stretch, Move, or Scale commands to edit any viewport border. The view within the viewport itself will remain at the same scale and location, while the viewport changes in size. You can move and stretch the viewports' borders with no effect on the size and location of the objects within the view. The only thing to remember is to be sure that the viewport border is visible (its layer should be on).

If you have a situation where you need to overlay one drawing on another, you can overlap viewports. Use the Osnap overrides to select geometry within each viewport, even while in Paper Space. This allows you to align one viewport on top of another at exact locations.

You can also add a title block in Paper Space at a 1:1 scale to frame your viewports, and then plot this drawing from Paper Space at a scale of 1:1. Your plot will appear just as it does in Paper Space, at the appropriate scale.

WARNING   While working in Paper Space, pay close attention to whether you're in Paper Space or floating Model Space mode. It is easy to accidentally perform a pan or zoom within a floating Model Space viewport when you intend to pan or zoom your Paper Space view. This can cause you to lose your viewport scaling or alignment with other parts of the drawing. It's a good idea to save viewport views using View ➤ Named Views in case you happen to change a viewport view accidentally.

Setting Layers in Individual Viewports

Another unique feature of Paper Space viewports is their ability to freeze layers independently. You could, for example, display the usual drawing and dimensions in the overall view and show only a large-scale detail in the other viewport.

You control viewport-layer visibility through the Layer Properties Manager dialog box. You may have noticed that once you're working in a layout, there are three sun icons for each layer listing. You're already familiar with the sun icon farthest to the left. This is the Freeze/Thaw icon that controls the freezing and thawing of layers globally. Just to the right of that icon is a sun icon with a transparent rectangle. This icon controls the freezing and thawing of layers in individual viewports. The icon on the left is for the current viewport. The next exercise shows you how it works.

1. Activate the lower viewport.

2. Open the Layer Properties Manager dialog box.

3. In the listing for the Hidden layer, move your cursor over a column heading for the icon that shows a transparent rectangle over a sun. You'll see a tool tip describing its purpose: Active Viewport Freeze.

NOTE   The Active VP Freeze and the New VP Freeze options in the Layer Properties Manager dialog box cannot be used while you're in tiled Model Space.

4. Click the sun with the transparent rectangle icon. The sun changes into a snowflake, telling you that the layer is now frozen for the current viewport.

5. Click OK. The active viewport will regenerate with the Hidden layer made invisible in the current viewport. However, the hidden lines remain visible in the other viewport (see Figure 9.16).

FIGURE 9.16:

The drawing editor with the Hidden layer frozen in the active viewport

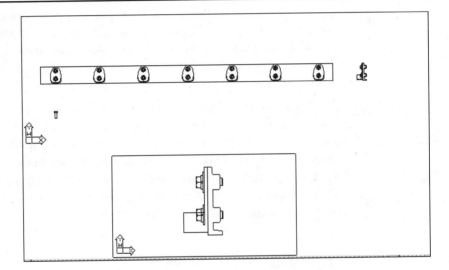

6. Save this drawing using the name Xref-1 and close it.

You may have noticed the other, similar-looking sun icon next to the one you used in the previous exercise. This icon shows an opaque rectangle over the sun. This icon controls layer visibility in any new viewports you might create next, rather than controlling existing viewports.

Linetype Scales and Paper Space

As you've seen from previous exercises, drawing scales have to be carefully controlled when creating viewports. Fortunately, this is easily done by typing **z↵** and then entering the scale factor followed by **xp↵**. While Paper Space offers the flexibility of combining different scale images in one display, it also adds to the complexity of your task in controlling that display. Your drawing's linetype scale, in particular, needs careful attention.

In Chapter 4, "Organizing Your Work," you saw how you had to set the linetype scale to the scale factor of the drawing in order to make the linetype visible. If you intend to plot that same drawing from Paper Space, you'll have to set the linetype scale back to one to get the linetypes to appear correctly. This is because AutoCAD faithfully scales linetypes to the current unit system. Remember that Paper Space units may differ from Model Space units. So when you scale a Model Space image down to fit within the smaller Paper Space area, the linetypes remain scaled to the increased linetype scale settings.

The Psltscale system variable allows you to determine how linetype scales are applied to Paper Space views. You can set Psltscale so that the linetypes will appear the same regardless of whether you view them directly in tiled Model Space, or through a viewport in Paper Space. By default, this system variable is set to one. This causes AutoCAD to scale all of the linetypes uniformly across every viewport in Paper Space. You can set Psltscale to zero to force the viewports to display linetypes exactly as they appear in Model Space.

Dimensioning in Paper Space

At times, you may find it more convenient to add dimensions to your drawing in Paper Space rather than directly on your objects in Model Space. There are several dimension settings you'll want to know about that will enable you to do this.

In order to have your dimensions produce the appropriate values in Paper Space, you need to have AutoCAD adjust the dimension text to the scale of the viewport from which you are dimensioning. You can have AutoCAD scale dimension values so that they correspond to a viewport zoom-scale factor. Here's how this setting is made.

1. Open the Dimension Style Manager dialog box.

2. Click the Modify button.

3. Click the Primary Units button.

4. In the Scale Factor input box of the Measurement Scale group, enter the inverse of the scale factor of the viewport you intend to dimension. For example, if the viewport is scaled to a 1/4xp scale, enter **4**.

NOTE   This change can be set to apply to layout viewport dimensions only, by checking that option.

---

5. Click OK, then click Close to exit the Dimension Style Manager.

You're ready to dimension in Paper Space. Remember that you can snap to objects in a floating viewport, so you can add dimensions as you normally would in Model Space.

WARNING   While AutoCAD offers the capability of adding dimensions in Paper Space, you may want to refrain from doing so until you've truly mastered AutoCAD. Because Paper Space dimensions won't be visible in Model Space, it is easy to forget to update your dimension when your drawing changes. Even with the variable Dimaso turned on, dimensions created on a layout in Paper Space will not be associative with the geometry in Model Space. However, you can use Osnaps through the viewport to set up the dimension. Dimensioning in Paper Space can also create confusion for others editing your drawing at a later date.

Other Uses for Paper Space

The exercises presented in this section should give you a sense of how to work in Paper Space. We've provided examples that reflect the more common uses of Paper Space. Remember that it is like a page-layout portion of AutoCAD, separate from Model Space, yet connected to Model Space through viewports.

You needn't limit your applications of Paper Space to drawing format. When used in conjunction with AutoCAD raster-import capabilities, Paper Space can be a powerful tool for creating presentations.

Advanced Tools—Selection Filter, Quick Select, and Calculator

Before finishing this chapter, you'll want to know about two other tools that are extremely useful in your day-to-day work with AutoCAD: Selection Filters, Quick Select and the Calculator. We've saved the discussion of these tools for this chapter because you won't really need them until you've become accustomed to the way AutoCAD works. Chances are you've already experimented with some of AutoCAD's menu options not yet discussed in the tutorial. Many of the pull-down menu options and their functions are self-explanatory. Selection Filters, Quick Select, and the Calculator. We'll start with Selection Filters.

Filtering Selections

Suppose you need to take just the circles in your drawing and isolate them in order to use the pattern in another version of the bar. This is a simple problem in our bar model, but suppose you have a plate like the one in Plate.dwg on the companion CD-ROM. One way to select only the circles would be to turn off all of the layers except the one that contains the circles; then you could use the Wblock command and select the remaining circles using a window to write the circle information to a file. Filters can simplify this operation by allowing you to select groups of objects based on their properties.

1. Open the Plate.dwg file from the companion CD-ROM.

2. Start the Wblock command by choosing File ➤ Export. In the Export Data dialog box, enter **plate2.dwg** in the File input box, choose Block (*.dwg) from the Save As Type drop-down list, and click Save.

3. Press ↵ at the prompt for a block name, and then enter **0,0** at the Specify insertion base point: prompt.

4. At the Object Selection prompt, type **'filter**↵. The Object Selection Filters dialog box appears.

5. Open the drop-down list in the Select Filter button group.

6. Scroll down the list; find and highlight the Circle option.

7.   In the Object Selection Filters dialog box, click the Add to List button toward the bottom of the Select Filter button group. Object = Circle is added to the list box.

8.   Click Apply. The dialog box closes.

9.   Type **all**⏎ to select everything in the drawing. Only circle objects are selected. You'll see a message in the Command window indicating how many objects were found and how many were filtered out.

10. Press ⏎ and you'll see the message `Exiting Filtered selection.` `<Selection set: a> 6 found.`

11. Press ⏎ again to complete the Wblock command. All of the circles are written out to a file called `Plate2`. You can type **oops**⏎ to get the circles back into this drawing.

TIP   In this exercise, you filtered out objects using the filter command. By adding the () in front of the typed command you were able to use filter "transparently," or while the Wblock command was active. Once a filter is designated, you then select the group of objects you want AutoCAD to filter through. AutoCAD finds the objects that match the filter requirements and passes those objects to the current command, in this case, Wblock. Filter can also be used by itself to create a selection set. This set can be accessed by later commands that will accept a selection set, simply by responding to the `Select Objects:` prompt, with a **p** (for previous) ⏎.

As you've seen from the previous exercise, there are many options to choose from in this utility. Let's take a closer look.

Working with the Object Selection Filters Dialog Box

To use the Object Selection Filters dialog box, first select the criteria for filtering from the drop-down list. If the criterion you select is a named item (layers, linetypes, colors, or blocks), you can then click the Select button to choose specific items from a list. If there is only one choice, the Select button is grayed out.

Once you've determined what to filter, you must add it to the list by clicking the Add to List button. The filter criteria then appear in the list box at the top of the dialog box. Once you have something in the list box, you can then apply it to your current command or to a later command. AutoCAD will remember your filter settings, so if you need to reselect a filtered selection set, you don't have to redefine your filter criteria.

Saving Filter Criteria

If you prefer, you can preselect a filter criterion. Then, at any `Select objects:` prompt, you can type **'filter↵**, highlight the appropriate filter criteria in the list box, and click Apply. The specifications in the Object Selection Filters dialog box remain in place for the duration of the current editing session.

You can also save a set of criteria by entering a name in the input box next to the Save As button and then clicking the button. The criteria-list data is saved in a file called `Filter.nfl`. You can then access the criteria list at any time by opening the Current drop-down list and choosing the name of the saved-criteria list.

Filtering Objects by Location

Notice the X, Y, and Z drop-down lists just below the main Select Filters drop-down list. These lists become accessible when you select a criterion that describes a geometry or a coordinate (such as an arc's radius or center point). You can use these lists to define filter selections even more specifically, using the *relational-operator* greater-than (>), less-than (<), equal-to (=), or not-equal-to (!=) comparisons.

For example, suppose you want to grab all of the circles whose radii are greater than 4.0 units. To do this, choose Circle Radius from the Select Filters drop-down list. Then, in the X list, select >. Enter **4.0** in the input box to the right of the X list, and click Add to List. You see the item

```
Circle Radius > 4.0000
```

added to the list box at the top of the dialog box. You've used > to indicate a circle radius greater than 4.0 units.

Creating Complex Selection Sets

There will be times when you'll want to create a very specific filter list. For instance, say you need to filter out all of the circles in the `Plate` drawing representing holes greater than 0.25 radius, and all arcs with a radius less than 0.5. To do this, use the *grouping operators* found at the bottom of the Select Filter drop-down list. Open the Object Selection Filters dialog box by typing **filter↵**. You'll need to build a list as follows:

```
** Begin OR
** Begin AND
Object = Circle
Circle Radius > 0.2500
** End AND
```

```
** Begin AND
Object = Arc
Arc Radius < 0.5000
** End AND
** End OR
```

Notice that the `Begin` and `End` operators are balanced; that is, for every `Begin` `OR` or `Begin` `AND`, there is an `End` `OR` or an `End` `AND`.

This list may look rather simple, but it can get confusing, mostly because of the way we normally think of the terms *and* and *or*. If criteria are bounded by the `AND` grouping operators, then the objects must fulfill all criteria before the objects are selected. If criteria are bounded by the `OR` grouping operators, then the objects fulfilling any one criterion will be selected.

Here are the steps to build the list shown just above.

1. In the Select Filter drop-down list, choose Begin OR, and click Add to List. Then do the same for Begin AND.

2. Click Circle in the Select Filters list, and then click Add to List.

3. Select Circle Radius from the Select Filter List, and the X section below becomes active. Drop the list box next to the X and click the greater-than symbol (>). Then enter **0.25** in the input box and click Add to List.

4. In the Select Filter list, choose End AND and click Add to List. Then do the same for Begin AND.

5. Select Arc from the Select Filter drop-down list and click Add to List.

6. Select Arc Radius from the Select Filter list, and enter **0.5** in the input box next to the X drop-down. Be sure the less-than symbol (<) shows in the X drop-down, and then click Add to List.

7. Choose End AND and click Add to List. Then do the same for End OR.

TIP   If you make an error in any of these steps, just highlight the item, select an item to replace it, and click the Substitute button instead of the Add to List button. If you only need to change a value, click Edit Item near the center of the dialog box.

---

8. When you're satisfied with your list, click the Apply button. If the operators are not balanced, the dialog box will let you know. If they can be applied to the drawing, the dialog box will close, and you'll see the `Applying filter to selection` prompt followed by the `Select objects:` prompt in the

command line. We hadn't selected anything to apply the filter to yet, so type **all**↵ to see the results of the filter on all objects in the drawing. The six circles are highlighted because they have a 0.256 radius, and the slots are highlighted because they are 0.4066 radius arcs. Another ↵ will activate grips on these for noun/verb editing.

Using Quick Select

The Filter command offers a lot of power in isolating specific types of objects. However, if you just need to filter your selection based on the object properties, you can use Quick Select (Qselect). Select Tools ➤ Quick Select from the menu bar, or right-click the drawing area when no command is active and choose Quick Select from the pop-up menu. Quick Select is also offered as an option in a few dialog box, such as the one for the Wblock command. After you start the command, the Quick Select dialog box appears.

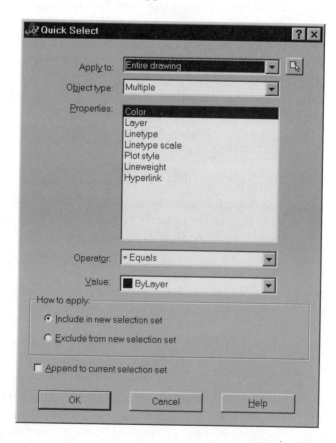

Quick Select selects objects based on the object's properties, as shown in the Properties list box. You can apply the selection criteria based on the entire drawing, or you can use the Select Objects button in the upper-right corner of the dialog box to isolate a set of objects to which you want to apply the selection criteria. The following options are available in the Quick Select dialog box:

Apply To   Lets you determine the set of objects you want to apply the Quick Select filters to. The default is the entire drawing, but you can use the Select Objects button to select a set of objects. If a set of objects has been selected before issuing the Quick Select command, you also see the Current Selection option in the Apply To drop-down list.

Object Type   Lets you limit the filter to specific types of objects, such as lines, arc, or circles. The Multiple option lets you filter your selection from all the objects in the drawing regardless of its type.

Properties   Once you select an object type, you can then select the property of the object type that you want to filter. The Properties list changes to reflect the properties that are available to be filtered.

Operator   Offers comparison (relational) operators to apply to the property you select in the Properties list. Depending on the property you select, your choices may include equal to (=), not equal to (!=), greater than (>) or less than (<) a given property value.

Value   Displays the different values of the property you select in the Properties list.

How to Apply   Lets you determine whether to include or exclude the filtered objects in a new selection set.

Append to Current Selection Set   Lets you append the filtered objects to an existing selection set or create an entirely new selection set.

Finding Geometry with the Calculator

Another useful AutoCAD tool is the Geometry Calculator. Like most calculators, it adds, subtracts, divides, and multiplies. If you enter an arithmetic expression such as 1+2, the calculator returns 3. This is useful for doing math on the fly, but the Calculator does much more than arithmetic, as you'll see in the next examples.

Finding the Midpoint between Two Points

One of the most common questions heard from AutoCAD users is, "How can I locate a point midway between two objects?" You can draw a construction line between the two objects, and then use the Midpoint Osnap override to select the midpoint of the construction line. The Calculator offers another method that doesn't require drawing additional objects.

In the following exercise, you'll start a line midway between the center of an arc and the endpoint of a line. Draw a line and an arc and try this out.

1. Start the Line command, and at the Specify first point: prompt, type 'cal↵.

2. At the >> Expression: prompt, enter **(end + cen)/2**↵.

3. At the >> Select entity for END snap: prompt, the cursor turns into a square. Place the square on the endpoint of a line and click it.

4. At the >> Select entity for CEN snap: prompt, click an arc. The line will start midway between the arc's center and the endpoint of the line.

TIP   Typing the Calculator expressions may seem a bit too cumbersome to use on a regular basis, but if you find you could use some of its features, you can create a toolbar macro to simplify the Calculator's use. See Chapter 21, "Integrating AutoCAD into Your Projects and Organization," for more on customizing toolbars.

Using Osnap Modes in Calculator Expressions

In the previous exercise, you used Osnap modes as part of arithmetic expressions. The Calculator treats them as temporary placeholders for point coordinates until you actually pick the points (at the prompts shown in steps 3 and 4 above).

The expression

```
(end + cen)/2
```

finds the average of two values. In this case, the values are coordinates, so the average is the midpoint between the two coordinates. You can take this one step further and find the centroid of a triangle using this expression

```
(end + end + end)/3
```

Notice that only the first three letters of the Osnap mode are entered in Calculator expressions. Table 9.1 shows what to enter in an expression for Osnap modes.

TABLE 9.1:   The Geometry Calculator's Osnap Modes

Calculator	Meaning
End	Endpoint
Mid	Midpoint
Int	Intersection
Cen	Center
Qua	Quadrant
Tan	Tangent
Per	Perpendicular
Nod	Node
Ins	Insert
Nea	Nearest
Rad	Radius
Cur	Cursor Pick

We've included two items in the table that are not really Osnap modes, although they work similarly when used in an expression. The first is Rad. When you include Rad in an expression, you get the prompt

```
Select circle, arc or polyline segment for RAD function:
```

You can then select an arc, polyline arc segment, or circle, and its radius is used in place of Rad in the expression.

The other item, Cur, prompts you for a point. Instead of looking for specific geometry on an object, it just locates a point. You could have used Cur in the previous exercise in place of the End and Cen modes, to create a more general-purpose midpoint locator, as in the following form:

```
(cur + cur)/2
```

Because AutoCAD doesn't provide a specific tool to select a point midway between two other points, the form shown here would be useful as a custom toolbar macro. You'll learn how to create macros in Chapter 21, "Integrating AutoCAD into Your Projects and Organization."

Finding a Point Relative to Another Point

Another common task in AutoCAD is starting a line at a relative distance from another line. The following steps describe how to use the Calculator to start a line from a point that is 2.5" in the X axis and 5.0" in the Y axis from the endpoint of another line.

1. Start the Line command. At the Specify first point: prompt, enter **'cal**↵.

2. At the >> Expression: prompt, enter **end + [2.5,5.0]**↵.

3. At the >> Select entity for END snap: prompt, pick the endpoint. The line starts from the desired location.

In this example, you used the Endpoint Osnap mode to indicate a point of reference. This is added to the Cartesian coordinates in square brackets, describing the distance and direction from the reference point. You could have entered any coordinate value within the square brackets. Or you could have entered a polar coordinate in place of the Cartesian coordinates, as in the following:

```
end + [5.59<63]
```

You don't have to include the @ because the Calculator assumes you want to add the coordinate to the one indicated by Endpoint Osnap mode. Also, it's not necessary to include every coordinate in the square brackets. For example, to indicate a displacement in only one axis, you can leave out a value for the other two coordinates, as in the following examples:

```
[4,5] = [4,5,0]
[,1] = [0,1,0]
[,,2] = [0,0,2]
[] = [0,0,0]
```

Adding Feet and Inch Distances on the Fly

One of the more frustrating situations you may have run across is having to stop in the middle of a command to find the sum of two or more distances. Say you start the Move command, select your objects, and pick a base point. Then you realize you don't know the distance for the move, but you do know that the distance is the sum of two values—unfortunately, one value is in millimeters and the other is in inches. Usually, in this situation, you would have to reach for pen and paper or use the Windows accessory called the Calculator, figure out the distance, then return to your computer to finish the task. AutoCAD's Geometry Calculator puts an end to this runaround.

The following steps show you what to do if you want to move a set of objects a distance that is the sum of 12' 6-5/8" and 115'-3/4".

1. Issue the Move command, select objects, and pick a base point.

2. At the `Second point:` prompt, start the Calculator.

3. At the `>> Expression:` prompt, enter [@12'6" + 115-3/4" < 45]. Press ↵, and the objects move into place at the proper distance.

WARNING   You must always enter an inch symbol (") when indicating inches in the Calculator.

In this example, you are mixing inches and feet, which under normal circumstances is a time-consuming calculation. Notice that the feet-and-inch format follows the standard AutoCAD syntax (no space between the feet and inch values). The coordinate value in square brackets can have any number of operators and values, as in the following:

```
[@4 * (22 + 15) - (23.3 / 12) + 1 < 13 + 17]
```

This expression demonstrates that you can also apply operators to angle values.

Guidelines for Working with the Calculator

You may be noticing some patterns in the way expressions are formatted for the Calculator. Here are some guidelines to remember.

- Coordinates are enclosed in square brackets.

- Nested or grouped expressions are enclosed in parentheses.

- Operators are placed between values, as in simple math equations.

- Object snaps can be used in place of coordinate values.

Table 9.2 lists all of the operators and functions available in the Calculator. You may want to experiment with these other functions on your own.

TABLE 9.2: The Geometry Calculator's Functions

Operator/ Function	What It Does	Example
+ or -	Adds or subtracts numbers or vectors	2 - 1 = 1 [a,b,c] + [x,y,z] = [a+x, b+y, c+z]
* or /	Multiplies or divides numbers or vectors	2 * 4.2 = 8.4 a*[x,y,z] = [a*x, a*y, a*z]
^	Exponentiation of a number	3^2 = 9
sin	Sine of angle	sin (45) = 0.707107
cos	Cosine of angle	cos (30) = 0.866025
tang	Tangent of angle	tang (30) = 0.57735
asin	Arcsine of a real number	asin (0.707107) = 45.0
acos	Arccosine of a real number	acos (0.866025) = 30.0
atan	Arctangent of a real number	atan (0.57735) = 30.0
ln	Natural log	ln (2) = 0.693147
log	Base-10 log	log (2) = 0.30103
exp	Natural exponent	exp (2) = 7.38906
exp10	Base-10 exponent	exp10 (2) = 100
sqr	Square of number	sqr (9) = 81.0
abs	Absolute value	abs (–3.4) = 3.4
round	Rounds to nearest integer	round (3.6) = 4
trunc	Drops decimal portion of real number	trunc (3.6) = 3
r2d	Converts radians to degrees	r2d (1.5708) = 90.0002
d2r	Converts degrees to radians	d2r (90) = 1.5708
pi	The constant pi	3.14159

TIP	The Geometry Calculator is capable of much more than the uses you've seen here. If you want to know more about the Calculator, consult the *AutoCAD Command Reference* and the *User's Guide* that come with AutoCAD 2000.

If You Want to Experiment...

Finish changing the objects in the end view of the bar to hidden lines. You'll find that the Fence and Match Properties commands continue to be very useful as you do this.

You might want to experiment further with Paper Space in order to become more familiar with it. You'll be using it again in Chapter 11, "Introducing 3D," and Chapter 12, "Mastering 3D Solids."

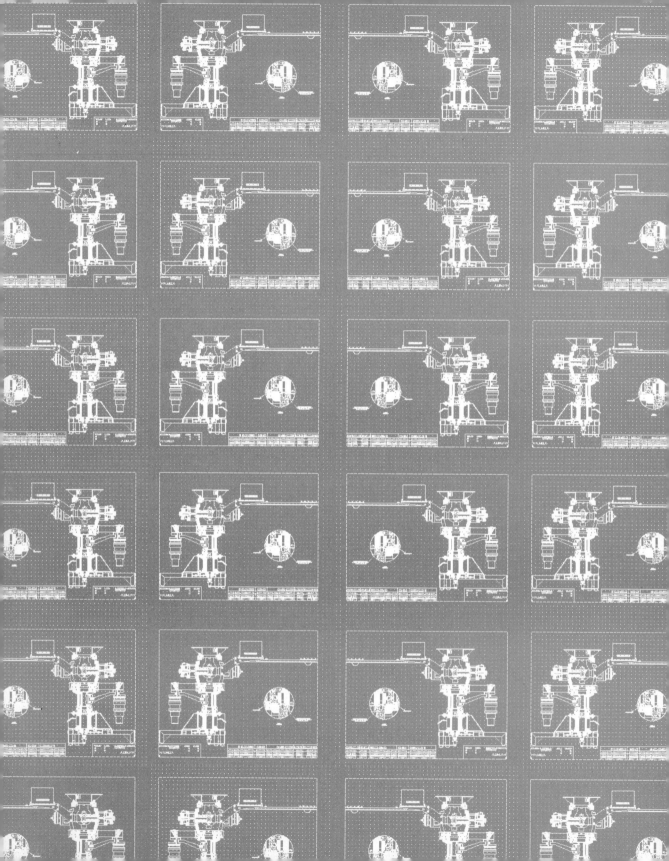

CHAPTER

TEN

Drawing Curves
and Solid Fills

- Editing Polylines

- Creating a Polyline Spline Curve

- Using True Spline Curves

- Marking Divisions on Curves

- Sketching with AutoCAD

- Filling In Solid Areas

- Using AutoCAD's Hatch Tools

10

So far in this book, you've been using basic lines, arcs, and circles to create your drawings. Now it's time to add polylines and spline curves to your repertoire. Polylines offer many options for creating objects. Spline curves are perfect for drawing smooth, nonlinear objects. (The spline curves are true *NURBS*, which stands for Non-Uniform Rational B-Splines.) We'll also look at several options for filling in the areas inside drawn objects, including hatching and solid fills.

Introducing Polylines

Polylines are similar in appearance to lines, but can have additional properties, such as constant or tapering width and several types of curvature. A polyline may look like a series of line segments, but it acts like a single object. This characteristic makes polylines useful for a variety of applications, as you'll see in the upcoming exercises. You've already created a few polylines in the form of rectangles.

Drawing a Polyline

First, to become acquainted with the polyline, you'll begin a drawing of the top view of the clevis in Figure 10.1.

FIGURE 10.1:

A sketch of a clevis

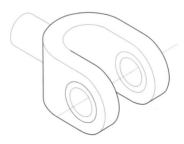

1. Start a new file and save it as Clevis2d. Don't bother to make any special setting changes, as you'll do this drawing with the default settings.

2. Click the Polyline tool in the Draw toolbar, or type **pl↵**.

3. At the `Specify startpoint:` prompt, enter a point at coordinate 3,3 to start your polyline.

4. At the `Specify next point or [Arc/Close/Halfwidth/Length/Undo/Width]:` prompt, enter **@3<0↵** to draw a horizontal line (see Figure 10.2).

FIGURE 10.2:

A polyline line and arc

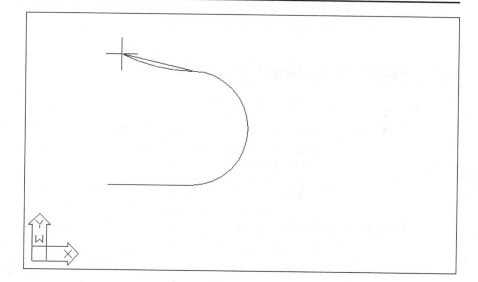

> **NOTE**
>
> You can draw polylines as you would with the Line command. Or, you can use the other Polyline tool options to enter a polyline arc, specify polyline width, or add a polyline segment in the same direction as the previously drawn line.

5. At the `Arc/Close/Halfwidth...` prompt, enter **a↵** to continue your polyline with an arc.

> **NOTE**
>
> The Arc option allows you to draw an arc that starts from the last point you selected. Once selected, the Arc option offers additional options. The default Save option is the endpoint of the arc. As you move your cursor, an arc follows it in a tangent direction from the first line segment you drew.

6. At the `Specify endpoint of arc or [Angle/CEnter/CLose/Direction/Halfwidth/Line/Radius/Second pt/Undo/Width]:` prompt, enter **@4<90↵** to draw a 180° arc from the last point you entered. Your drawing should look like Figure 10.2.

7. Continue the polyline with another line segment. To do this, enter **l↵**.

8. At the `Arc\Close\Halfwidth…` prompt, enter **@3<180↵**. Another line segment continues from the end of the arc.

9. Press **↵** to exit the Polyline command.

You now have a sideways U-shaped polyline that you'll use in the next exercise to complete the top view of the clevis.

Polyline Options

Let's pause from the tutorial to look at some of the Polyline options you didn't use.

Close   Draws a line segment from the last endpoint of a sequence of lines to the first point picked in that sequence. This works exactly like the Close option of the Line command.

Length   Enables you to specify the length of a polyline segment. It is also used when changing from Polyline Arc mode back to Polyline Line mode. When changing from Arc to Line mode, this option also forces the line segment to be tangent to the arc.

Halfwidth   Creates a tapered line segment or arc by specifying half of its beginning and ending widths (see Figure 10.3).

FIGURE 10.3:

Tapered line segment and arc created with Halfwidth

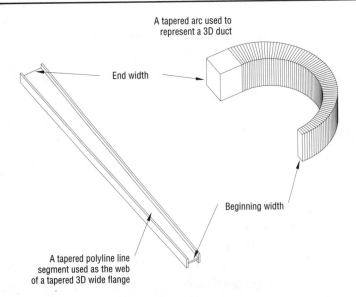

A tapered arc used to represent a 3D duct

End width

Beginning width

A tapered polyline line segment used as the web of a tapered 3D wide flange

Width   Allows the polyline width to be something more than the width of the plotter pen. You may also create a tapered line segment or arc by specifying a different full width for the segment's beginning and ending points. (Polyline width should not be confused with line weight, although they are very similar.)

Undo   Deletes the last line segment drawn.

If you want to break down a polyline into simple lines and arcs, you can use the Explode option in the Modify toolbar, just as you would with blocks. Once a polyline is exploded, it becomes a set of individual line segments or arcs, and loses all width settings.

To turn off the filling of solid polylines type **fill↵off↵**. You'll have to regenerate the drawing to see the effect. (The Display options are explained in detail in the section "Toggling Solid Fills On and Off," later in this chapter.)

NOTE   The Fillet tool in the Modify toolbar can be used to fillet all of the nonparallel line segments of a polyline. To do this, click Fillet, and then set your fillet radius. Click Fillet again, type **p↵** to select the Polyline option, and then pick the polyline you want to fillet.

Editing Polylines

You can edit polylines with many of the standard editing commands. To change the properties of a polyline, click the Properties tool in the Standard toolbar, or type **ddchprop↵**. The Stretch command in the Modify toolbar can be used to displace vertices of a polyline; the Trim, Extend, and Break commands in the Modify toolbar also work with polylines. In addition, there are many editing capabilities offered only for polylines.

In the following exercise, you'll use the Offset command in the Modify toolbar to add to the clevis.

1.   Click the Offset tool in the Modify toolbar or type **o↵**.

2. At the `Offset distance` prompt, enter **1**↵.

3. At the `Select object to offset or <exit>:` prompt, pick the U-shaped polyline you just drew.

4. At the `Specify point on side to offset:` prompt, pick a point toward the inside of the U. A concentric copy of the polyline appears (see Figure 10.4).

5. Press ↵ to exit the Offset command.

FIGURE 10.4:

The offset polyline

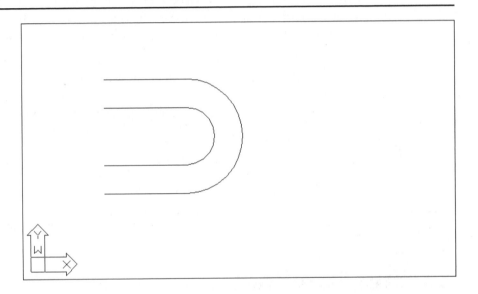

The concentric copy of a polyline made with Modify ➤ Offset can be very useful when you need to draw complex parallel curves like the ones in Figure 10.5.

Next, complete the top view of the clevis.

1. Connect the two top ends of the polylines with a short line segment (see Figure 10.6).

WARNING   The objects to be joined must touch the existing polyline exactly endpoint to endpoint, or they won't join. To ensure that you place the endpoints of the lines exactly on the endpoints of the polylines, use the Endpoint Osnap override to select each polyline endpoint.

FIGURE 10.5:

Sample complex curves
drawn using offset polylines

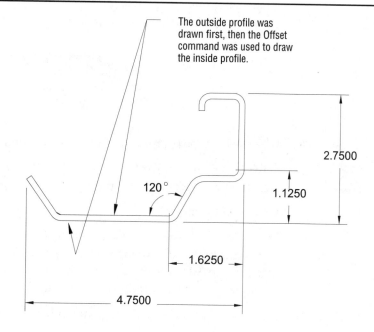

The outside profile was
drawn first, then the Offset
command was used to draw
the inside profile.

FIGURE 10.6:

The joined polyline

2. Choose Modify ➤ Object ➤ Polyline, or type **pedit**.↵. You can also click the Edit Polyline tool in the Modify II toolbar.

3. At the `Select polyline:` prompt, pick the outermost polyline.

4. At the `Enter an option Close/Join/Width/Edit Vertex/Fit/Spline/ Decurve/Ltype gen/Undo]:` prompt, enter **j**↵ for the Join option.

5. At the `Select objects:` prompt, select all of the objects you've drawn so far.

6. Once all of the objects are selected, press ↵ to join them into one polyline. It appears that nothing has happened, though you'll see the message `4 segments added to polyline` in the Command window. The segments referred to by the message are the objects in your drawing.

7. Press ↵ again to exit the Pedit command.

8. Click the drawing to expose its grips. The entire object is highlighted, telling you that all of the lines have been joined into a single polyline.

By using the Width option under Edit vertex, you can change the width of all segments of a polyline. Let's change the width of your polyline to give some thickness to the outline of the joint.

1. Click the Edit Polyline tool in the Modify II toolbar.

2. Click the polyline.

3. At the `Close/Join/Width...:` prompt, enter **w**↵ for the Width option.

4. At the `Specify new width for all segments:` prompt, enter **.03**↵ for the new width of the polyline. The line changes to the new width (see Figure 10.7), and you now have a top view of your joint.

5. Press ↵ to exit the Pedit command.

6. Save this file.

FIGURE 10.7:

The polyline with a new thickness

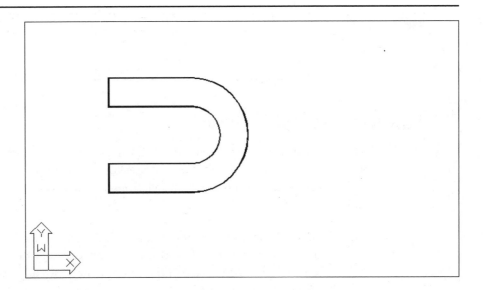

Now here's a brief look at a few of the Pedit options you didn't try yet.

Close   Connects the two endpoints of a polyline with a line or arc segment. If the polyline you selected to be edited is already closed, this option displays as Open.

Open   Removes the last segment added to a closed polyline.

Spline/Decurve   Smooths a polyline into a spline curve (discussed in detail later in this chapter in the section "Smoothing Polylines").

Edit Vertex   Lets you edit each vertex of a polyline individually (discussed in detail in the section "Edit Vertex Suboptions").

Fit   Turns polyline line segments into a series of arcs.

Ltype Gen   Controls the way noncontinuous linetypes pass through the vertices of a polyline. If you have a fitted or spline curve with a noncontinuous linetype, you'll want to turn on this option.

TIP You can change the plotted width of regular lines and arcs by using Pedit to change them into polylines, and then using the Pedit Width option to change their width.

Smoothing Polylines

There are many ways to create a curve in AutoCAD. If you don't need the representation of a curve to be exactly accurate, you can use a polyline curve. In the following exercise, you'll draw a polyline curve to represent a contour on a topographical map.

1. Open the Topo.dwg drawing that is included on the CD-ROM that comes with this book. You'll see the drawing of survey data shown in the top image of Figure 10.8. Some of the contours have already been drawn in between the data points.

2. Zoom into the upper-right corner of the drawing, so that your screen looks like the middle image of Figure 10.8.

3. Click the Polyline tool in the Draw toolbar. Using the Center Osnap, draw a polyline that connects the points labeled 254.00. Your drawing should look like the bottom image of Figure 10.8.

FIGURE 10.8:

The Topo.dwg drawing shows survey data portrayed in an AutoCAD drawing. Notice the dots indicating where elevations were taken. The actual elevation value is shown with a diagonal line from the point.

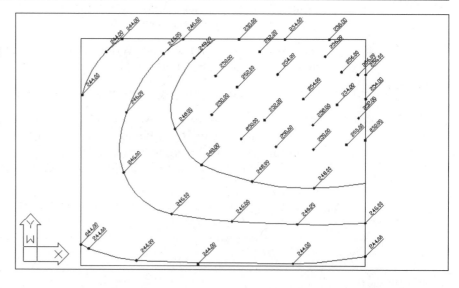

**FIGURE 10.8
CONTINUED:**

The Topo.dwg drawing shows survey data portrayed in an AutoCAD drawing. Notice the dots indicating where elevations were taken. The actual elevation value is shown with a diagonal line from the point.

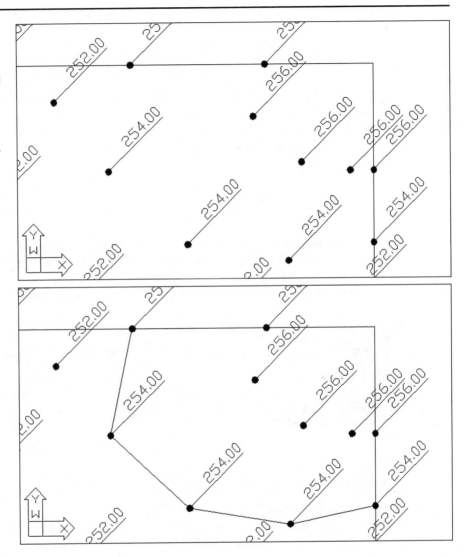

4. When you've drawn the polyline, press ↵.

If Running Osnaps aren't set, you can right-click the OSNAP label in the status bar to open the Osnaps Settings tab of the Drafting Settings dialog box. From there, you can select Center to set the Center Running Osnap.

Next you'll convert the polyline you just drew into a smooth contour line.

1. Choose Modify ➤ Polyline, or type **pedit**↵.

2. At the PEDIT Select polyline: prompt, pick the contour line you just drew.

3. At the Enter an option [Close/Join/Width/Edit vertex/Fit/Spline/ Decurve/Ltype gen/Undo]: prompt, press **f**↵ to select the Fit option. This causes the polyline to smooth out into a series of connected arcs that pass through the data points.

4. Press ↵ to end the Pedit command.

Your contour is now complete. The Fit curve option under the Pedit command causes AutoCAD to convert the straight-line segments of the polyline into arcs. The endpoints of the arcs pass through the endpoints of the line segments, and the curve of each arc depends on the direction of the adjacent arc. This gives the effect of a smooth curve. Next, you'll use this polyline curve to experiment with some of the editing options unique to the Pedit command.

Turning Objects into Polylines and Polylines into Splines

There may be times when you'll want to convert regular lines, arcs, or even circles into polylines. You may want to change the width of lines, or join together lines to form a single object such as a boundary. Here are the steps for converting lines, arcs, and circles into polylines.

1. Choose Modify ➤ Polyline. You can also type **pedit**↵ at the command prompt.

2. At the Select polyline: prompt, pick the object you wish to convert. If you want to convert a circle to a polyline, you must first break the circle (use the Break tool in the Modify toolbar) so that it becomes an arc of approximately 359°.

3. At the prompt

        Object selected is not a polyline. Do you want to turn it
        into one? <Y>

    press ↵ twice. The Pedit command ends, and the object is converted into a polyline.

Continued on next page

If you have a polyline that you would like to turn into a true spline curve, do the following:

1. Choose Modify ➤ Polyline, or type **pedit**↵. Select the polyline you want to convert.

2. Type **s**↵ to turn it into a polyline spline; then press ↵ to exit the Pedit command.

3. Click the Spline tool in the Draw toolbar or type **spline**↵. You can also select Draw ➤ Spline from the menu bar.

4. At the `Specify first point or [Object]:` prompt, type **o**↵ for the Object option.

5. At the `Select object:` prompt, click the polyline spline. Though it may not be apparent at first, the polyline is converted into a true spline.

You can also use the Spline Edit tool (Modify ➤ Spline or **spe**↵) on a polyline spline. If you do, the polyline spline is automatically converted into a true spline. If you select an object that is not a polyline spline you'll see the following warning: `Only spline fitted polylines can be converted to splines. Unable to convert the selected object.`

Editing Vertices

One of the Pedit options, Edit Vertex, has numerous suboptions that allow you to fine-tune your polyline by giving you control over its individual vertices. To access the Edit Vertex options, follow these steps.

1. First, turn off the Data and Border layers to hide the data points and border.

2. Issue the Pedit command again. Then select the polyline you just drew.

3. Type **e**↵ to enter the Edit Vertex mode. An X appears at the beginning of the polyline, indicating the vertex that will be affected by the Edit Vertex options.

WARNING   When using Edit Vertex, you must be careful about selecting the vertex to be edited. Edit Vertex has six options, and you often have to exit the Edit Vertex operation and use Pedit's Fit option to see the effect of Edit Vertex's options on a curved polyline.

Edit Vertex Suboptions

Once you've entered the Edit Vertex mode of the Pedit command, you have the option to perform the following functions:

- Break the polyline between two vertices.

- Insert a new vertex.

- Move an existing vertex.

- Straighten a polyline between two vertices.

- Change the Tangent direction of a vertex.

- Change the width of the polyline at a vertex.

These functions are presented in the form of the prompt

```
[Next/Previous/Break/Insert/Move/Regen/Straighten/Tangent/Width/eXit] <N>:
```

We'll examine each of the options presented in this prompt, starting with the Next and Previous options.

The Next and Previous Suboptions The Next and Previous suboptions enable you to select a vertex for editing. When you started the Edit Vertex option, an X appeared on the selected polyline to designate its beginning. As you select Next or Previous, the X moves from vertex to vertex to show which one is being edited. Let's try this out.

1. Press ↵ a couple of times to move the X along the polyline. (Because Next is the default option, you only need to press ↵ to move the X.)

2. Type **p**↵ for Previous. The X moves in the opposite direction. Notice that now the default option becomes P.

TIP   To determine the direction of a polyline, note the direction the X moves in when you use the Next option. Knowing the direction of a polyline is important for some of the other Edit Vertex suboptions discussed below.

The Break Suboption   The Break suboption breaks the polyline between two vertices.

1. Position the X on one end of the segment you want to break.

2. Enter **b↵** at the command prompt.

3. At the `Enter an option [Next/Previous/Go/eXit] <N>:` prompt, use Next or Previous to move the X to the other end of the segment to be broken.

4. When the X is in the right position, select Go from the Edit Vertex menu or enter **g↵**, and the polyline will be broken (see Figure 10.9).

FIGURE 10.9:

How the Break option works

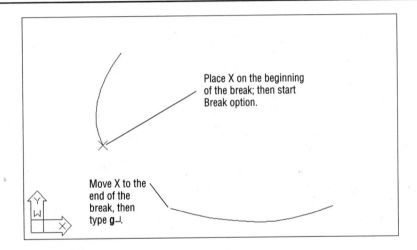

Place X on the beginning of the break; then start Break option.

Move X to the end of the break, then type **g↵**.

NOTE   You can also use the Break and Trim commands in the Modify toolbar to break a polyline anywhere.

The Insert Suboption   Next, try the Insert option, which inserts a new vertex.

1. Type **x↵** to temporarily exit the Edit Vertex option. Then type **u↵** to undo the break.

2. Type **e↵** to return to the Edit Vertex option, and position the X before the new vertex.

3. Press ↵ to advance the X marker to the next point.

4. Enter **i↵** to select the Insert option.

5. When the `Specify location of new vertex:` prompt appears, along with a rubber-banding line originating from the current X position (see Figure 10.10), pick a point indicating the new vertex location. The polyline is redrawn with the new vertex.

FIGURE 10.10:

The new vertex location

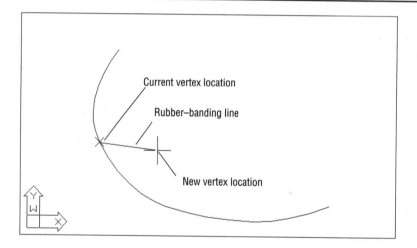

Notice that the inserted vertex appears between the currently marked vertex and the next one, so this Insert option is sensitive to the direction of the polyline. If the polyline is curved, the new vertex won't immediately appear as curved (see the top image of Figure 10.11). You must smooth it out by exiting the Edit Vertex option and then using the Fit option, as you did to edit the site plan (see the bottom image of Figure 10.11). You can also use the Stretch command (in the Modify toolbar) to move a polyline vertex.

FIGURE 10.11:

The polyline before and after the curve is fitted

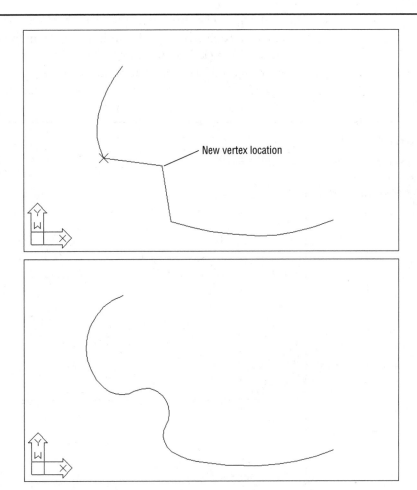

New vertex location

The Move Suboption   In this brief exercise, you'll use the Move option to relocate a vertex.

1. Undo the inserted vertex by exiting the Edit Vertex option (x↵) and typing **u**↵.

2. Restart the Edit Vertex option, and use the Next or Previous option to place the X on the vertex you wish to move.

3. Enter **m**↵ for the Move option.

4. When the Specify new location for marked vertex: prompt appears along with a rubber-banding line originating from the X (see the top image of Figure 10.12), pick the new vertex. The polyline is redrawn (see the middle image of Figure 10.12). Again, if the line is curved, the new vertex appears as a sharp angle until you use the Fit option (see the bottom image of Figure 10.12).

FIGURE 10.12:

Picking a new location for a vertex with the polyline before and after the curve is fitted

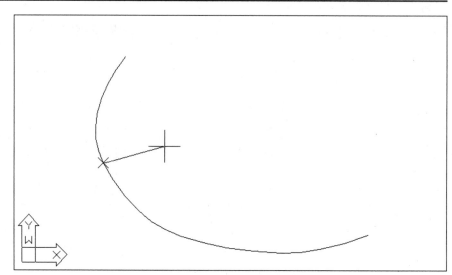

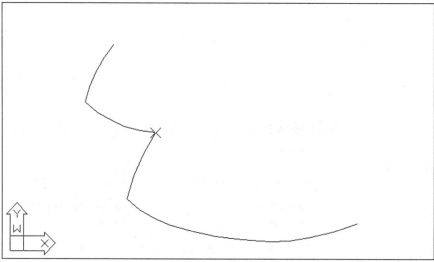

**FIGURE 10.12
CONTINUED:**

Picking a new location for a vertex with the polyline before and after the curve is fitted

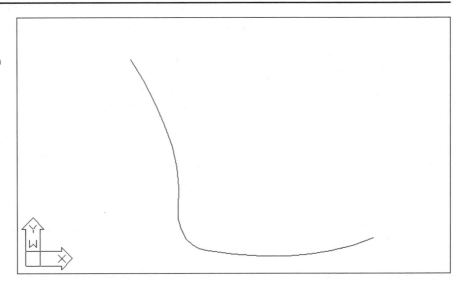

TIP   You can also move a polyline vertex using its grip. This is usually much easier than using Pedit.

The Regen Suboption   This option is available any time you need to correct the image. Regen will rebuild the view.

The Straighten Suboption   The Straighten suboption straightens all of the vertices between two selected vertices, as shown in the following exercise.

1. Undo the moved vertex from the previous exercise.

2. Start the Edit Vertex option again, and select the starting vertex for the straight line.

3. Enter s↵ for the Straighten option.

4. At the [Next/Previous/Go/eXit]: prompt, move the X to the location for the other end of the straight-line segment.

5. Once the X is in the proper position, enter g↵ for the Go option. The polyline straightens between the two selected vertices (see Figure 10.13).

FIGURE 10.13:

A polyline before and after straightening

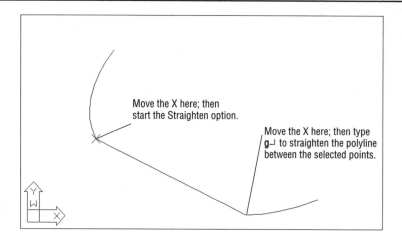

Move the X here; then start the Straighten option.

Move the X here; then type **g**↵ to straighten the polyline between the selected points.

TIP   The Straighten option offers a quick way to delete vertices from a polyline.

The Tangent Suboption   The Tangent option alters the direction of a curve on a curve-fitted polyline.

1. Undo the straightened segment from the previous exercise.

2. Restart the Edit Vertex option, and position the X on the vertex you wish to alter.

3. Enter **t**↵ for the Tangent option. A rubber-banding line appears (see the top image of Figure 10.14).

4. Point the rubber-banding line in the direction for the new tangent, and click the mouse. An arrow appears, indicating the new tangent direction (see the middle image of Figure 10.14).

Don't worry if the polyline shape doesn't change. You must use Fit to see the effect of Tangent (see the bottom image of Figure 10.14).

FIGURE 10.14:

Picking a new tangent direction

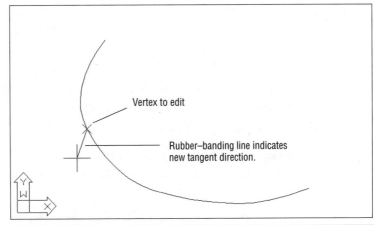

Vertex to edit

Rubber-banding line indicates new tangent direction.

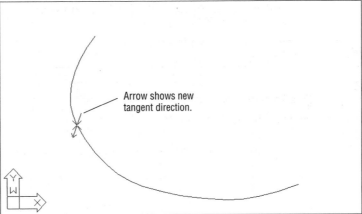

Arrow shows new tangent direction.

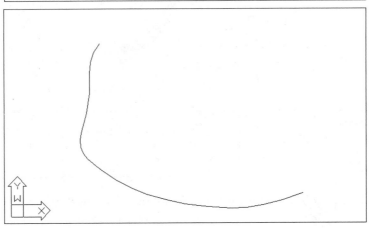

The Width Suboption Finally, try out the Width suboption. Unlike the Pedit command's Width option, Edit Vertex/Width enables you to alter the width of the polyline at any vertex. Thus you can taper or otherwise vary polyline thickness.

1. Undo the tangent arc from the previous exercise.

2. Return to the Edit Vertex option, and place the X at the beginning vertex of a polyline segment you want to change.

3. Type **w**↵ to issue the Width option.

4. At the `Specify starting width for next segment <0'-0">:` prompt, enter a value, **1'** for example, indicating the polyline width desired at this vertex.

5. At the `Specify ending width for next segment <1'-0">:` prompt, enter the width, **2'** for example, for the next vertex.

The result is shown in Figure 10.15.

FIGURE 10.15:

A polyline with the width of one segment increased

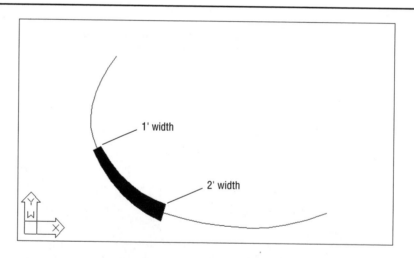

NOTE The Edit Vertex/Width option is useful when you want to create an irregular or curved area in your drawing that is to be filled in solid. This is another option that is sensitive to the polyline direction.

As you've seen throughout these exercises, you can use the Undo option to reverse the last Edit Vertex option used. You can also use the Edit option to leave Edit Vertex at any time. Just enter **x**↵, and this brings you back to the Pedit `Close/ Join/Width...` prompt.

Before we move on, note that the Pedit tool has, as you've seen, the ability to change the width of a polyline. However, at times you may not wish to see the polyline filled in. The Fill tool will take care of that. Simply type **fill**↵, **off**↵, followed by **regen**↵. To reverse the command, type **fill**↵, **on**↵, followed by **regen**↵.

Please note that to exit the Pedit command Edit Vertex suboptions, you must use the eXit option. Simply type **x**↵ to exit the suboptions. Press ↵ once more to complete the command.

Creating a Polyline Spline Curve

The Pedit command's Spline option (named after the Spline tool used in manual drafting) offers you a way to draw smoother and more controllable curves than those produced by the Fit option. A polyline spline doesn't pass through the vertex points as does a fitted curve. Instead, the vertex points act as weights pulling the curve in their direction. The polyline spline only touches its beginning and end vertices. Figure 10.16 illustrates this concept.

FIGURE 10.16:

The polyline spline curve
pulled toward its vertices

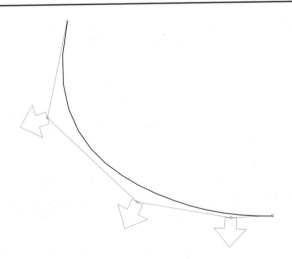

A polyline spline curve doesn't represent a mathematically true curve. See "Using True Spline Curves" later in this chapter to learn how to draw a more accurate spline curve.

Let's see how using a polyline spline curve might influence the way you edit a curve.

1. Undo the width changes you made in the previous exercise.

2. To change the contour into a polyline spline curve, choose Modify ➤ Polyline.

3. Pick the polyline to be curved.

4. At the ...Close\Join\Width...: prompt, enter s↵. Your curve will change to look like Figure 10.17.

FIGURE 10.17:

A spline curve

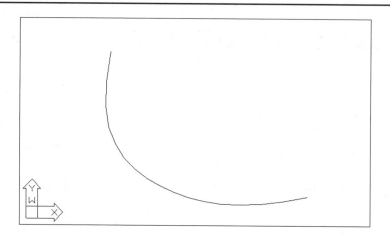

5. Press ↵ to exit Pedit.

The curve takes on a smoother, more graceful appearance. It no longer passes through the points you used to define it. To see where the points went and to find out how spline curves act, do the following:

1. Make sure the Noun/Verb Selection mode and the Grips feature are turned on.

2. Click the curve. You'll see the original vertices appear as grips (the top image of Figure 10.18).

3. Click the grip that is second from the top of the curve, as shown in the middle image of Figure 10.18, and move the grip around. Notice how the curve follows, giving you immediate feedback on how the curve will look.

4. Pick a point. The curve is fixed in its new position, as shown in the bottom image of Figure 10.18.

FIGURE 10.18:

The fitted curve changed to a spline curve, with the location of the second vertex and the new curve

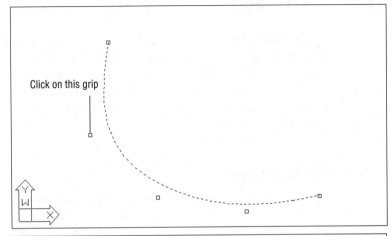

Click on this grip

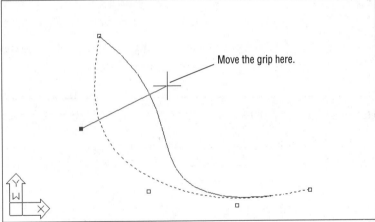

Move the grip here.

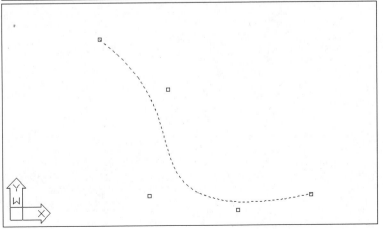

Using True Spline Curves

So far, you've been working with polylines to generate spline curves. The advantage to using polylines for curves is that they can be enhanced in other ways. You can modify their width, for instance, or join together several curves. But at times you'll need a more exact representation of a curve. The spline object, created with Draw ➤ Spline, offers a more accurate model of a spline curve, as well as more control over its shape.

Drawing a Spline

The next exercise demonstrates the creation of a spline curve.

1. Undo the changes made in the last two exercises.

2. Turn on the Data layer so that you can view the data points.

3. Adjust your view so that you can see all of the data points with the elevation of 250.00 (see Figure 10.19).

FIGURE 10.19:

Starting the spline curve at the first data point

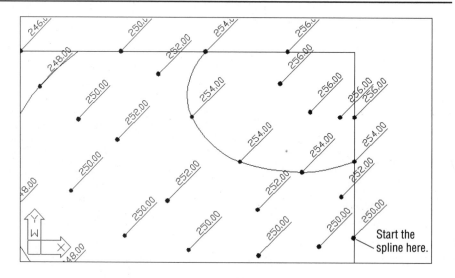

4. Click the Spline tool in the Draw toolbar or type **spl**↵.

5. At the Specify first point or [Object]: prompt, use the Center Osnap to start the curve on the lowest-right data point (see Figure 10.19). At the Specify next point: prompt, pick the next data point. The prompt changes to Specify next point or [Close/Fit tolerance] <start tangent>:.

6. Continue to select the 250.00 data points until you reach the last one. Notice that as you pick points, a curve appears, and bends and flows as you move your cursor.

7. Once you've selected the last point, press ↵. Notice that the prompt changes to Specify start tangent:. Also, a rubber-banding line appears from the first point of the curve to the cursor. As you move the cursor, the curve adjusts to the direction of the rubber-banding line. Here, you can set the tangency of the first point of the curve (see the top image of Figure 10.20).

8. Press ↵. This causes AutoCAD to determine the first point's tangency based on the current shape of the curve. A rubber-banding line appears from the last point of the curve. As with the first point, you can indicate a tangent direction for the last point of the curve (see the bottom image of Figure 10.20).

FIGURE 10.20:

The last two prompts of the Spline command let you determine the tangent direction of the spline.

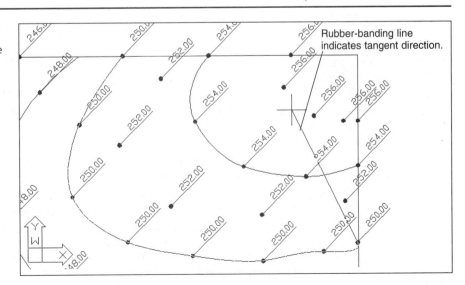

FIGURE 10.20 CONTINUED:

The last two prompts of the Spline command let you determine the tangent direction of the spline.

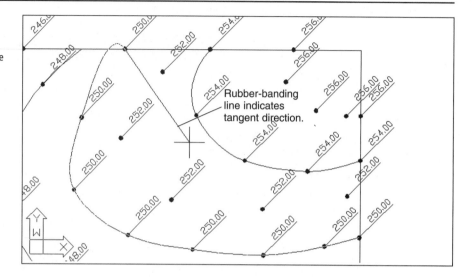

Rubber-banding line indicates tangent direction.

9. Press ↵ to exit the Spline command without changing the endpoint tangent direction.

You now have a smooth curve that passes through the points you selected. These points are called the *control points*. If you click the curve, you'll see the grips appear at the location of these control points, and you can adjust the curve simply by clicking the grip points and moving them. (You may need to turn off the Data layer to see the grips clearly.)

TIP See Chapter 2, "Creating Your First Drawing," for more detailed information on grip editing.

You may have noticed two other options—Fit Tolerance and Close—as you were selecting points for the spline in the last exercise. Here is a description of these options.

Fit Tolerance Lets you change the curve so that it doesn't actually pass through the points you pick. When you select this option, you get the prompt `Specify fit tolerance <0.0000>:`. Any value greater than 0 will cause the curve to pass close to, but not through the points. A value of 0 causes the curve to pass through the points. (You'll see how this works in the next exercise.)

Close Lets you close the curve into a loop. If you choose this option, you're prompted to indicate a tangent direction for the closing point.

Fine-Tuning Spline Curves

Spline curves are different from other types of objects, and many of the standard editing commands won't work on splines. AutoCAD offers the Modify ➤ Spline tool (Splinedit command) for making changes to splines. The following exercise will give you some practice with this command. You'll start by focusing on Splinedit's Fit Data option, which lets you fine-tune the spline curve.

Controlling the Fit Data of a Spline

The following exercise will demonstrate how the Fit Data option lets you control some of the general characteristics of the curve.

1. Choose Modify ➤ Spline, or type **splinedit**↵ at the command prompt.

2. At the `Select Spline:` prompt, select the spline you drew in the previous exercise.

3. At the prompt

   `Enter an option [Fit data/Close/Move vertex/Refine/rEverse/Undo]:`

   type **f**↵ to select the Fit Data option.

> **NOTE**   The Fit Data option is similar to the Edit Vertex option of the Pedit command in that Fit Data offers a subset of options that let you edit certain properties of the spline.

4. To control tangency at the beginning and end points, at the next prompt

   `Enter a fit data option [Add/Close/Delete/Move/Purge/Tangents/ toLerance/eXit] <eXit>:`

   type **t**↵ to select the Tangents option. Move the cursor, and notice that the curve changes tangency through the first point, just as it did when you first created the spline (see Figure 10.20).

5. Press ↵. Now you can edit the other endpoint tangency.

6. Press ↵ again. You return to the `Add/Close/Delete…` prompt.

7. Now we'll add another control point to the spline curve. At the `Add/ Close/Delete…` prompt, type **a**↵ to select the Add option.

8. At the `Specify control point <exit>:` prompt, click the second grip point from the bottom end of the spline (see the top image of Figure 10.21). A rubber-banding line appears from the point you selected. That point and

the next point are highlighted. The two highlighted points tell you that the next point you select will fall between these two points. You also see the `Specify new point <exit>:` prompt.

9. Click a new point. The curve changes to include that point. In addition, the new point becomes the highlighted point, indicating that you can continue to add more points between it and the other highlighted point (see the bottom image of Figure 10.21).

FIGURE 10.21:

Adding a new control point to a spline

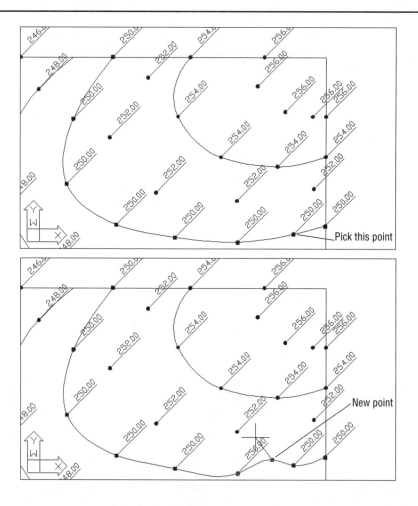

10. Press ↵. The `Specify control point <exit>:` prompt appears, allowing you to select another point if you so desire.

11. Press ↵ again to return to the `Add/Close/Delete...` prompt.

Before we end our examination of the Fit Data options, let's see how the Tolerance option works.

1. At the Add/Close/Delete… prompt, type l↵ to select the Tolerance option. This option sets the tolerance between the control point and the curve.

2. At the Specify fit tolerance <0.0000>: prompt, type **30**↵. Notice how the curve no longer passes through the control points, except for the beginning and end points (see Figure 10.22). The fit tolerance value you enter determines the maximum distance from any control point the spline can be.

FIGURE 10.22:

The spline after setting the control point tolerance to 30

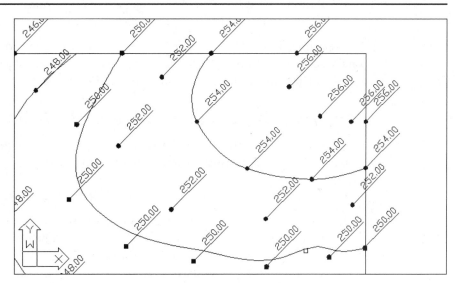

3. Type **x**↵ to exit the Fit Data option.

You've seen how you can control many of the shape properties of a spline through the Fit Data option. Here are descriptions of the other Fit Data options you didn't try in these exercises.

Delete   Removes a control point in the spline.

Close   Lets you close the spline into a loop.

Move   Lets you move a control point.

Purge   Deletes the fit data of the spline, thereby eliminating the Fit Data option for the purged spline. (See the next section, "When *Can't* You Use Fit Data?")

When *Can't* You Use Fit Data?   The Fit Data option of the Splinedit command offers many ways to edit a spline; however, this option is not available to all spline curves. When you invoke certain of the other Splinedit options, a spline curve will lose its fit data, thereby disabling the Fit Data option. These operations are as follows:

- Fitting a spline to a tolerance and moving its control vertices.

- Fitting a spline to a tolerance and opening or closing it.

- Refining the spline.

- Purging the spline of its fit data using the Purge option when you select when you select Modify ➤ Fit Data ➤ Purge or type **splinedit** ↵).

Also note that the Fit Data option is not available when you edit spline curves that have been created from polyline splines. See the "Turning Objects into Polylines and Polylines into Splines" sidebar earlier in this chapter.

Adjusting the Control Points with the Refine Option

While you're still in the Splinedit command, let's look at another one of its options, Refine, with which you can fine-tune the curve.

1. Type **u**↵ to undo the changes you made in the previous exercise.

2. At the Fit Data/Close/Move Vertex/Refine… prompt, type **r**↵. The Refine option lets you control the "pull" exerted on a spline by an individual control point. This isn't quite the same effect as the Fit Tolerance option you used in the previous exercise.

3. At the prompt

    ```
    Enter a refine option [Add control point/Elevate order/
    Weight/eXit] <eXit>:
    ```

    type **w**↵. The first control point is highlighted.

4. At the next prompt

    ```
    Spline is not rational. Will make it so.
    Enter new weight (current = 1.0000) or [Next/Previous/Select
    point/eXit] <N>:
    ```

    press ↵ three times to move the highlight to the fourth control point.

5. Type **25**↵. The curve not only moves closer to the control point, it also bends around the control point in a tighter arc (see Figure 10.23).

FIGURE 10.23:

The spline after increasing the Weight value of a control point

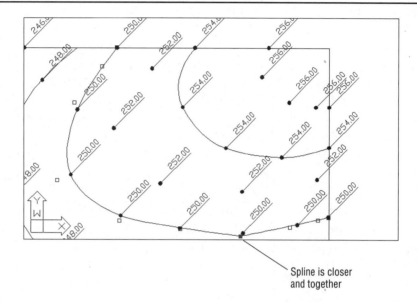

Spline is closer
and together

You can use the Weight value of Splinedit's Refine option to pull the spline in tighter. Think of it as a way to increase the "gravity" of the control point, causing the curve to be pulled closer and tighter to the control point.

Continue your look at the Splinedit command by adding more control points without actually changing the shape of the curve. You do this by using Refine's Add Control Point and Elevate Order options.

1. Type **1**↵ to return the spline to its former shape.

2. Type **x**↵ to exit the Weight option; then type **a**↵ to select the Add Control Point option.

3. At the `Specify a point on the spline <exit>:` prompt, click the second-to-last control point toward the top end of the spline (see the top image of Figure 10.24). The point you select disappears and is replaced by two control points roughly equidistant from the one you selected (see the bottom image of Figure 10.24). The curve remains unchanged. Two new control points now replace the one control point you selected.

FIGURE 10.24:

Adding a single control point using the Refine option

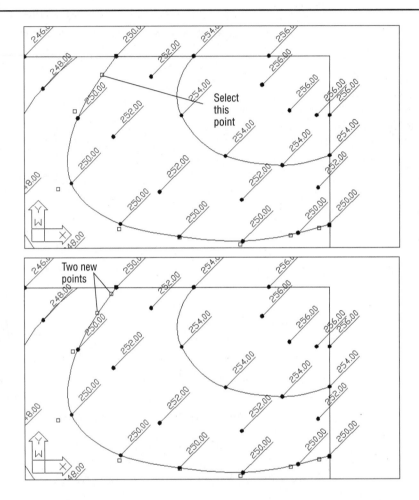

4. Press ↵ to exit the Add Control Point option.

5. Now type **e**↵ to select the Elevate Order option.

6. At the Enter new order <4>: prompt, type **6**↵. The number of control points increases, leaving the curve itself untouched.

7. Type **x=**↵ twice to exit the Refine option and then the Splinedit command.

You would probably never edit contour lines of a topographical map in quite the way these exercises have shown. However, by following this tutorial you've explored all the potential of AutoCAD's spline object. Aside from its usefulness for drawing contours, it can be a great tool for drawing free-form illustrations. Overall, it is an excellent tool for mechanical applications, where precise, nonuniform curves are required, such as drawings of cams, horsepower curves, or sheet metal work.

Marking Divisions on a Curve

One of the most difficult things to do in manual drafting is to mark regular intervals on a curve. AutoCAD offers the Divide and Measure commands to help you perform this task with speed and accuracy.

NOTE   Divide and Measure are discussed here in conjunction with polylines, but you can use these commands on any objects except blocks and text.

Dividing Objects into Segments of Equal Length

The Divide command can be used to divide an object into a specific number of equal segments. For example, suppose you needed to mark off the contour you've been working on in this chapter into nine equal segments. One way to do this is to first find the length of the contour by using the List command, and then sit down with a pencil and paper to figure out the exact distances between the marks. But there is another, easier way.

Divide will place a set of point objects on a line, arc, circle, or polyline, marking off exact divisions. The next exercise shows how it works.

1. Open the file called 10a-divd.dwg from the companion CD-ROM.

2. Choose Draw ➢ Point ➢ Divide or type **div**↵.

3. At the `Select object to divide:` prompt, pick the spline contour line.

4. The `Enter the number of segments or [Block]:` prompt that appears next is asking for the number of divisions you want on the selected object. Enter **9**↵. The command prompt now returns, and it appears that nothing has happened. But AutoCAD has placed several points on the contour that indicate the locations of the nine divisions you've requested. To see these points more clearly, continue with step 5.

5. Click Format ➢ Point Style or type **ddptype.**↵. The Point Style dialog box appears.

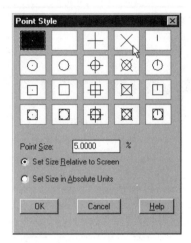

6. Click the X point style in the upper-right side of the dialog box, click the Set Size Relative to Screen radio button, and then click OK.

7. A set of Xs appears, showing the nine divisions (see Figure 10.25).

FIGURE 10.25:

Using the Divide commands on a polyline

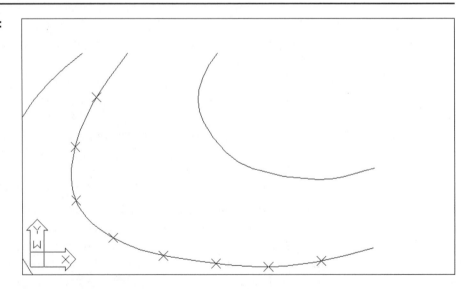

TIP   You can also change the point style by changing the Pdmode system variable. When Pdmode is set to 3, the point appears as an X. See Appendix D, "System Variables," for more on Pdmode.

The Divide command uses point objects to indicate the division points. Point objects are created by using the Point command; they usually appear as dots. Unfortunately, such points are nearly invisible when placed on top of other objects. However, remember that you can alter their shape using the Point Style dialog box. You can use these X points to place objects or as references along the object being divided. (Divide doesn't actually cut the object into smaller divisions.)

TIP   If you're in a hurry, and you don't want to bother changing the shape of the point objects, you can set the Running Osnaps to Node. Then when you're in Point Selection mode, move the cursor over the divided curve. When the cursor gets close to a point object, the Node Osnap marker will appear.

Dividing Objects into Specified Lengths

The Measure command acts just like Divide. However, instead of dividing an object into segments of equal length, Measure marks intervals of a specified distance along an object. For example, suppose you need to mark some segments exactly 5' apart along the contour. Try the following exercise to see how Measure is used to accomplish this task.

1. Erase the X-shaped point objects.

2. Choose Draw ➤ Point ➤ Measure or type **measure**↵.

3. At the `Select object to measure:` prompt, pick the contour at a point closest to its lower endpoint. We'll explain shortly why this is important.

4. At the `Specify length of segment or [Block]:` prompt, enter **5'**↵. The X points appear at the specified distance.

5. Now close this file.

NOTE   Measure is AutoCAD's equivalent of the Divider tool in manual drafting. A divider is a V-shaped instrument, similar to a compass, used to mark off regular intervals along a curve or line.

Bear in mind that the point you pick on the object to be measured will determine where Measure begins measuring. In the last exercise, for example, you picked the contour near its bottom endpoint. If you had picked the top of the contour, the results would have been different because the measurement would have started at the top, not at the bottom.

Marking Off Intervals Using Blocks Instead of Points

You can also use the Block option under the Divide and Measure commands to place blocks at regular intervals along a line, polyline, or arc. Here's how to use blocks as markers.

1. First, be sure the block you want to use is part of the current drawing file.

2. Start either the Divide or Measure command.

3. At the `Enter the number of segments or [Block]:` prompt, enter **b**↵.

4. At the `Enter name of block to insert:` prompt, enter the name of a block and press ↵.

5. At the `Align Block with Object? [Yes/No] <Y> :` prompt, press ↵ if you wish the blocks to follow the alignment of the selected object. (Entering **n**↵ causes each block to be inserted at a 0 angle.)

6. At the `Enter the number of Segments:` prompt, enter the number of segments and press ↵. The blocks appear at regular intervals on the selected object.

One example of using the Divide or Measure Block option is to place a row of brackets equally spaced along a bar. Or, you might use this technique to make multiple copies of an object along an irregular path defined by a polyline.

Sketching with AutoCAD

Although AutoCAD isn't a sketch program, you can draw freehand using the Sketch command. Sketch allows you to rough in ideas in a free-form way, and later overlay a more formal drawing using the usual lines, arcs, and circles. You can use Sketch with a mouse, but it makes more sense to use this command with a digitizing tablet that has a stylus. The stylus affords a more natural way of sketching.

Freehand Sketching with AutoCAD

Here's a step-by-step description of how to use Sketch.

1. Make sure the Ortho and Snap modes are turned off. Then type **skpoly↵1↵**. This sets the Sketch command to draw using polylines.

2. Type **sketch↵** at the command prompt.

3. At the `Record increment:` prompt, enter a value that represents the smallest line segment you'll want Sketch to draw. This command approximates a sketch line by drawing a series of short line segments; the value you enter here determines the length of those line segments.

4. At the `Sketch. Pen eXit Quit Record Erase Connect:` prompt, press the pick button and then start your sketch line. Notice that the message `<Pen down>` appears, telling you that AutoCAD is recording your cursor's motion.

NOTE   You can also start and stop the sketch line by pressing P on the keyboard.

---

5. Press the pick button to stop drawing. The message `<Pen up>` tells you AutoCAD has stopped recording your cursor motion. As you draw, notice that the line is green. This indicates that you have drawn a temporary sketch line and have not committed the line to the drawing.

6. A line drawn with Sketch is temporary until you use Record to save it, so turn the sketch line into a polyline now by typing **r↵**.

7. Type **x↵** to exit the Sketch command.

Filling in Solid Areas

You've learned how to create a solid area by increasing the width of a polyline segment. But suppose you want to create a simple solid shape or a very thick line. AutoCAD provides the Trace and Donut commands to help you draw simple filled areas. The Trace command acts just like the Line command (with the added feature of drawing wide line segments). You can also use the Hatch command to fill in areas, either with a pattern or a solid color.

Adding Hatch Patterns

Hatch patterns can represent types of materials, special regions, or textures in your drawings. You also can create free-form solid filled areas using the solid hatch pattern. Create an enclosed area using any set of objects, then use the Hatch tool, as described in this section, to apply a solid hatch pattern to the area.

For this example, you'll create a sectional view of the bracket and bar drawing you worked on in Chapter 6, "Enhancing Your Drawing Skills."

1. Open the Bar_9.dwg file from the companion CD-ROM. Your goal is to modify the view on the left side of Figure 10.26 to look like the view on the right side of the figure. First, edit the drawing by trimming the vertical lines between the horizontal lines to break them into several short segments. Then add the missing short line segments using the Endpoint Osnap. Use the Match Properties tool in the Standard toolbar to "paint" the linetype properties you require on the different line segments.

TIP   You can use the Match Properties tool in the Standard toolbar to transfer selected properties, such as layer, color, linetype, dimension style, and text settings. Click the tool, click the item you want to copy, and then type **s[cr]**. This opens the Property Settings box. Here, you can select the properties you want to transfer. By default, all the properties are selected.

---

2. After you modified the lines in the drawing, click the Hatch tool in the Draw toolbar, type **h.⏎**, or select Draw ➢ Hatch.

FIGURE 10.26:

The Bar_9 drawing (left),
modified, to create a sectional view of the bracket
and bar (right)

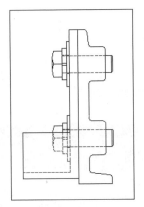
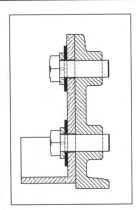

The Boundary Hatch dialog box appears.

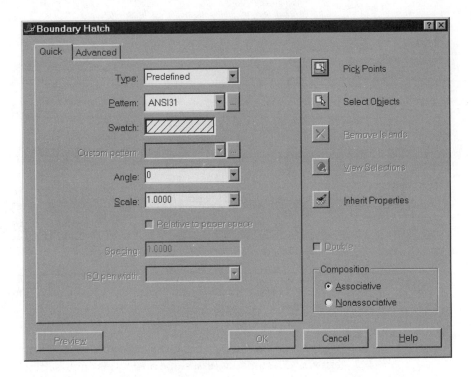

3. Click the Pick Points button in the top-right area of the dialog box. The
    dialog box temporarily closes to allow you to pick points inside the areas
    you want to hatch. Click a point inside the top area of the bracket and bar

assembly (see Figure 10.26). Then press [↵] to return to the Boundary Hatch dialog box.

4. Click the ellipsis (…) button to the right of the Pattern drop-down list. The Hatch Pattern Palette dialog box appears. Click the ANSI tab, then click the pattern labeled ANSI31 (the pattern for cast iron and aluminum). Click OK to close the dialog box.

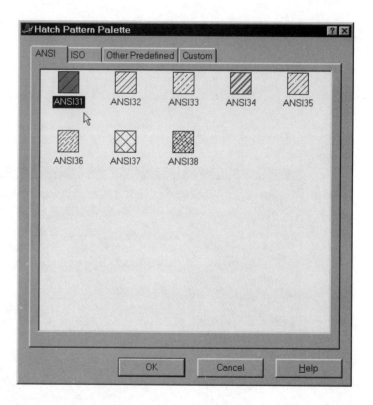

NOTE

If you know the name of the pattern you want, you can select it from the Pattern drop-down list.

5. In the Angle drop-down list, choose the 90 setting. Because the ANSI31 pattern was developed with a slope of 45°, this angle setting will display the hatch pattern at a 135° angle. Alternatively, you could leave the 0 setting,

which sets the hatch pattern at a 45° angle (The required angle will vary with different patterns.)

6. Double-click the Spacing input box and enter **1**.

7. Click the Preview button in the lower-left corner of the dialog box. The drawing appears with the hatching you've selected. Right-click to return to the dialog box.

8. Double-click the Spacing input box and enter **2**. This will make the hatching look less crowded.

9. Click the Preview button to check the hatching.

10. Right-click to return to the dialog box. Adjust the hatch settings as necessary (the options in the dialog box are discussed after this exercise). Then click OK to close the Boundary Hatch dialog box.

Once you've completed this first hatch pattern, try adding hatch patterns to the other parts. The pattern for the large washer is the one labeled Solid in the Other Predefined tab of the Hatch Pattern Palette. This pattern creates solid fills (and it is a vast improvement over the Solid command that was used in earlier versions of AutoCAD).

Understanding the Boundary Hatch Options

The Boundary Hatch dialog box offers many other options that you didn't explore in the previous exercise. The Quick tab of the dialog box offers the following selections:

Type Offers the choices User-Defined and Predefined. You can use the User-Defined choice to create your own hatch pattern, as described in Chapter 21, "Integrating AutoCAD into Your Projects and Organization."

Swatch Shows a sample view of the selected hatch pattern. Click this button to see a graphical representation of the predefined hatch patterns.

Pattern Lists the predefined patterns by name. Click the ellipsis button to see the Hatch Pattern Palette dialog box, with four tabs: ANSI, ISO, Other Predefined, and Custom. The Custom tab is empty until you create your own set of custom hatch patterns.

Angle   Sets the slant of the hatch pattern. As you saw in the previous exercise, the angle will be added to any existing slant of the pattern.

Scale   Allows you to scale the hatch pattern.

ISO Pen Width   Shows a drop-down list of pen widths that conform to the ISO standard (when you pick a predefined pattern from the ISO tab of the Hatch Pattern Palette dialog box).

Pick Points   Allows you to select the area to be hatched by clicking a point within it. AutoCAD automatically traces around the boundary of the zone you choose.

Select Objects   Allows you to define the area to be hatched by selecting the objects that bound that area. (This method can be a bit more difficult than picking points, because overlapping geometry may confuse the Hatch command.)

Remove Islands   Lets you add hatching to an area that would otherwise be left unhatched (like an island). This option is available when you use the Pick Points option and an island has been detected.

View Selections   Temporarily closes the dialog box and highlights the objects that have been selected as the hatch boundary by AutoCAD.

Inherit Properties   Lets you select a hatch pattern from an existing one in the drawing and use its properties (such as its scale and rotation).

Double   Sets the hatch pattern to run both vertically and horizontally.

Composition   Contains the choices Associative and Nonassociative. An associative hatch pattern automatically changes to fill its boundary whenever that boundary is stretched or edited. A nonassociative hatch pattern doesn't do this.

WARNING   A hatch pattern can lose its associativity when you erase or explode a hatch boundary or a block that forms part of the boundary. It also can lose its associativity when you move a hatch pattern away from its boundary.

Using the Advanced Hatch Options

The Advanced tab of the Boundary Hatch dialog box lets you control how AutoCAD treats islands within patterns and provides other advanced tools.

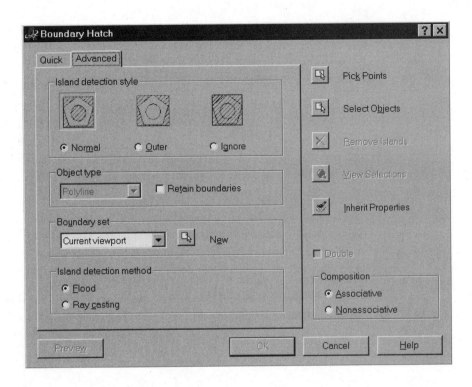

Island Detection Style Options

The Island Detection Style options control how nested boundaries affect the hatch pattern:

Normal   Causes the hatch pattern to alternate between nested boundaries. The outer boundary is hatched; if there is a closed object within the boundary, it is not hatched. (This is the default setting.)

Outer   Applies the hatch pattern to an area defined by the outermost boundary and by any boundaries nested within the outermost boundary. Any boundaries nested within the nested boundaries are ignored.

Ignore   Supplies the hatch pattern to the entire area within the outermost boundary, ignoring any nested boundaries.

Object Type Options

The Hatch command creates a temporary boundary (polyline by default) to establish the hatch area. You can check the Retain Boundaries check box to retain the boundaries in the drawing. The drop-down list in this area lets you choose between Polyline and Region (similar to a 2D plane) for the boundary.

Boundary Set Options

The options in this area can help if you have problems selecting objects to create hatch boundaries. The Hatch command is view dependent, meaning that it locates boundaries based on what is visible in the current view. If the current view contains a lot of graphic data, AutoCAD may have difficulty finding a boundary.

The Current Viewport setting shows that AutoCAD will use all of the current view to determine the hatch boundary. The New button lets you limit the area by selecting the objects from which you want AutoCAD to determine the hatch boundary. After you've selected a set of objects using the New button, Existing Set will appear as an option in the drop-down list in this area.

NOTE   The Boundary Set options are designed to give you more control over the way a point selection boundary is created (when you've chosen Pick Points); they have no effect when you use the Select Objects button.

Island Detection Method Options

The options in this area let you control the way that islands are hatched.

Flood   Includes islands as boundary objects and hatches them.

Ray Casting   Looks for the object nearest to your pick point and then traces the boundary in a counterclockwise direction, thus excluding islands.

Tips for Adding Hatch Patterns

Here are some tips on using the Hatch command:

- It's a good idea to set up a separate layer for your hatch patterns. That way, you will be able to turn them on and off.

- To help AutoCAD locate boundaries, zoom in on the area to be hatched.

- If the area to be hatched is large yet requires fine detail, first outline the hatch using a polyline (as described earlier in this chapter). Then use the Select Object option in the Boundary Hatch dialog box to select the polyline boundary manually.

- Consider turning off layers that might interfere with AutoCAD's ability to find a boundary.

- The Hatch command works on nested blocks as long as the nested block entities are parallel to the current UCS and are uniformly scaled in the X and Y axes.

- Hatch patterns are like blocks in that they act like single objects. You can explode a hatch pattern to edit its individual lines.

- You can modify hatch patterns by using the Properties tool. The Properties dialog box includes the settings for the pattern, type, and other hatch properties.

Drawing Filled Circles

If you need to draw a thick circle or a solid filled circle, take the following steps:

1. Choose Draw ➤ Donut or type **donut**↵ at the command prompt.

2. At the `Specify inside diameter of donut <0.5000>:` prompt, enter the desired diameter of the donut "hole." This value determines the opening at the center of your circle.

3. At the `Specify outside diameter of donut <1.0000>:` prompt, enter the overall diameter of the circle.

4. At the `Specify center of donut or <exit>:` prompt, click the desired location for the filled circle. You can continue to select points to place multiple donuts (see Figure 10.27).

5. Press ↵ to exit this process.

If you need to fill only a part of a circle, such as a pie slice, you can use the Donut command to draw a full, filled circle. Then use the Trim or Break tools in the Modify toolbar to cut out the portion of the donut you don't need.

FIGURE 10.27:

Drawing wide circles using the Donut command

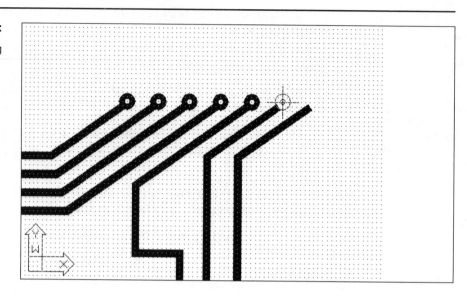

Toggling Solid Fills On and Off

Once you've drawn a solid area with the Polyline, Solid, Trace, or Donut commands, you can control whether the solid area is actually displayed as filled in. Open the Options dialog box and choose Display ➤ Display Performance ➤ Apply Solid Fill. If the Apply Solid Fill check box doesn't show a check mark, thick polylines, solids, traces, and donuts appear as outlines of the solid areas (see Figure 10.28).

NOTE   You can shorten regeneration and plotting time if solids are not filled in.

WARNING   You'll have to issue the Regen command to display the effects of the Fill command.

The Apply Solid Fill check box option is an easy-to-remember way to control the display of solid fills. Or, you can enter **fill↵** at the command prompt; then, at the [ON/OFF] <ON>: prompt, enter your choice of **on** or **off**.

FIGURE 10.28:

Two polylines with the Fill option turned on (top) and turned off (bottom)

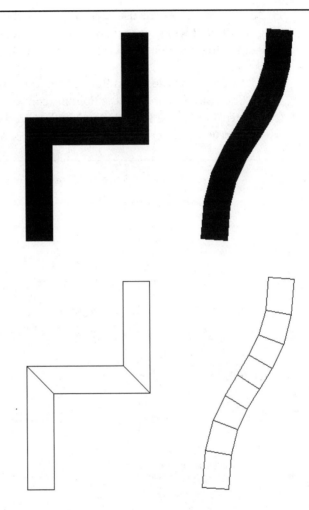

If You Want to Experiment...

There are many valuable uses for polylines beyond those covered in this chapter. We encourage you to become familiar with the polyline so that you can take full advantage of AutoCAD.

To further explore the use of polylines, try the following exercise illustrated in Figure 10.29. It will give you an opportunity to try out some of the options discussed in this chapter that weren't included in exercises.

1. Open a new file called Part13. Set the Snap mode to 0.25 and be sure the Snap mode is on. Use the Polyline command to draw the object shown at step 1 of Figure 10.29. Draw it in the direction indicated by the arrows, and start at the upper-left corner. Use the Close option to add the last line segment.

FIGURE 10.29:

Drawing a simple plate with curved edges

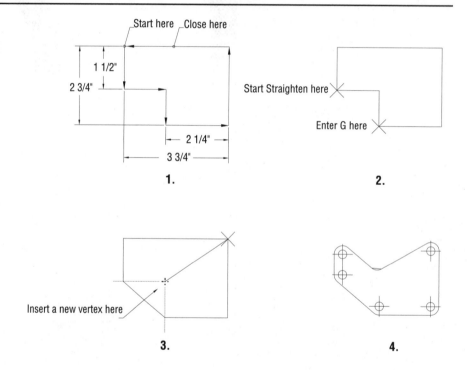

2. Start the Pedit command, select the polyline, and then type e↵ to issue the Vertex option. At the Next/Previous/Break: prompt, press ↵ until the X mark moves to the first corner shown in the figure to the right. Enter s↵ for the Straighten option. At the Next/Previous/Go: prompt, press ↵ twice to move the X to the other corner shown in the figure. Press g↵ for Go to straighten the polyline between the two selected corners.

3. Press ↵ twice to move the X to the upper-right corner, then enter **i**↵ for Insert. Pick a point as shown in the figure. The polyline changes to reflect the new vertex. Enter an **x**↵ to exit the Edit Vertex option, and then press ↵ to exit the Pedit command.

4. Start the Fillet command and use the Radius option to set the fillet radius to 0.30. Press ↵ to start the Fillet command again, but this time use the Polyline option and pick the polyline you just edited. All of the corners fillet to the 0.30 radius. Add the 0.15 radius circles as shown in the figure and exit the file with the End command.

PART III

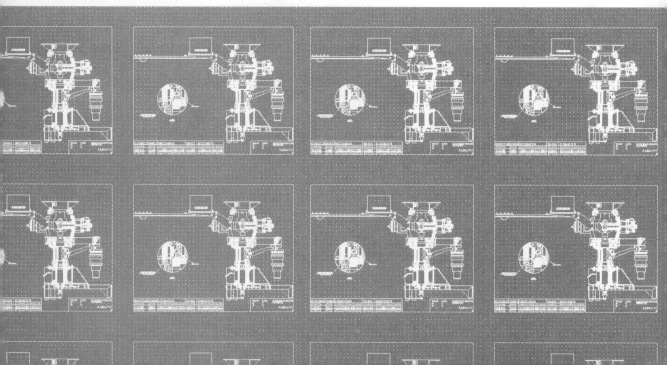

Modeling and Imaging in 3D

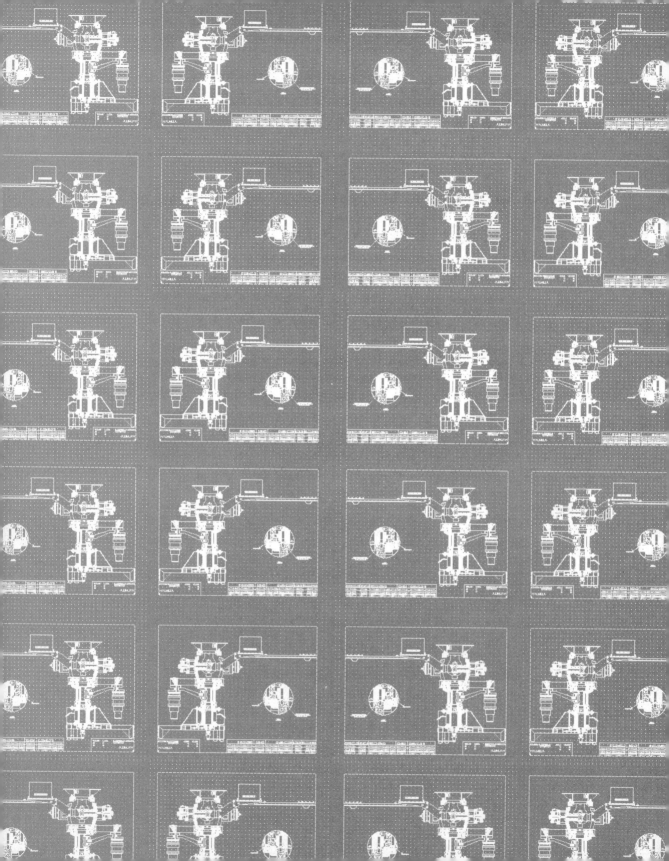

CHAPTER

ELEVEN

11

Introducing 3D

- Creating a 3D Drawing

- Understanding the UCS

- Viewing a 3D Drawing

- Visualizing Your Model

- Creating a Solid Model from a 2D Drawing

- Building an Assembly of Parts

Viewing an object in three dimensions gives you a sense of its true shape and form. It also helps you conceptualize the design, which will ultimately result in better design decisions. Using three-dimensional objects also helps you visually communicate your ideas to others on your design team.

Another advantage to creating parts and assemblies in three dimensions is that you derive 2D drawings from your 3D model. For example, you could model a mechanical part in 3D and then quickly derive top, front, and right-side views using the techniques discussed in this chapter.

AutoCAD offers three methods for creating 3D models: wire-frame modeling, surface modeling, and solid modeling. All three methods will look pretty much the same as you view them in AutoCAD.

A *wire-frame model* represents an object using lines, circles, and arcs drawn along the boundaries of the surfaces of the object. The wire-frame model cannot be rendered or shaded because only the hollow outline of the model is displayed. You won't learn about wire-frame model making in this book. It is an old technique, and you'll only use it to revise old drawings. The methods used to draw and edit new solids and surfaces will suffice to provide examples for editing old wire-frame files.

A *surface model* is drawn by creating the model's surfaces. So a box that has six flat sides requires six flat surfaces. If the box is hollow and the sides have thickness, six more surfaces are required to represent the interior. A surface model can be shaded and rendered by bouncing light off the surfaces. You'll get a taste of surface modeling in Chapter 13, "Using 3D Surfaces."

NOTE Many 3D designs were created in AutoCAD Releases 10–12 and are still used in production today. The only method available to create 3D models in the earlier releases was wire frame, and you may be called on to edit or update some of these wire-frame models. Wire frame isn't a good way to learn 3D modeling, because a wire-frame model is just what it looks like—line, arc, and circle objects assembled in 3D space to represent the boundary of a 3D model. To edit wire frame, you use many of the same tools that you use to edit the properties of 2D objects (with a few additional options). As you proceed through the following 3D chapters, note the methods used to manipulate the User Coordinate System (UCS), the way in which objects are created in each UCS, and the visualization tools. Once mastered, these techniques will allow you to revise existing 3D models.

A *solid model* has both surfaces and volume. You may have difficulty working with solids in that you're actually modeling in 3D. You're creating objects that AutoCAD will treat as if they have volume and mass. They can be rendered like the surface model, but unlike the surface model they can be analyzed for mass properties. Let's explore solids and learn about 3D.

Creating a 3D Drawing

Let's start by drawing a 3D 3/8" socket-head machine screw. You'll not draw every nuance of the screw because it's often a purchased part. (Of course, the manufacturer of the screw would want it drawn complete with every detail.)

First you'll use the Cylinder command to start the head of the screw. Then you'll use the Polygon command to draw the hexagon shape of the screw-head drive, which you'll make into a solid with the Extrude tool. Next you'll edit the cylinder using the Boolean operation of subtracting one solid from another, and you'll learn that you can use the Chamfer command with solids by taking the sharp edge off the top of the screw head.

It sounds like a lot to do, but you'll see that these are pretty simple operations, following a pretty simple logic.

Preparing to Draw Solid Models

First, begin a new drawing. Then tell AutoCAD which drawing tools you'll need.

1. Right-click any toolbar. The Toolbars dialog box appears.

2. Scroll down the toolbar choices until you find the Modify II toolbar; click the check box to the left to activate it.

3. Use the same method from steps 1 and 2 to select the Solids, UCS, and View toolbars.

NOTE   If you find that the toolbars are in your way, remember that you can turn them on and off by right-clicking any toolbar. You can also dock a toolbar by dragging it to the side of your screen.

Drawing a Solid Cylinder

Let's start with the head of the screw.

1. Click the Cylinder tool in the Solids toolbar, choose Draw ➤ Solids ➤ Cylinder, or type **cylinder.**↵.

You'll see the following information:

```
Current wire frame density: ISOLINES=4
```

2. At the `Specify center point for base of cylinder or [Elliptical]` `<0,0,0>:` prompt, pick a point in the lower-left corner of the screen.

3. At the `Specify radius for base of cylinder or [Diameter]:` prompt, enter **d.**↵. At the `Specify diameter for base of cylinder:` prompt, enter **.562**↵. This is the head diameter of a standard 3/8 socket-head machine screw. So far, this has been just like drawing a circle.

4. At the `Specify height of cylinder or [Center of other end]:` prompt, type **.375.**↵. A circle appears.

5. You need to change your point of view if you want to see the cylinder. Right now, you're looking straight down on the cylinder. Drag the mouse across the View toolbar while looking at the tool tips. Choose the S(outh) E(ast) Isometric View option.

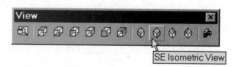

NOTE There are a couple of important things to note about what happened when you chose Isometric view. The cylinder looks transparent, or as though it is composed of lines and ellipses. This is the wire-frame representation of the solid form. The X-Y icon, called the UCS icon, has changed in appearance as though it, too, is drawn in an Isometric view. However, it has not actually changed—only your view of it has altered. A Zoom All command was issued at the end of the View command.

Next, draw the hexagon-shaped drive socket.

6. Click the Polygon tool in the Draw toolbar, choose Draw ➤ Polygon from the menu bar, or type **polygon**⏎.

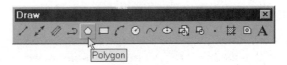

At the `Number of sides <4>:` prompt, type **6**⏎. At the `Specify center of polygon or [Edge]:` prompt, pick the top center of the cylinder.

7. At the `Enter an option [Inscribed in circle/Circumscribed about circle] <I>:` prompt, type **c**⏎. At the `Specify radius of circle:` prompt, type **5/32**⏎.

Your drawing is starting to take on the shape of the screw head, even though it is still basically a hexagon and a cylinder.

Let's take a closer look at the prompts generated by the Cylinder command to understand them better. The first message reads:

`Current wire frame density: ISOLINES=4`

This message indicates the current setting for the Isolines system variable. The Isolines system variable controls the way that curved objects, such as cylinders and holes, are displayed. A setting of 4 means that cylinders are represented by four lines with circles at each end. You can change the Isolines setting by entering **Isoline**⏎ at the command prompt, followed by the number of lines to use to represent surfaces. This setting is also controlled by the Contour Lines Per Surface option in the Display tab of the Options dialog box (choose Tools ➤ Options).

The next prompt offers two options.

`Specify center point for base of cylinder or [Elliptical] <0,0,0>`
You can specify the point for the base of the cylinder, or you can choose the Elliptical option. Selecting Elliptical allows you to draw an ellipse, rather than a circle, as the base of your cylinder.

The third prompt is the same one you see when you use the Circle tool:

`Specify radius for base of cylinder or [Diameter]:`

Finally, you see this prompt:

```
Specify height of cylinder or [Center of other end]:
```

If you enter a positive number, the command creates the cylinder along the positive Z axis. Entering a negative number creates the cylinder along the negative Z axis.

Turning a Polyline into a Solid

Now you've got the basic shape outlined as a polyline. In order to make it a solid, you'll make the 2D object into a 3D solid object with the Extrude tool.

1. To turn the polygon into a solid, click the Extrude tool in the Solids toolbar, choose Draw ➤ Solids ➤ Extrude from the menu bar, or type **extrude**↵.

2. At the `Select objects:` prompt, type l↵ to select the last object drawn (the polygon), and press ↵ again to complete the selection set.

3. At the `Specify height of extrusion or [Path]:` prompt, type **.182**↵. The normal extrusion direction is in the positive Z direction, unless you tell AutoCAD otherwise. The minus sign (–) tells AutoCAD to extrude in the negative Z direction (away from you).

4. At the `Specify angle of taper for extrusion <0>:` prompt, hit ↵ to accept the default, no taper angle.

Now you've taken another step in creating 3D objects—you've converted a polyline into a solid with an extrusion thickness of .182 units.

Before continuing, let's take a closer look at the Extrude tool. At the `Select objects:` prompt, you selected a closed polyline. In another drawing, you might use this tool with other shapes, such as a closed spline, a region, a circle, or an ellipse.

NOTE   You cannot extrude text or dimensions using the Extrude tool because you have to use an object that can become a solid.

The next prompt was `Specify height of extrusion or [Path]:`. You established the height of an extrusion using the same sort of prompt that you used in the Cylinder tool (defining positive or negative measurements along the Z axis).

The Path option is something different. The path of an extrusion could follow a line, an arc, a 2D polyline, or a spline. If you're careful and don't draw a polyline that causes the extrusion to curl inside itself, you can sweep your extrusion along interesting shapes such as an ellipse or a spiral. This can be very handy for representing piping or the plumbing on a pneumatic layout, for example.

Removing the Volume of One Solid from Another

Next, you'll remove the hexagon solid from the head of the screw to create the socket-drive feature.

1. Click the Subtract tool in the Modify II toolbar, choose Modify ➤ Solids Editing ➤ Subtract from the menu bar, or type **subtract**↵ in the command line.

2. At the Select solids and regions to subtract from: prompt, choose the cylinder. When the cylinder is highlighted, press ↵ to complete the selection set.

3. At the Select solids and regions to subtract: prompt, pick the extruded polygon, or type 1↵ to use the last object created. Press ↵ again to complete the selection set. Your drawing should look like Figure 11.1.

FIGURE 11.1:

The head of the hex-drive socket screw

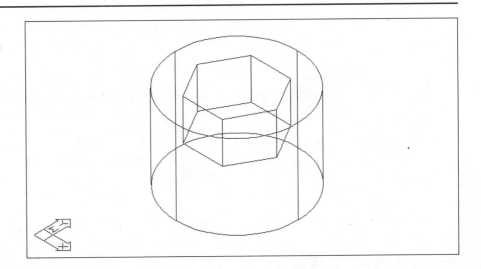

The hexagon shape was subtracted from the cylinder that represents the head of the screw.

Using the Hide Tool to Verify Your Design

Your model looks the same after the subtraction of the hexagon solid as it looked before. You'll use the Hide tool to see the difference. The Hide tool evaluates any solids or surfaces and displays each one relative to its position and your viewpoint. If one shape is closer to you than another, the Hide command obscures the lines of the shape that is farther away, like a house might block your view of a shrub. So external surfaces hide internal surfaces, and near surfaces hide far surfaces.

> **TIP**
>
> The Hide tool can be very useful when your drawing gets more complex. This tool can clarify the model by eliminating extra lines that are hidden.

1. Click the Hide tool in the Render toolbar, choose View ➢ Hide, or type **hi**↵ in the command line. The surfaces of the cylinder (and all non-planar surfaces in any model) are shown as triangles. This solid-looking object, with triangles for non-planar surfaces, is the AutoCAD default view. Your drawing should look like the top image of Figure 11.2.

2. Change the Hide view default to show the silhouette of the surfaces with the system variable Dispsilh (for display silhouette). Type **dispsilh**↵ in the command line. At the New value for DISPSILH <0>: prompt, type **1**↵.

3. Type **hi**↵ to see the results of changing the system variable. Your drawing should look like the bottom image of Figure 11.2.

4. When you've finished looking at the model in Hidden view, type **regen**↵. This will regenerate the view and return the view to wire frame.

> **TIP**
>
> As a matter of personal preference, you can continue to work with the Hidden view. This view will only allow you to select objects that you can see; this can be limiting. A number of normal view regenerations occur during solid modeling. For example, a Regen occurs when you edit a model and when you change viewpoint. At each Regen, the hidden line view is redrawn as a wire-frame view, and you'll have to use the Hide tool to restore your hidden-line view.

FIGURE 11.2:

The chamfered head of the hex-drive socket screw with Dispsilh set to 0 in the top view and set to 1 in the bottom view

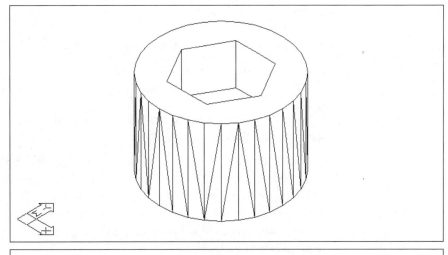

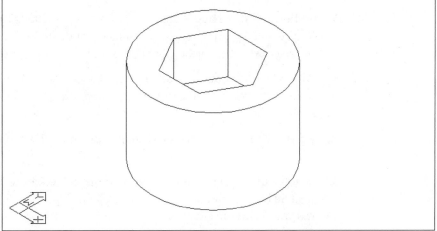

Editing a Solid Model with the Chamfer Tool

The Chamfer tool offers a way of breaking sharp edges. In a hand tool where sharp edges are a problem, or in some other part where a certain elegance of design is desired, chamfering will allow you to remove material using a few simple steps.

1. Click the Chamfer tool in the Modify toolbar.

2. The prompt will show the status and option

   (TRIM mode) Current chamfer Dist1 = 0.5000, Dist2 = 0.5000

3. At the Select first line or [Polyline/Distance/Angle/Trim/Method]: prompt, pick the circle that is the top of the screw head's cylinder.

4. At the Base surface selection Enter surface selection option [Next/OK (current)] <OK>: prompt, look at the cylinder. If the top appears to be highlighted, press ↵; if not, press **n**↵ until it is. If you pick the edge of a solid, there will be more than one surface that you could have intended for the base surface, so the Next option cycles through the available surfaces, allowing you to select the one that you want.

5. At the Specify base surface distance <0.5000>: prompt, type .035↵. You're telling AutoCAD the dimension of the top edge of the chamfer for the selected solid object (the highlighted surface) of the screw head.

6. At the Specify other surface distance <0.5000>: prompt, type .035↵ to form the second edge of the chamfer a little distance down the side of the screw head.

7. At the Select an edge or [Loop]: prompt, pick the top edge of the cylinder again.

8. To tell AutoCAD that you're finished picking edges to be chamfered, at the next Select an edge or [Loop]: prompt, press ↵. The screw head is redrawn with the chamfer.

9. Your drawing should look like Figure 11.3. To better visualize the results, use the Hide command to see your work.

Next, you'll add the body of the screw. AutoCAD can't draw the helical threads of the screw, so you'll represent the threads of the screw with a simple cylinder.

FIGURE 11.3:

The chamfered head of the hex-drive socket screw

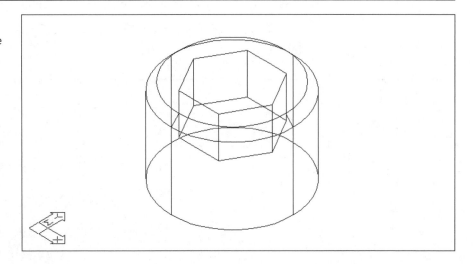

Adding the Volumes of Two Solids

You've drawn the cylindrical head of the screw. Next you'll need to draw the threaded portion of the screw. New solids are always drawn as separate parts, so you'll then have to join the head and threads together.

1. Click the Cylinder tool in the Solids toolbar.

2. When you're prompted for <center point>, use the Center Osnap tool to pick the center of the bottom of the head of the screw.

3. At the Specify radius for base of cylinder or [Diameter]: prompt, type **d**↵, and at the Diameter: prompt, enter **.375**↵ (because you've been drawing a 3/8" socket-head screw).

4. At the Specify height of cylinder or [Center of other end]: prompt, type **–1**↵ to draw a 1-unit-long cylinder down the negative Z axis to make the threads 1 unit long.

5. Click the Union tool in the Solids Editing toolbar, choose Modify ➤ Solids Editing ➤ Union, or type **union**↵ at the command line.

6. At the `Select objects:` prompt, pick the head and the cylinder, and press ↵.

7. Choose Zoom All to see your handiwork. Use the Hide tool, and your drawing should look like Figure 11.4.

FIGURE 11.4:

The completed socket-head screw

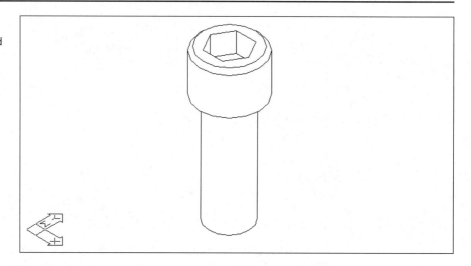

8. Save your work.

Using the Box Primitive

We've looked at some basic shapes—the line, the polyline, spline curves, and circles. Let's look at how we can manipulate another basic shape. Because we're now talking about more than two dimensions, we're not going to talk about a simple rectangle. We'll use the box shape to build the base and the post of an arbor press.

1. Click the Box tool in the Solids toolbar, choose Draw ➢ Solids ➢ Box, or type **box**↵.

2.   At the Specify corner of box or [CEnter] <0,0,0>: prompt, type **1,1**↵.

3.   At the Specify corner or [Cube/Length]: prompt, type **1**↵ to choose the Length option.

4.   Respond to the rest of the prompts as indicated below.

     Specify length: **.98**↵

     Specify width: **3.5**↵

     Specify height: **6**↵

The post is going to be L-shaped. You need to create more objects to use as tools to edit the post. These new objects will need to be created in other 3D planes. All new objects are created relative to the current User Coordinate System (UCS).

Let's take a closer look at the coordinate system.

The UCS and the WCS

The *World Coordinate System* (WCS) is the default when you start any new drawing (unless you've created a special template file with other settings).

The UCS Icon

The *User Coordinate System* (UCS) icon is the figure in the lower-left corner of every drawing that you've made since the beginning of this book. The purpose of the icon is to show you the current UCS settings. The *W* means that the icon is showing the WCS X and Y axes. The arrows point in the positive X and Y directions. Take a look at Figure 11.5 to see the UCS icon showing the WCS from a southeast isometric viewpoint.

FIGURE 11.5:

The World UCS icon

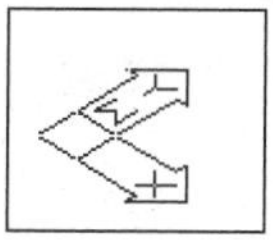

The UCS icon doesn't plot or print and it remains the same size regardless of the zoom scale. In Chapter 8, "Using Dimensions," you created a new UCS. The UCS icon stayed in place, but the W was turned off, demonstrating that you were using a system other than the default **wcs**. If you wanted to, you could have set the UCS icon to follow the origin.

The View ➤ Display ➤ UCS Icon cascading menu offers options for turning on and off the UCS icon and changing its location. When the View ➤ Display ➤ UCS Icon ➤ On option is checked, UCS is on, and the icon is displayed. To turn off the display of the icon, deselect the On option.

When the View ➤ Display ➤ UCS Icon ➤ Origin option is checked, the UCS icon appears at the location of the current 0,0,0 origin point. When Origin is not checked, the UCS icon appears in the lower-left corner of the AutoCAD window.

NOTE   If the origin of the UCS is off the screen while the Origin option is active (shown by a check mark in the menu), the UCS icon appears in the screen's lower-left corner.

In addition, there are two other settings that control the UCS icon's appearance.

- The All option lets you set the UCS icon's appearance in all of the viewports on your screen at once. To use All, type **ucsicon↵a↵**.

- The UCSFollow system variable, when set to 1, causes the display in a given viewport to always show a plan view of the current UCS. To use this option, highlight a viewport, type **ucsfollow↵**. At the New value for UCSFOLLOW <0>: prompt, type **1↵**.

NOTE   The Shademode tool includes an additional option. New in this version of AutoCAD is a 3D wire-frame icon with the Z axis also displayed. If you wish, you can turn on this icon. Choose View ➤ Shade ➤ 3D Wireframe, or type **shademode↵3d↵**. Immediately, a 3D icon will replace the 2D icon. The 2D Wireframe option of Shademode will reverse the process.

Coordinate Systems for UCS and WCS

You've been working with the three-axis coordinate system since the beginning of this book. You know how to specify the two main X and Y coordinates for 2D drawings.

You may be familiar with the *right-hand rule,* which is useful when you're looking at Cartesian coordinates and trying to recall which directions are positive. Hold your right hand palm up and stretch out your thumb. Then align your thumb so that it points in the positive X-axis direction; your index finger automatically points 90° from your thumb in the positive Y axis. You've been using these axes to define locations in 2D space since your first drawing in Chapter 2, "Creating Your First Drawing." With your right hand still stretched out, curl your middle finger until it is perpendicular to your thumb and index fingers. Your middle finger is pointing in the positive Z-axis direction.

With the X, Y, and Z axes you can define any point in 3D space, just as the X and Y axes alone allowed you to define any point in 2D space. Both the WCS and the UCS allow you to enter values (both the positive and negative) for the Z-axis location of a point.

NOTE   If you'll never use 3D, all you need to know about the Z axis is that when you rotate anything, the point you pick about which those objects are to be rotated is a point on the Z axis. Also, many (but not all) AutoCAD commands will accept 2D or 3D input. This is defined as the Z coordinate. If you don't supply a Z value, AutoCAD assumes it is zero.

By now, without realizing it, you've become an expert in the use of the WCS because you've been using it since you started this book. Each object that you've created has been located on, or relative to, the X,Y plane of the WCS. When you created a circle, the plane of the circle was parallel to the X,Y plane of the WCS. You might think of WCS as a sort of cube of 3D space with a plane located in the middle. One point on the plane is the origin—0,0,0. So far, you've been viewing the plane from a position normal to the origin.

But you're not limited to using the plane of the WCS. You can create a coordinate system of your own, relative to the WCS, called a UCS. A 2D UCS might be rotated to ease drawing and dimensioning of a view that is projected from an angled surface. If you've used a drafting machine, creating a rotated UCS is akin to rotating the head and scales to some angle, as if you were rotating the scales on the X, Y plane about the Z axis.

All objects are created on, or relative to, the X,Y plane of the current UCS. You could draw every solid object on, or relative to, the WCS and move and rotate the object into position using the Move, 3D Rotate, and Align commands. The UCS allows you to create, save, and restore any number of planes to draw on. The most difficult concept to master is the use of the UCS in 3D, and yet, it's also the simplest. If you draw on a table and then draw on the wall you can tell yourself that you've just moved the UCS from the table to the wall. See how simple?

Controlling the UCS

Have a look at the UCS controls by clicking the Tools menu so that the UCS menus cascade. Or, if you wish, pull up the UCS and UCS II toolbars. You've got lots of options.

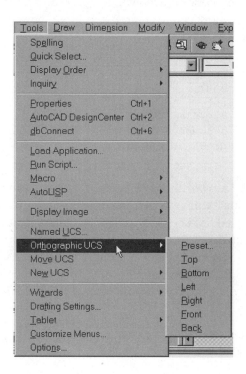

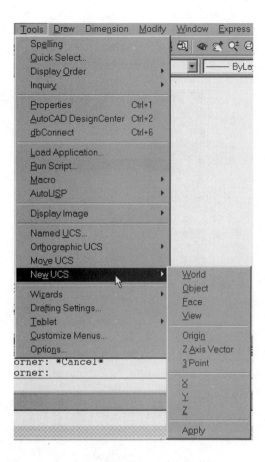

Right-click any toolbar and select UCS and UCSII from the list of toolbars. The UCS toolbar provides the tools you'll use in the next exercises.

The first tool is the UCS tool, which manages the UCS. It gives you access to all of the UCS options.

The next tool is the Display UCS Dialog tool. Click this tool to see the UCS dialog box, which has three tabs. The Named UCSs tab alllows you to save UCSs and lists the UCSs that you have saved previously.

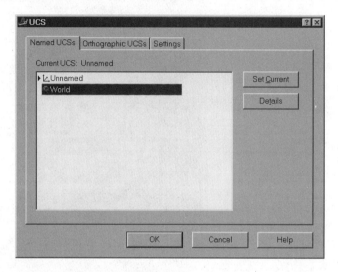

The Orthographic UCSs tab of the UCS dialog box lists a set of predefined UCSs (supplied by AutoCAD), such as Top, Bottom, Front, and Back.

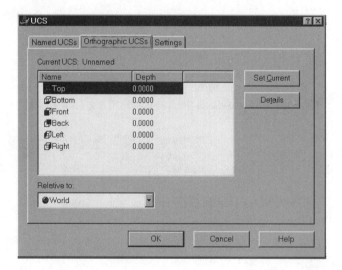

The Settings tab of the UCS dialog box includes UCS icon and UCS settings. The three UCS Icon settings work the same as those described in the "The UCS Icon" section earlier in this chapter. The Save UCS with Viewport setting allows you to save a UCS with a viewport. The Update View to Plan View When UCS Is Changed setting changes the view in the viewport to a plan view every time you reset the UCS (which can be confusing if you aren't prepared for it).

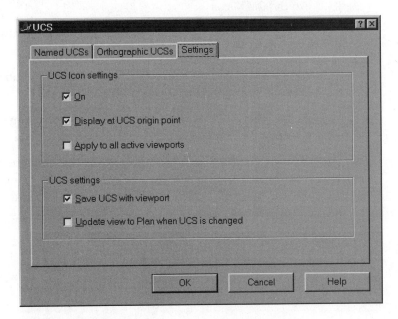

The remaining toolbar buttons work as follows:

Previous UCS   Returns to the UCS used previously.

World UCS   Returns to the WCS.

Object UCS   Uses an object to define the X, Y plane of a UCS. See Table 11.1 for the description of how objects determine the UCS orientation.

Face UCS   Sets the UCS parallel to the face of a solid object.

View UCS   Uses the current viewpoint to define a plane parallel to the screen (an isometric plane).

Origin UCS   Sets an origin point relative to the current UCS.

Z Axis Vector UCS Lets you to define a UCS by picking the point through which the new Z axis will extend.

3 point UCS Lets you to define three points for a UCS: the origin point, a positive point on the X axis, and a positive point on the Y axis.

X Axis Rotate UCS Lets you define a plane by rotating the current UCS around the X axis.

Y Axis Rotate UCS Lets you define a plane by rotating the current UCS around the Y axis.

Z Axis Rotate UCS Lets you define a plane by rotating the current UCS around the Z axis.

Apply UCS Applies the current UCS setting to a specified viewport or to all active viewports.

The UCS II toolbar includes a drop-down list that contains all of the saved UCSs in a drawing. You can use this list as a quick way to move between UCSs that you've set up, as well as to move between the predefined orthogonal UCSs.

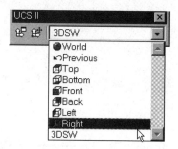

This toolbar also contains the Display UCS Dialog tool, which is the same as the one on the UCS toolbar, and the Move UCS Origin tool, which moves an existing UCS to another location.

TABLE 11.1:   Effects of Objects on the Orientation of a UCS

Object Type	UCS Orientation
2D Polyline	The starting point of the polyline establishes the UCS origin. The X axis is determined by the direction from the first point to the next vertex.
3D Face	The first point of the 3D face segment establishes the origin. The first and second points establish the X axis. The plane defined by the face determines the orientation of the UCS.
Arc	The center of the arc establishes the UCS origin. The X axis of the UCS passes through the pick point on the arc.
Circle	The center of the circle establishes the UCS origin. The X axis of the UCS passes through the pick point on the circle.
Dimension	The midpoint of the dimension text establishes the origin of the UCS origin. The X axis of the UCS is parallel to the X axis that was active when the dimension was drawn.
Line	The endpoint nearest the pick point establishes the origin of the UCS, and the X, Z plane of the UCS contains the line.
Point	The point location establishes the UCS origin. The UCS orientation is arbitrary.
Shapes, Text, Blocks, Attributes, and Attribute Definitions	The insertion point establishes the origin of the UCS. The object's rotation angle establishes the X axis.
Solid	The first point of the solid establishes the origin of the UCS. The second point of the solid establishes the X axis.
Trace	The direction of the trace segment establishes the X axis of the UCS with the beginning point setting the origin.

Setting the UCS Origin

When you create a UCS, you begin by defining the origin. You can change the origin location of the current UCS with the UCS Origin tool. First let's set the UCS icon to follow the UCS origin so we can view a change in the origin location as it happens. The origin will change to the new setting whether or not the UCS icon is set to follow. It's usually bad practice to set the UCS icon to follow the UCS origin because the UCS icon can be confused with object lines around the origin. Let's set the UCS icon to follow the origin here just so you can see what happens. Go back to the box you were drawing for the base and the post of the press.

1. Select Tools ➤ New UCS ➤ Origin. The UCS icon changes to include a plus sign (+) near the apex and moves to the current WCS origin. The UCS will look like the left image of Figure 11.6.

2. To move the UCS origin to the corner of the box, select the Origin tool from the UCS toolbar. Use the Endpoint Osnap to select the lower-left corner of the box. Now the icon should have moved to look like the right image of Figure 11.6.

Note that the X and Y axes remain pointing in the same direction.

FIGURE 11.6:

Left: The UCS icon changed to follow the UCS origin and moved to the WCS point 0,0. Right: The UCS icon moved again to the new origin location.

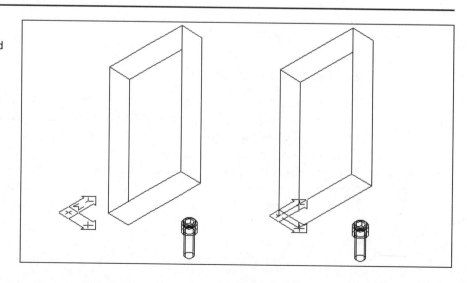

A UCS Defined with Three Points

You can easily define a UCS by locating three points in 3D space. The first point will be the origin, the second point will be a point on the positive X axis, and the third point will a point somewhere in the positive Y direction. Use this method when you know three points on a plane and you need the plane to define some objects.

1. Use the right side of the box to define a new UCS plane. From the UCS tool-bar, click the 3-Point UCS tool. At the `Origin point <0,0,0>:` prompt, pick the bottom-front corner of the right side of the box.

2. At the `Specify point on positive portion of the X-axis <1.9800, 0.0000,0.0000>:` prompt, pick the point at the bottom-back corner of the right side of the box.

3. At the `specify point on positive Y portion of the UCS XY plane <-0.0200,0.0000, 0.0000>:` prompt, pick the corner of the box above the first corner that you selected. (This sets the Y axis to the direction of the former Z axis.) The UCS should look like the one shown in Figure 11.7.

FIGURE 11.7:

The circle drawn on the vertical plane of the block

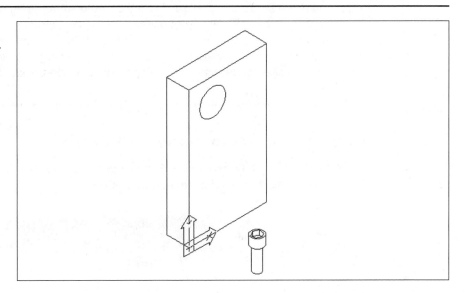

As mentioned in the section "Coordinate Systems for UCS and WCS" earlier in this chapter, each object is drawn on the current UCS plane. You've now created your first UCS plane that is not parallel to the WCS. Draw a circle and watch how it is positioned relative to the solid box. You'll use this circle later to make a hole in the part for the rack gear.

Locate a circle 1.062 units horizontal and 5 units vertical from your new UCS origin.

1. From the Draw toolbar, select the Circle tool and at the `CIRCLE Specify center point for circle or [3P/2P/Ttr (tan tan radius)]:` prompt, type **1.062,5**↵ to locate the circle in the X and Y axes of the current UCS.

2. At the `Specify radius of circle or [Diameter]:` prompt, type **d**↵. Finish the command by typing **1.17**↵ at the `Specify diameter of circle:` prompt.

A 1.17-unit diameter circle is drawn in on the side of the block. It looks like an ellipse, but this is an illusion due to your Isometric view of the object.

NOTE If you choose to select points with the pointing device but without the benefit of Osnaps while in a nonorthogonal viewpoint, you may not get the results that you expect. The pointing device works only on the X, Y plane. Any point you pick that is not attached to an object (by an Osnap) offers AutoCAD the choice of picking a point anywhere along an axis formed by the point receding directly into the screen. When you look at your object from another perspective, you might find that the object is far from where you expected it. The UCS tool can prevent problems of this type, because you're locking your geometry onto a defined plane. Using absolute coordinates as we did in the previous exercise can also save a lot of confusion.

Using an Object to Define a UCS Plane

The circle in the previous exercise can be used to define a UCS. Use the Object UCS tool on the UCS toolbar to set a definition.

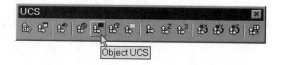

At the `Select object to align UCS:` prompt, pick the circle that you just created. The UCS origin has moved to the center of the circle and the UCS has rotated about –45°. See Table 11.1 to identify objects that can be used to establish a UCS origin and plane. See Figure 11.8 for what the current UCS icon looks like.

FIGURE 11.8:

Setting the UCS using an object

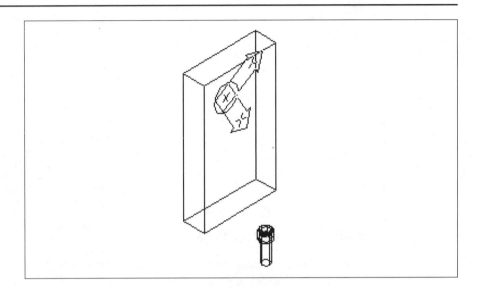

Rotating the UCS Plane about the Z Axis

Suppose you want to rotate the current X, Y plane about the Z axis. You can accomplish this by using the Z Axis Rotate option of the UCS command. Let's try rotating the UCS about the Z axis to see how this works.

1. Click the Z Axis Rotate UCS tool in the UCS toolbar.

2. At the `specify rotation angle about Z axis <90>:` prompt, enter **35** for 35°. The UCS icon rotates to reflect the new orientation of the current UCS (see Figure 11.9).

FIGURE 11.9:

Rotating the UCS about the Z axis

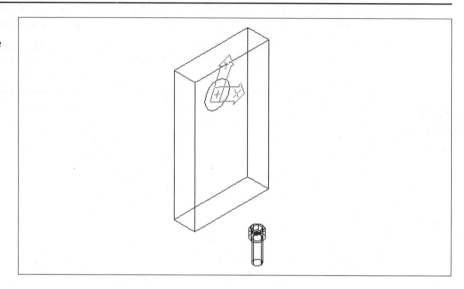

Returning to the WCS

Regardless of the location or plane of the current UCS, you can always restore the WCS. This can be a boon—especially for those times when you would rather start over to define a new UCS plane. From the UCS toolbar, select the World UCS tool. You can see that the UCS icon has returned to the original position and the *W* once more appears in the Y leg (see the left image of Figure 11.6).

TIP

If you prefer, merely type **ucs.⏎⏎**. This will accept the default setting of UCS, which is World.

Rotating the UCS Plane about the X or Y Axis

Next you'll move the origin back to the bottom of the box again and rotate the UCS about the Y axis.

1. Move the UCS origin back to the previous location by typing **UCS↵p↵**. Then, on the UCS toolbar, click the Origin UCS tool. Select the bottom, rear corner of the box.

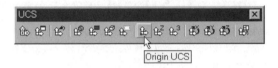

2. Rotate the UCS about the Y axis using the Y Axis Rotate UCS tool in the UCS toolbar.

At the `specify rotation angle about Y axis <0>:` prompt, enter **-90↵**. Your drawing should look like Figure 11.10. The UCS is now rotated 90° and appears aligned with the southwest plane of the box.

FIGURE 11.10:

The UCS rotated about the X axis

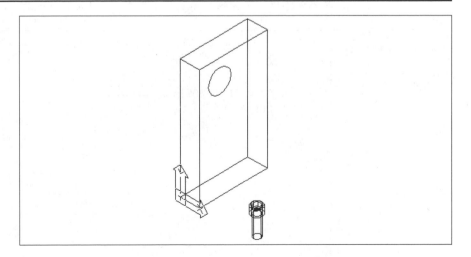

Similarly, the X Axis Rotate UCS tool allows you to rotate the UCS about the current X axis.

Defining a UCS Origin and Plane with the Z Axis

You can align a UCS perpendicular to two points and set the origin in one simple command. Use this option when you can select two points on the Z axis of the plane that you want to define.

1. From the UCS toolbar, select the Z Axis Vector UCS tool.

2. At the `Origin point <0,0,0>:` prompt, use the Endpoint Osnap marker to find the upper-left front corner of the box, and pick this point.

3. At the next prompt, `Point on positive portion of Z-axis <0'-0",0'-0", 0'-1">:`, use the Endpoint Osnap to pick the upper-right corner of the front of the box, as shown in Figure 11.11. The UCS twists to reflect the new Z axis of the UCS. See Figure 11.12 for the final result.

FIGURE 11.11:

Picking points for the Z Axis Vector option

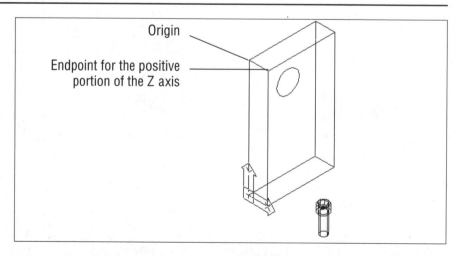

Origin

Endpoint for the positive portion of the Z axis

FIGURE 11.12:

The new UCS after assigning the Z axis points

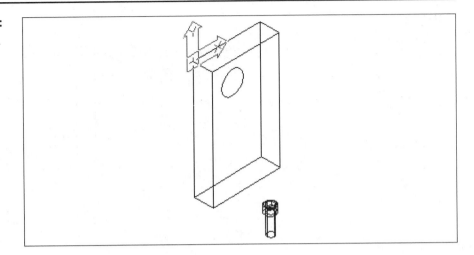

TIP   The result of this last exercise might be more clearly shown by changing the UCS icon's appearance. Type **shademode**↵ and then choose 3D↵. The icon will change to the new 3D icon.

Orienting a UCS to the View Plane

You can define a UCS in the current view plane. This is useful if you want to quickly switch to the current view plane for editing or for adding text to a 3D view. To try this, let's create text in the current UCS, switch to the WCS, and then change the UCS to be parallel to the screen.

1. Select Draw ➢ Text ➢ Single Line Text.

2. At the Start point>: prompt, use the current origin by typing **0,0**↵.

3. At the Height <0.2000>: prompt, press ↵ (to accept the default value), and at the Rotation angle <0>: prompt, press ↵ to accept the default.

4. At the Text: prompt, type **MASTERING AUTOCAD**↵. It is not necessary to use capital letters. For this exercise, capital letters are a little bit easier to see. The top image of Figure 11.13 demonstrates how the text should look.

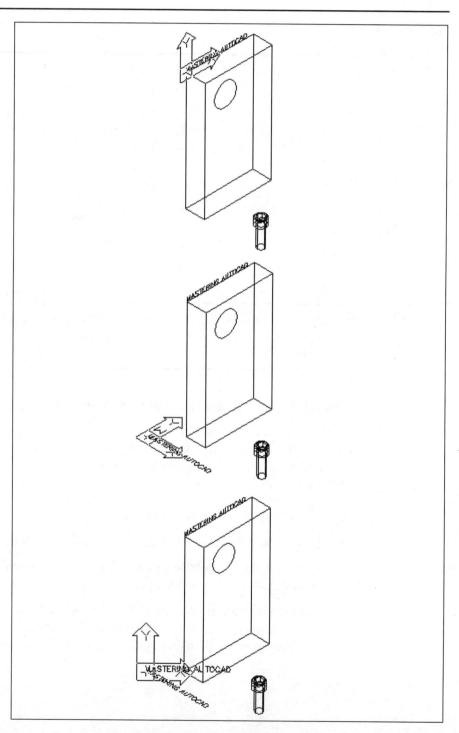

5. Use the World UCS tool in the UCS toolbar to return the UCS to the original position.

6. Use the Single Line Text command again to write **MASTERING AUTO-CAD** at the origin of the WCS. The middle image of Figure 11.13 shows the text at the WCS center.

7. Change the UCS to be parallel to the screen. Click the View UCS tool in the UCS toolbar, or choose View ➢ Set UCS ➢ View. You can also type **ucs↵v↵**. The UCS icon changes to show that the UCS is aligned with the current view.

8. Use the Single Line Text command again to write **MASTERING AUTO-CAD**. You can also cut and paste the words from one of the previous lines. Your work will look like the bottom image of Figure 11.13.

AutoCAD uses the current UCS origin point for the origin of the new UCS. By defining a UCS as parallel to your monitor's screen (the view), you can enter text to label your drawing, as you would in a technical illustration. Text entered in a plane created in this way will appear normal (that is, it will be legible and organized relative to the plane on which it is typed).

Saving a Named UCS

The UCS Save option is designed for speed. You could type **ucs↵s↵i↵** very quickly if you wanted to restore a saved UCS named **i** that could stand for *a UCS parallel to the screen when the object is viewed from the southeast corner*. But first, you must save a UCS named **i**.

1. Click the UCS tool from the UCS toolbar. The prompt is the same as if you had typed **ucs↵**. At the Enter an option [New/Move/orthoGraphic/Prev/Restore/Save/Del/Apply/?/World] <World>: prompt, type s↵ to use the UCS Save option.

2. You'll see the prompt Enter name to save current UCS or [?]:. The ? allows you to list the currently saved UCS, if any. At the prompt, type i↵. You've just saved your first UCS.

Restoring and Deleting a Saved UCS

You could use the UCS toolbar or the menu bar to issue the Restore command, but these are very slow compared to using the command line. If you find any of the commands hard to remember, the toolbars and pull-down menus are there to help (AutoCAD's Help system can also be very useful).

In the following exercise, you'll change the UCS back to the WCS again, recall the **i** UCS, and then you'll delete the **i** UCS.

1. Use the World UCS tool in the UCS toolbar to return to the WCS. You'll see the icon return to the origin and orientation of the WCS.

2. Type **ucs⏎**, and at the prompt, type **r⏎** to use the Restore option. At the Enter name of UCS to restore or [?]: prompt, type **i⏎**. You can see that the UCS is restored to normal on the screen.

You could also have used the Named UCS tool in the UCS toolbar or chosen Tools ➢ UCS ➢ Named UCS from the menu bar. Either of these options brings up the UCS dialog box, so you have complete control over restoring (making current) a UCS.

The UCS Presets Menu

You may recall that another way to define a UCS is with the help of the Orthographic UCSs tab of the UCS dialog box. This tool gives you names and visual references for a set of orthogonal planes that you can select when you need them, or you can use the ÜCS dialog box to save a set of planes to use with the UCS Restore command. Let's try it.

1. Select the World UCS option from the UCS toolbar. This is a good starting point for the new or occasional user to define a new UCS.

NOTE
The preset uses the current UCS plane to generate the new UCS. This can be quite disconcerting. You expect the preset to look like the one in your drawing, but it doesn't—it looks as if it were viewed from the previous UCS perspective.

---

2. Click the Display UCS dialog tool in the UCS toolbar, or choose Tools ➢ Named UCS. The UCS dialog box appears.

3. Click Orthographic UCSs. Double-click the UCS labeled Right. Click OK. The dialog box closes and the UCS icon is set parallel to the southeast face of the block, as shown in Figure 11.14.

4. Click the UCS tool or type **ucs⏎** and type **s⏎** for Save. You'll want to use this UCS tool to draw other geometry.

5. At the Enter name to save current UCS or [?]: prompt, type **s⏎** to save a UCS named **s**. The **s** stands for *side*—it's faster to type and less prone to error.

FIGURE 11.14:

The UCS icon after the Right preset was selected

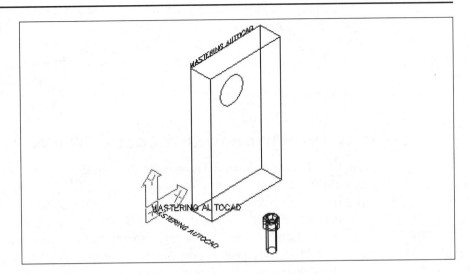

Viewing Your Model

Earlier in this chapter, you used the View toolbar to select the SE Isometric view. It's time to take a closer look at the View toolbar and the options it provides. You're about to take a virtual walk around your model.

As you go through the tutorial, be sure to notice the UCS icon. The UCS icon will remain attached and pointing in the direction of the positive X and Y axes. The only thing that changes is your position as you view the model.

You won't see the progress of the change in viewport, only the end result as Auto-CAD regenerates the screen. There is a practical reason for this. The current graphics are not fast enough to show it "live"—the effect would be to slow you down.

The Viewpoint command allows you to view your model from any point in 3D space. You may select a point that is relative to the World UCS.

AutoCAD 2000 includes the Camera command, which acts like a camera, setting up a perspective view of your 3D part. (Experienced AutoCAD users may recognize the similarity to CAmera from the Dview command.)

Your first 3D view of a model is a wire-frame view. This means that your model appears to be made of wire and none of the sides appear solid. This section describes how to manipulate this wire-frame view so you can see your drawing from any angle. You'll also learn how to view the 3D drawing as a solid object with the hidden lines removed. (This is done once you've selected your view.) Additionally, you'll learn methods for saving views.

Finding Isometric and Orthogonal Views

First, let's start by looking at some of the viewing options available. You used one option already to get the current 3D view. That option, View ➤ 3D Views ➤ SE Isometric, brings up an Isometric view from a southeast direction, where north is the same direction as the Y axis. Figure 11.15 illustrates the "camera" locations for the three other Isometric view options: SE Isometric, NE Isometric, and NW Isometric. In Figure 11.15, the cameras represent the different viewpoint locations. You can get an idea of their locations by looking at the UCS icon shown in the figure.

FIGURE 11.15:

This diagram shows the viewpoints for the four Isometric views available from the View ➤ 3D Views pull-down menu.

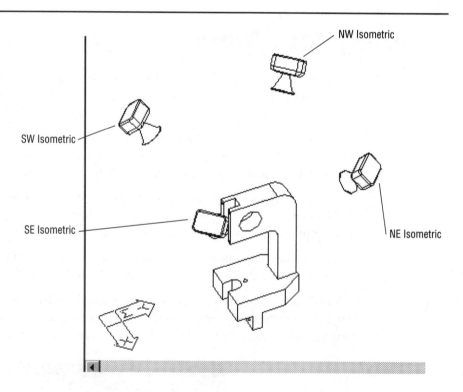

Another set of options available on the View ➤ 3D Views menu contains Top, Bottom, Left, Right, Front, and Back. These are Orthogonal views that show the top, bottom, and sides of the model, as shown in Figure 11.16. In this figure, the cameras once again show the points of view.

To give you a better idea of what an Orthogonal view looks like, Figure 11.17 shows the view that you see when you choose View ➤ 3D Views ➤ Right. This is a side view of the unit.

FIGURE 11.16:

The six viewpoints of the Orthogonal view options on the View ➤ 3D Views menu

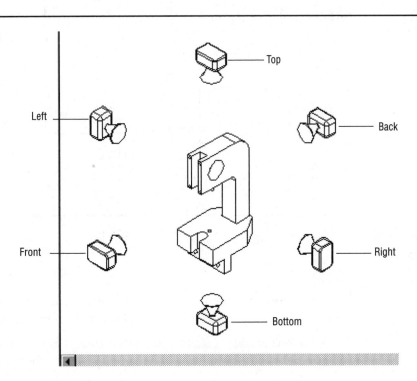

From time to time you'll see the UCS icon change to a broken pencil. This will only happen if your view is aligned to the current X, Y plane. The icon will look like the one in Figure 11.18. The simplest way out of this situation is to type **ucs**↵↵. This will ensure that the WCS is being used. Then type **plan**↵↵. This will establish a plan view of the WCS. You could also try one of the other views illustrated in Figures 11.15 and 11.16.

NOTE   The pencil is broken to discourage you from drawing in this mode. It's not broken because it won't work—it will, but you won't have control of object placement.

FIGURE 11.17:

The view of the model you
see when you choose
View ➤ 3D Views
Right

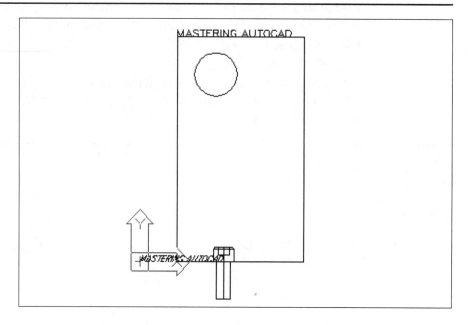

FIGURE 11.18:

The UCS icon showing a
broken pencil

When you use any of the View options described here, AutoCAD will display the
extents of the drawing. You can then use the Pan and Zoom tools to adjust your view.

The View toolbar offers quick access to all of the options discussed in this section.
To open it, right-click any toolbar, and then click View in the Toolbars dialog box.

Using a Dialog Box to Select 3D Views

You now know that you can select from a variety of "canned" viewpoints to view
your 3D model. You can also fine-tune your view by indicating an angle from the

drawing's X axis and from the floor plane using the Viewpoint Presets dialog box. Let's try it.

1. Choose View ➢ 3D Views ➢ Viewpoint Presets, or type **vp**↵ in the command line. The Viewpoint Presets dialog box appears (see Figure 11.19). The square dial to the left lets you select a viewpoint location in degrees relative to the X axis. The semicircle to the right lets you select an elevation in degrees for your viewpoint above the X, Y plane.

FIGURE 11.19:

The Viewpoint Presets dialog box

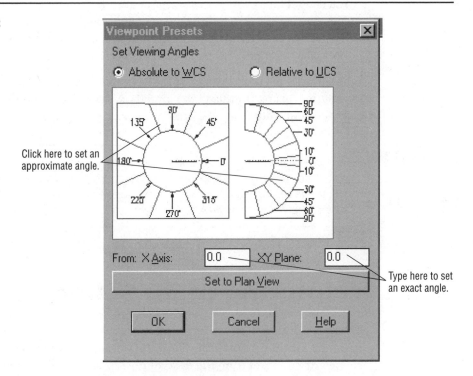

2. Click the angle labeled 135° in the upper-left section of the square dial. Then click the angle labeled 60° in the right-hand semicircle. Notice that the pointer moves to the angle you've selected and the input boxes below the graphics change to reflect the new settings.

3. Click OK. Your view changes according to the new settings you just made.

Other settings in the Viewpoint Presets dialog box let you determine whether the selected view angles are relative to the WCS or to the current UCS. You can also go directly to a plan view by clicking the Set to Plan View button.

There are a couple of features of the Viewpoint Presets dialog box that aren't readily apparent. First of all, you can select the exact angle indicated by the label of either graphic by clicking anywhere inside the outlined regions around the pointer (see Figure 11.19). For example, in the graphic to the left, click anywhere in the region labeled 90° to set the pointer to 90° exactly.

You can set the pointers to smaller degree increments by clicking within the pointer area, inside the circle on the left, or inside the arc on the right. For example, if you click in the area just below the 90° region in the left graphic, the pointer will move to that location. The angle will be slightly greater than 90°.

If you want to enter an exact value from the X axis or X, Y plane, you can do so by entering an angle value in the input boxes provided. You can obtain virtually any view you want using the options offered in this dialog box.

NOTE Three other options—Rotate, Tripod, and Vector—are also available from the View pull-down menu. These options of the Vpoint command are somewhat difficult to use and duplicate the functions described here. If you'd really like to learn more about these options, consult the AutoCAD Help system and look up the Vpoint command.

More Sculpturing of 3D Solids

The socket-head screw model required the use of two interesting commands to construct a solid. You used the Subtract Boolean operation to remove the hexagon-shaped drive socket from the head of the screw, and you used the Union Boolean operation to add the cylinder for the threads. This is the normal method used in AutoCAD to construct solids. This process would be similar to one you would use if you were sculpting in clay. You would gather a lump of clay that is about the right size, and you might find that you need to add some more here or to cut away some over there. You might need to fashion a special tool to remove material or carefully create a piece to be added.

In the following exercise, you'll continue to work with the Subtract and Union commands. You'll create the base and post of the press by the subtraction and union of material. You don't need the text *Mastering AutoCAD* any more, so erase it by clicking the Erase tool in the Modify toolbar and selecting each line of text. Right-click to complete the command.

1. You should still have the UCS set to the right side of the block. If not, type **ucs↵r↵s↵** to restore the saved UCS called *s*, click in the drop-down list on the UCSII toolbar and restore the *s* UCS, or choose Tools ➢ Named UCS. Select the UCS named *s*, and then select the Set Current button, followed by OK. The UCS toolbar also offers the Display UCS Dialog button which will open the Named UCS dialog box.

2. Be sure that the UCS origin is located at the southwest corner of the right side of the block. Your drawing should look like Figure 11.20.

FIGURE 11.20:

The box and screw ready to use

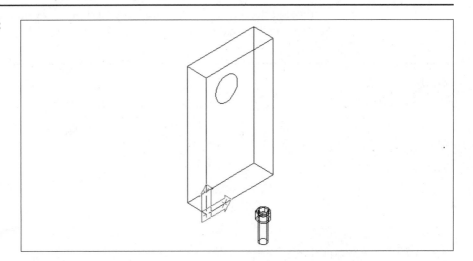

3. Draw a new box from the corner of the existing box. You'll use the new box to remove material from the existing box. Type **box** [↵] or click the Box tool in the Solids toolbar. At the Specify corner of box or [CEnter] <0,0,0>: prompt, press [↵r] to accept the default. At the Specify corner or [Cube/Length]: prompt, enter l to indicate that you wish to create a noncubic object. The rest of the prompts and required dimensions are

Specify length: **2.5**

Specify width: **4**

Specify height: **–2**

NOTE The negative value for the height is used to make a box that has thickness in the –Z direction. The thickness of the new box is not the same as the large box. Because the space occupied by the small box is going to be subtracted from the large box, you need only ensure that the small box is large enough to remove all of the material that you want to remove.

4. Type **su** ↵ to use the Subtract tool to remove the small box from the larger one and to create the L shape of the post.

5. At the `Select solids and regions to subtract from:` prompt, select the taller box. Press ↵ to complete the selection. This should be the box area that you want to keep. At the `Select objects: Select solids and regions to subtract:` prompt, select the small box and press ↵ to complete the selection. You should have a drawing that looks like Figure 11.21.

FIGURE 11.21:

The box after the subtraction

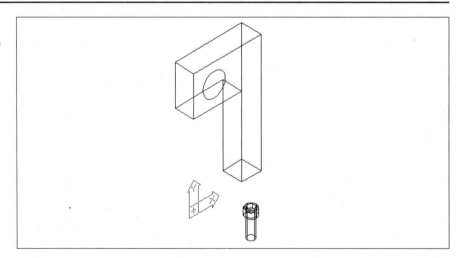

Using the Fillet Tool to Edit a Solid Model

You used the Chamfer tool earlier in this chapter to modify the screw. The post of the press needs an inside and an outside fillet. Although you can use the same

Fillet tool that you first learned about in Chapter 3, "Learning the Tools of the Trade," there will be a change in the way it works because you've selected a solid.

1. Type **f↵**, click the Fillet tool in the Modify toolbar, or choose Modify ➤ Fillet from the menu bar.

2. At the `Current settings: Mode = TRIM, Radius = 0.3800 Select first object or [Polyline/Radius/Trim]:` prompt, pick the line that represents the corner between the top and back side. (Your default fillet radius may be different from the 0.3800 listed here.)

3. This edge requires a 1-unit radius. At the `Enter radius <0.3800>:` prompt, enter **1↵**.

4. At the `Select an edge or [Chain/Radius]:` prompt, press ↵ to tell Auto-CAD that this is the only edge that you want to fillet. This prompt is asking whether you want to change the radius default, define a chain of consecutive edges, or select more edges to fillet. Your drawing should look like the top of Figure 11.22.

FIGURE 11.22:

The post before and after filleting

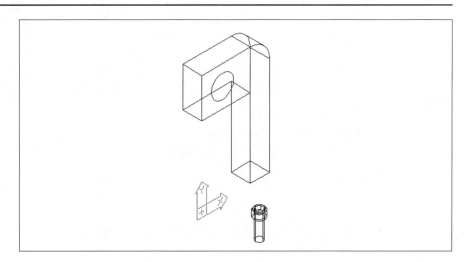

FIGURE 11.22 CONTINUED:

The post before and after filleting

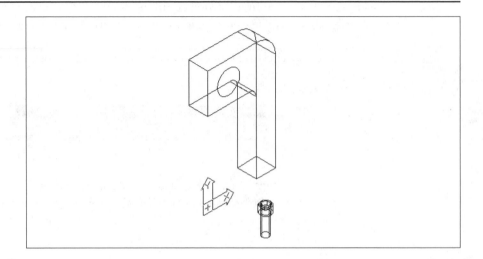

5. Fillet the inside corner the same way. Type **f.⏎** for the Fillet command. Pick the inside corner, type **0.38.⏎** for the radius, and press ⏎ again to complete the command. Now your drawing should look like the bottom image of Figure 11.22.

Next, let's see how fast we can complete the post. If you haven't been working along with the text in developing the post, it is recommended that you return to the beginning of this chapter and work through the section "Using the Box Primitive." Or, you may open the file named `Figure 11.22` on the CD, and then continue from here. We're going to use very abbreviated commands, much the way that you'll actually use AutoCAD to build and edit models. If you don't understand a command or sequence, be sure to pay attention to the prompts on the AutoCAD command line as you execute them.

Our first task is to create a slot on the top, front face of the post (see Figure 11.23, Panel 1). To help as you work through this part of the exercise, turn on the 3D UCS icon by typing **Shademode**, then **3D**.

1. Type **ucs.⏎⏎** to return to the World UCS.

2. Type **ucs.⏎x.⏎90.⏎** to rotate the UCS plane 90° about the X axis.

3. Type **ucs.⏎s.⏎f.⏎** to save a UCS named **f** for *front*.

4. See Panel 1 of Figure 11.23 to locate the lower-left corner of the front of the post. Type **box.⏎from⏎**, then use the Endpoint Osnap to pick the lower-left corner of the front of the post. Type **@.24<0⏎** for the offset. Type **@.5,3⏎** for the other corner, and then type **−.625⏎** for the height.

5. Type **su⏎** to subtract the new solid from the post. Pick the post, press ⏎, pick the new box, and then press ⏎. To see what this should look like, see Panel 1 in Figure 11.23.

6. Type **cylinder⏎** to draw the tapped holes in the front of the post. Use the From Osnap and pick the lower-left corner of the front. Type **@.125,.125⏎**, and then type **d⏎.138⏎** to give the diameter value. Type **−.375⏎** to give the depth of the threads.

TIP

If you need to show the tap-drill depth, you could draw the tap drill diameter for the thread. You could even draw a cone to represent the point of the drill and subtract this from your solid. If you need to show the actual thread depth, you could draw another cylinder the diameter and length of the threads.

---

7. Type **ar⏎** to use the Array command. There are four threaded holes in the front of the post spaced 0.75 in X and 1.75 in Y. Type **l⏎⏎** to use the last object. Type **r⏎** for a rectangular array, **2⏎** for the rows, and **2⏎** for the columns. Type **1.75⏎** for the distance between the rows and **.75⏎** for the distance between the columns.

8. Type **su⏎** to subtract the array from the post. Pick the post, press ⏎, and then use a window to select the four cylinders. Press ⏎ to complete the selection set. Your drawing should look like Panel 2 in Figure 11.23.

9. Type **ucs⏎r⏎s⏎** to restore the UCS named **s** for *side*.

NOTE

If you didn't create this saved UCS, review the options discussed in "Saving a Named UCS" earlier in this chapter, and save one now.

---

10. Type **ext⏎** to begin the Extrude command and pick the 1.17 diameter circle that you drew in the section "A UCS Defined with Three Points" earlier in this chapter. Type **−.778⏎** for the depth of the extrusion and press ⏎ for no taper angle.

11. Type **cylinder.⏎** to draw another cylinder for the drill-through. Use the Center Osnap to use the center of the original circle. Type **d.⏎.5.⏎** for the diameter and **−2.⏎** for the depth.

12. Type **su.⏎** to subtract the last two cylinders from the post. Pick the post and ⏎, and then the two cylinders and ⏎. Your drawing should look like Panel 3 in Figure 11.23.

13. Type **ucs.⏎r.⏎f.⏎** to restore the UCS named **f** for *front*.

14. Type **ucs.⏎or.⏎** to pick a new origin for the UCS. Use the Endpoint Osnap to pick the bottom back-left corner.

15. To draw the counterbored hole, type **cylinder.⏎**. Type **.5,.5.⏎** for the distance from the new UCS origin. Type **d.⏎.38.⏎** for the diameter, and **2.⏎** for the height. To draw the counterbore, type **cylinder.⏎** and select the center of the cylinder that you just drew. Type **d.⏎.562.⏎** to give the diameter value, and then type **.375.⏎** for the height.

16. Type **su.⏎** to subtract the last two cylinders from the post. Pick the post and ⏎ and then the two cylinders and ⏎. Your drawing should look like Panel 4 in Figure 11.23.

17. Save your work.

FIGURE 11.23:

Panel 1: The box subtracted from the post.

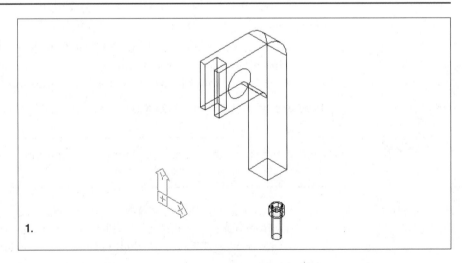

1.

FIGURE 11.23 CONTINUED:

Panel 2: The array of tapped holes subtracted from the post. Panel 3: The large counterbore subtracted from the side of the post. Panel 4: The small counterbore subtracted from the back of the post.

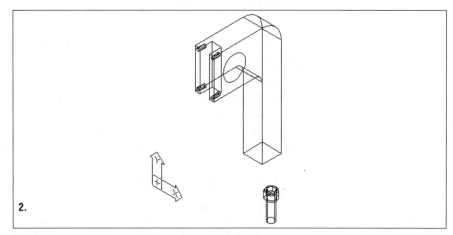

2.

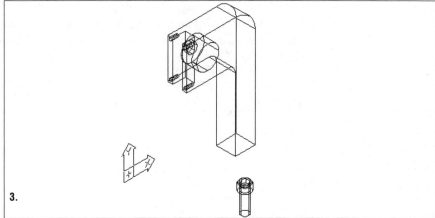

3.

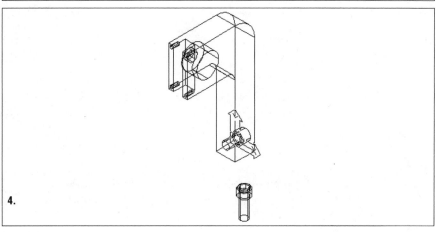

4.

Making a Solid Model from a 2D Drawing

As you'll see, making a solid model from a 2D drawing is not an automatic process. However, it is not a very difficult problem if you think your way through it. You learned about polylines in Chapter 10, "Drawing Curves and Solid Fills." Now you can put that knowledge to work to create closed polylines from the 2D objects. Next you'll extrude the polylines into solid objects, align them, and use a new tool that will remove any material that is not common to both solids. A 2D drawing has been created of the base of the press, called Press Base, on the companion CD-ROM.

1. Type **ucs**↵ to return to the World UCS.

2. Click the Insert tool in the Draw toolbar to open the Insert dialog box. Click the Browse button, and find and select the file named Press Base.dwg on the CD-ROM. Be sure the Specify Parameters on Screen check boxes for Insertion Point, Scale, and Rotation are not checked. The Explode check box normally does not have a check mark. Click the Explode check box to add a check mark. Click OK. (See Chapter 4 if you want to review the block-insertion process.)

3. Zoom All to see all of the new block and your solids. Your drawing should look like Figure 11.24.

FIGURE 11.24:

The base block inserted

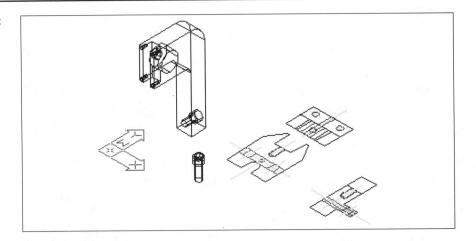

4. Type **pe**↵ to begin the Polyline Edit command. At the PEDIT Select polyline: prompt, pick any outside object line in the plan view of the base. At

the `Object selected is not a polyline Do you want to turn it into one? <Y>` prompt, press ↵ to accept the default and turn your selection into a polyline.

5. Join the other object lines to the first one to create a closed polyline. At the `Enter an option [Close/Join/Width/Edit vertex/Fit/Spline/Decurve/Ltype gen/Undo]:` prompt, enter j↵ to use the Join option.

6. At the `Select objects:` prompt, use a window to select all of the objects in the plan view. You don't need to be careful when you make this window because AutoCAD is only looking for lines that join the first line that you selected at the endpoint. Any others will be ignored, and AutoCAD will continue to test each line that it adds to the polyline to see if another line in your selection set has the same endpoint. The program will continue this way until all objects have been tested. If a closed polyline is formed, the first option in the prompt line changes from `Close` to `Open`, to see if you now want to open a closed polyline.

7. You should see the statement `13 segments added to polyline` and the prompt `Enter an option [Open/Join/Width/Edit vertex/Fit/Spline/Decurve/Ltype gen/Undo]:`. Press ↵ to exit the command.

8. To do the same thing to the side view, press ↵ again to reenter the Pedit command and at the `Select polyline:` prompt, pick any outside object line in the side view. Repeat steps 5–7 to complete the process. (If you aren't sure what to pick, check Figure 11.25 for the desired results.) You'll see the polyline close with seven segments added.

9. Save your work.

FIGURE 11.25:

The two views of the base after extrusion

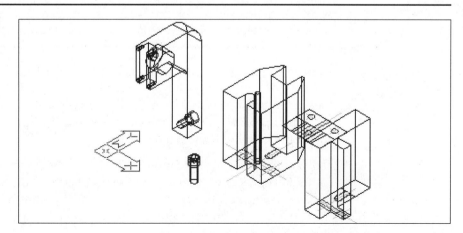

Next, you'll extrude to create solids and align the solids to take the interference.

1. Type **ext**↵ to use the Extrude command. You're going to extrude both views and the hole that you see as a circle in the plan view. At the `Select objects:` prompt, pick the two polylines and window the circle.

2. At the `Specify height of extrusion or [Path]:` prompt, enter **4**↵. Once again, there is no need for precision. The extrusion height only needs to be greater than the width and thickness of the actual part. Your drawing should look like Figure 11.25.

NOTE   If you see the message `Unable to extrude 1 selected object`, a polyline you tried to extrude probably was not a closed boundary. Try moving the polyline to get a better look. If there are segments that divide it into sections, fix them. Then try using the Extrude tool again.

Aligning Two 3D Objects

The Align tool allows you to define three points on two objects. The points will be used to define one plane on each object and the orientation of the objects to each other. In our example, the plan view extrusion will remain where it is and the side view extrusion will be aligned to the plan view.

TIP   The solid extruded from the side view will rotate 90° to the right, so that its left side will be on top and its right will be pointed down. It will also end up aligned to the top of the part extruded from the plan view. (See the top panel of Figure 11.26 to preview the points that you'll be selecting.) You could also use a series of discrete Move and Rotate commands to align the two extrusions.

The purpose of this alignment will become clear in just a few more steps.

1. Select Modify ➤ 3D Operation ➤ Align or type **align**↵. At the `Select objects:` prompt, pick the side view extrusion. The prompt is asking for the object that you want to move.

2. Look at the top image of Figure 11.26 to find the points in the correct order to align the two views. To select the objects, at the `Specify 1st source point:` prompt, pick the corresponding point; at the `Specify 1st destination point:` prompt, pick the corresponding point; and so on, through all three sets of points.

FIGURE 11.26:

Aligning the two extrusions

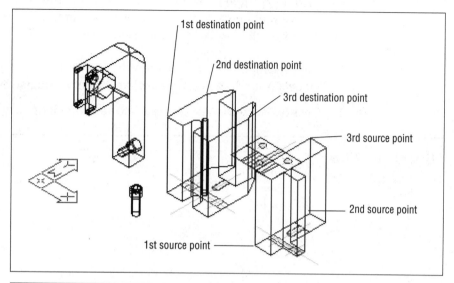

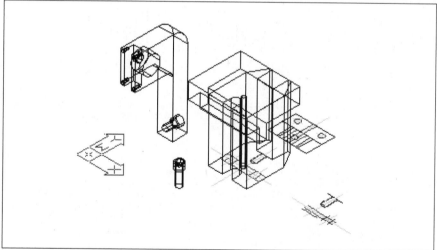

The first pair of points is the most important; they must be right on. The other points may not match because of the physical size of each feature, but AutoCAD will try to make them align.

You can use the Undo command or click the Undo tool in the Standard toolbar if you pick points out of order, or if the result is not as shown in the bottom image of Figure 11.26.

This would be a good time to look at your work from a few different views. Use the Front View option to look at your models from the front orthogonal view. Your drawing will look like the top image of Figure 11.27.

1. Click the Right View tool in the View toolbar to look at the right side of your models. Your drawing will look like the bottom image of Figure 11.27.

2. Click SE Isometric View to return to your previous viewpoint.

FIGURE 11.27:

The front and side views of the models

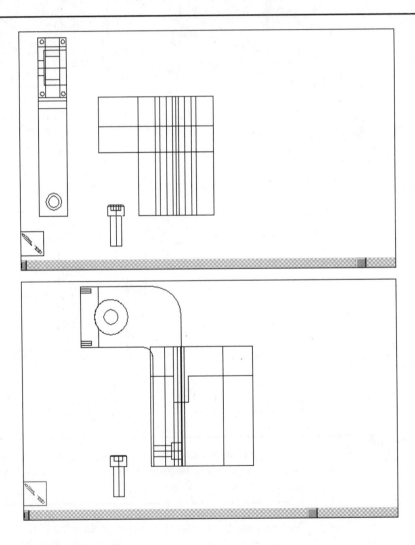

3. Click the Intersect tool in the Solids Editing toolbar, choose Modify ➤ Solids Editing ➤ Intersect from the menu bar, or type **intersect**⏎ in the command line.

4. At the `Select objects:` prompt, pick both extrusions and press ⏎ to complete the selection set. Your drawing should look like Figure 11.28. This Boolean operation evaluated the two models for all common volumes and discarded any portion of either model not found to be common to both.

FIGURE 11.28:

The two extrusions after the Intersection Boolean operation

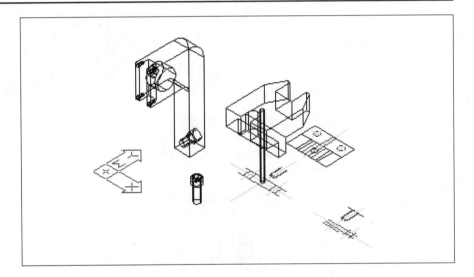

To complete the base, you still need to add the mounting holes to the model. If your drawings are created to scale, you can use these objects to make solids. The cylinder that you need is the extruded circle that is a hole in the 2D model. First you'll subtract the cylinder from the solid. Three holes are shown in the remaining view and you'll use them to finish editing your model. As you'll see, you only need the objects in the remaining view to finish editing the solid, so the second operation will be to change the viewpoint and make it easier to erase these objects.

1. Use the Subtract command to make a hole by subtracting the cylinder from the base.

2. Click Top View in the View toolbar to change your viewpoint to the top view.

3. Type **e↵w↵** or use your favorite method to issue the Erase command and the Window option to select objects. Figure 11.29 illustrates what objects shouldn't be erased.

FIGURE 11.29:

Top: The line to select to be the Z axis of rotation
Bottom: The rotated view

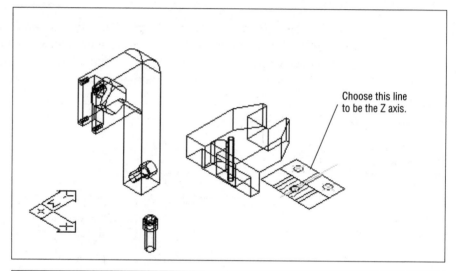

Choose this line to be the Z axis.

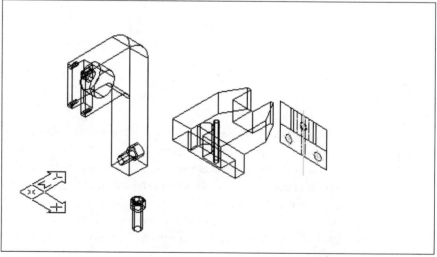

TIP The three circles and the top and bottom horizontal lines that will be references for the next operation should not be erased.

4. Click the SE Isometric View in the View toolbar to change your viewpoint to the previous view.

5. The remaining 2D end view would be much easier to use if it were parallel to the correct plane of the solid. Type **rotate3d**↵, or select Modify ➢ 3D Operation ➢ Rotate 3D to rotate the end view without changing the UCS.

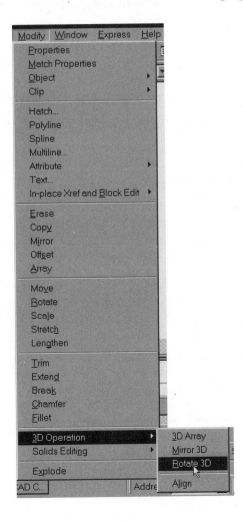

6. At the `Select objects:` prompt, type **w** to use a window to capture all of the objects in the remaining end view. Remember, if you accidentally get any objects that you don't want, type **r** for Remove and select the objects that you want to take out of the selection set. When you're satisfied with your choices, press ↵ to go to the next step.

7. You're going to pick a line to be the axis and rotate a set of objects about the line. At the `Specify first point on axis or define axis by Object/ Last/View/Xaxis/Yaxis/2points]:` prompt, type **o** to pick an object. At the `Pick Select a line, circle, arc or 2D-polyline segment:` prompt, pick the top line in the view. The line is shown in the top image of Figure 11.29.

8. At the `Specify rotation angle or [Reference]:` prompt, type **–90** to rotate the object clockwise 90°. See the bottom image of Figure 11.29.

The Rotate 3D Options

The following list shows several ways to rotate an object:

Axis by Object   Allows you to align the axis of rotation with an existing object (line, circle, arc, or 2D polylines).

Last   Uses the last axis of rotation.

View   Aligns the axis of rotation with the viewing direction of the current viewport that passes through a selected point. At the `specify a point on the view direction axis <0,0,0>:` prompt, select a point.

X/Y/Z Axis   Aligns the axis of rotation on the axes (X, Y, Z) that pass through the selected point. At the `Specify a point on (X, Y, Z) axis <0,0,0>:` prompt, select a point.

2 Points <the default>   Allows you to specify two points to define the axis of rotation. You'll be prompted for `Specify first point on axis` and `Specify second point on axis`. Select two points.

Tidying Up

You must align the view with the solid before you move the view into place.

1. Type **m.↵** to use the Move command. Type **p.↵** to use the previous selection set and use the Endpoint Osnaps, as shown in the top panel of Figure 11.30, to move the end view. The middle panel of Figure 11.30 shows the view after the move.

FIGURE 11.30:

Top panel: Osnap points used to move the view. Middle panel: The moved view.

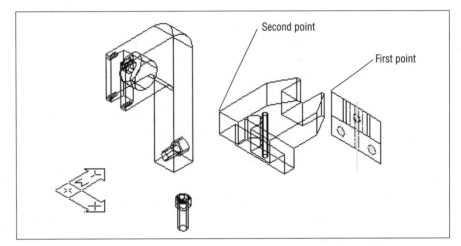

Second point

First point

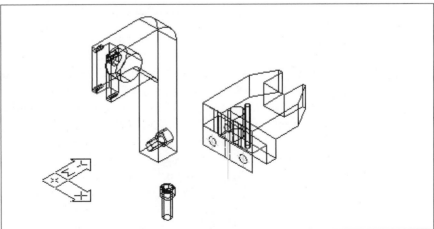

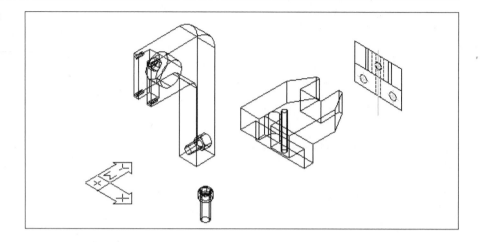

2. Be sure that Ortho is on and move the view by dragging it close to the final position, as shown in the bottom panel of Figure 11.30.

3. Save your work.

TIP You could have also used the Align tool we studied in the previous section to position the holes for this operation. If you have time, save your work, undo, and try it again.

The 3/8 tapped hole is 0.625 deep, and the other two base mounting holes go all the way through. You still need to extrude the circles and use them to subtract the holes. We won't draw the tap drill and point, but you can try this if you like. The tap drill is nearly always drilled 0.06 to 0.10 deeper than the threads (if you can't drill through) and has a 120° conical point. You could extrude the tap drill circle, create a 120° cone at the end of the extrusion, and subtract both the cylinder and the cone from the base, then begin step 1.

1. Type **ext↵** and pick the 0.375 diameter circle in the view that represents the outside diameter of the threads. You'll probably want to zoom in close enough to clearly see these objects. You can use the Transparent Zoom by typing **'z↵**. Window is the default, so just drag a small window around the view and pick the circle. (The circle is drawn with a hidden line.) Type **–.625↵** at the

Specify height of extrusion or [Path]: prompt. There is no taper, so press ↵ again.

The cylinder was created on the current layer. You can change it to the Layer 0 if you like.

2. Press ↵ again to reenter the Extrude command. This time you want to select the two mounting holes at the bottom of the view. At the Specify height of extrusion or [Path]: prompt, type −8↵↵ to create two cylinders −8 units long without taper. Your drawing should look like Figure 11.31.

FIGURE 11.31:

The screw with extruded thread hole and mounting holes

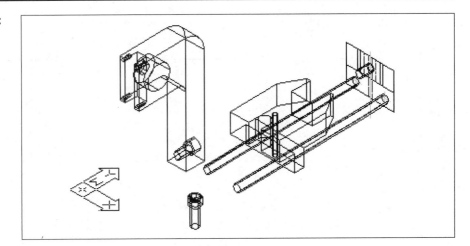

3. You made the mounting hole extrusions long enough to simply subtract them from the solid. Type **su**↵ to start the Subtract command, select the base, and press ↵. At the Select solids and regions to subtract: prompt, select the cylinders and press ↵ to complete the selection and the command. Your drawing will look like Figure 11.32.

4. You need to move the cylinder that represents the tapped hole into place in the cutout on the base. Use two Osnap points to move the cylinder the exact distance. See Figure 11.33 for these points. Type **m**↵, select the cylinder, press ↵, and use your Endpoint Osnap markers to find the points.

FIGURE 11.32:

The mounting holes subtracted from the base

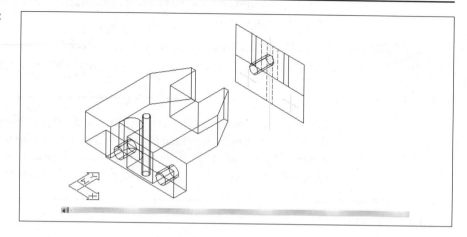

FIGURE 11.33:

The points needed to move the cylinder

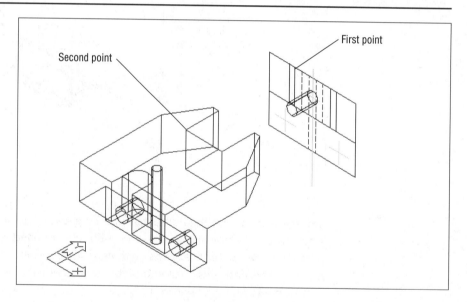

5. Type **su**↵, select the base and press ↵, and select the cylinder and press ↵. You've issued the Subtract command, selected an object to subtract from, and then selected the object to be subtracted. Your drawing should look like Figure 11.34.

6. You no longer need the lines of the view, so you can erase them. Drag a window around the remaining view to turn on the grips, and type **e**↲.

7. Save your work.

FIGURE 11.34:

The threaded hole subtracted from the base

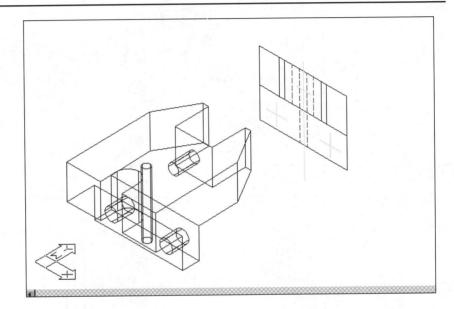

You can begin to see what the assembly will look like. The base and post are oriented correctly, but the screw will have to be rotated and moved into position. The easiest way to align the base and post is to find a common point, like the hole for the screw in the post and base. Use Osnaps to move the base to the post.

1. Type **m**↲ to start the Move command. Select the base, and press ↲. At the `base point` prompt, select the center of the threaded hole, and at the `Specify second point of displacement:` prompt, select the mating hole in the post. See the top image of Figure 11.35 for help in finding these points. See the bottom image of Figure 11.35 for the results of the move.

2. Use the 3DRotate command to rotate the screw. Type **rotate3d**↲, select the screw, and press ↲. At the `Specify first point on axis or define axis by [Object/Last/View/Xaxis/Yaxis/Zaxis/ 2points]:` prompt, pick the bottom of the threads using the Center Osnap. At the `Specify second point on axis:` prompt, and with Ortho on, drag the mouse to the right and pick any point. You've now defined an axis through the end of the screw.

FIGURE 11.35:

The points required to move the base, and the base after it has been moved

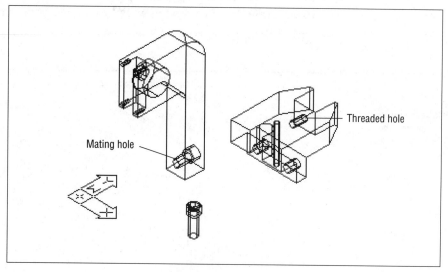

Mating hole

Threaded hole

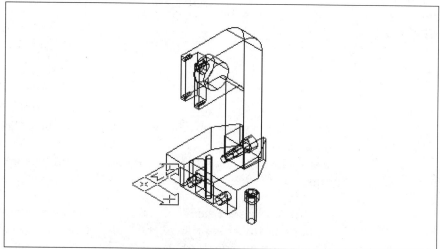

3. At the Specify rotation angle or [Reference]: prompt, type **–90**↲. The screw should look like Figure 11.36.

FIGURE 11.36:

The rotated screw

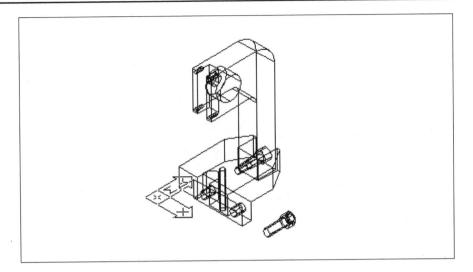

4. Zoom in so that you can easily see what you'll select, pick the screw to high-light the grips, and pick the grip that is at the center of the base of the head of screw; this grip is at the center of the cylinder shape. It is now the hot grip. Use the Transparent Zoom to Zoom Previous, and use it again to zoom into the counterbore on the post. Drag the cursor over the counterbore until you find the center of its depth; pick this point. Your model is assembled.

To best see the results of your work, use the Hide tool. Alternatively, you can use the Shade tool to do same thing. The Shade tool is not often used, but you should know it's there if you need it.

1. Type **shademode↵h↵**, choose View ➤ Shade ➤ Hidden, or click the Hidden button on the Shade toolbar.

2. Type **shademode↵f↵**, choose View ➤ Shade ➤ Flat Shaded, or click the Flat Shaded button on the Shade toolbar.

Figure 11.37 shows the results of the Hidden and Flat Shaded options.

FIGURE 11.37:

The assembly using Hidden in the top panel, and Flat Shaded in the bottom panel

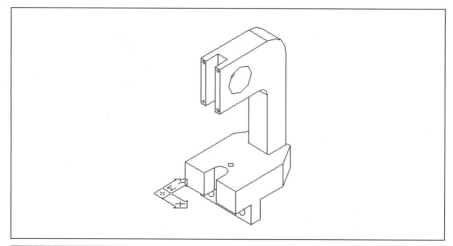

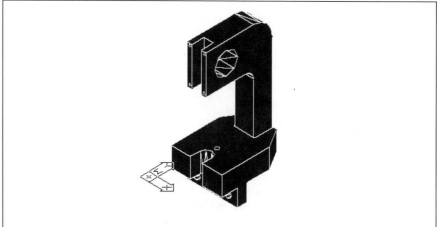

If You Want to Experiment...

- Dig out some of your old drafting books and try using Boolean addition and subtraction to construct some interesting parts.

- Use the View toolbar to walk around your model. View it from each of the orthogonal viewpoints. Then view it from each of the isometric angles.

- At each isometric viewpoint, use both Hide and Shade to see your model more clearly.

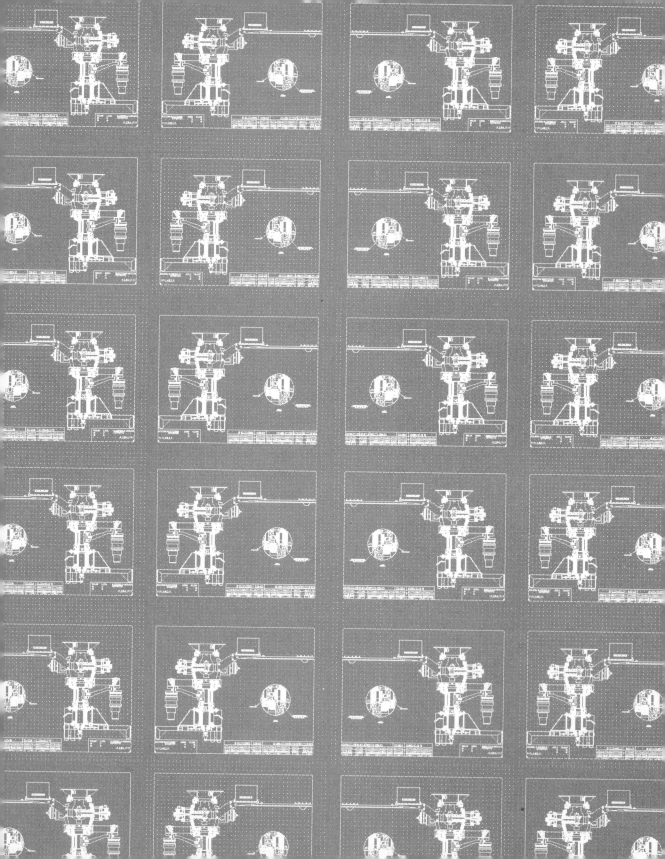

CHAPTER

TWELVE

Mastering 3D Solids

- The Union and Subtract Commands for Complex Models

- Primitive Solids—Box, Cylinder, Torus, Sphere, Cone, and Wedge

- Creating a Solid with Revolve

- Creating and Viewing Slides

- Using Slice to Edit a Solid and Section to Copy a Cross Section

- Regions and Boolean Operations

- Editing Solids

- Using the Clipboard with Solid Models

- 3D Solid Model to 2D Conversion Tools—Solview, Soldraw, and Solprof

You should now have a basic understanding of the power of 3D solids. The parts that you constructed in Chapter 11, "Introducing 3D," can be used to create an arbor press. In this chapter, you'll generate more parts for the press. Chapter 11 included the box and the cylinder primitives, and this chapter will illustrate the rest of the primitive solids. In Chapter 11, you learned about Shade and Hide as techniques for making your 3D drawing appear more realistic. Next, you'll learn some more subtle 3D effects.

First you'll put together two solids. Then you'll try revolving and sweeping closed polylines to create solids. You'll learn to make two solids from one using the Slice command; you'll see how to find out if two solids are occupying the same space by doing some interference checking; and you'll try rendering to see the work that you've done. Along the way, you'll learn about making slides, saving and viewing slides, and how to make and use other important snapshots of your work. You'll make a useful 2D drawing from a 3D solid image, and you'll learn how to manage these drawings. Finally, we'll look at exporting these files to in-house and out-of-house fabrication shops.

Putting Together Two Solids

You've already used the Union command to add material as you built a model. Now you'll use the Union command to edit two existing models. The base and the post from Chapter 11, "Introducing 3D," need to be joined because the stress on the bolt would be enormous if the press were used for heavy-duty work. The hole for the bolt needs to be removed, the clearance between the base and the post needs to be filled, and the screw should be moved away from the assembly.

1. Open the file Press.dwg that you saved in Chapter 11. If you've left it hidden or shaded, type **shademode↵3↵** to see the 3D wire frame. If you don't have this file or if you aren't sure that it's correct, you can start with the file named Ch12pres.dwg from the companion CD-ROM.

2. Type **m↵** to begin the Move command. At the Select objects: prompt, pick the screw. You can use selection cycling, or zoom in, to make it easier to pick the screw.

3. At the Select objects: prompt, press ↵ to complete the selection. Then at the Specify base point or displacement: prompt, type **4<90↵↵** to move the bolt 4 units on the Y axis. You've moved the bolt away from the base.

4. You'll join the post and base next. Type **union**⏎ to begin the Union command. At the `Select objects:` prompt, pick the post. At the next `Select objects:` prompt, pick the base, and at the next `Select objects:` prompt, press ⏎ to complete the selection set. The two parts are now one. Notice that there are some gaps on either side of the post, and that the holes for the screw are still there. Your drawing should look like Figure 12.1.

FIGURE 12.1:

The union of the post and base

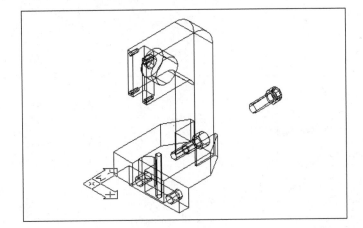

5. Use the Zoom Window tool to get a better look at the area between the base and the post.

6. The spaces to either side of the post need to be filled in. You can use the Box tool to create a solid that will fill in the spaces, and then union them to the new base/post. Type **box**⏎ and at the `Specify corner of box or [CEnter]` `<0,0,0>:` prompt, use the Endpoint Osnap to find the lower-left corner of the cutout in the base. At the `Specify corner or [Cube/Length]:` prompt, pick the diagonal corner. See Figure 12.2 for these points. You can barely see any change in your drawing because you just drew a box the exact same size as the gap between the parts. Copy the new filler piece to the other side of the base to fill it in also.

TIP   If you override the layer color with a contrasting color before you create either solid they will be easier to see and manage.

FIGURE 12.2:

Picking the point for the box on the base/post, and the base/post after the union

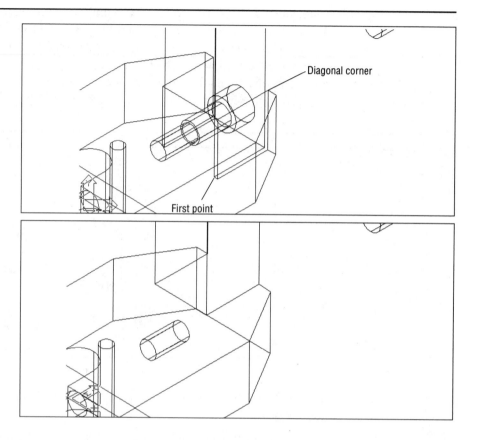

7. Type **union**↵ to begin the Union command. At the Select objects: prompt, pick the post/base, and at the next Select objects: prompt, type l↵ to use the last object created. At the next Select objects: prompt, press ↵ to complete the selection set.

Your drawing should look like the bottom of Figure 12.2. The spaces, counterbore, and the through hole that were in the post are gone. Part of the cylinder that was the tapped hole in the base still remains, and you'll need to fill it. AutoCAD doesn't check to see if it's possible to make a part, so you must be aware of the feasibility of what you draw; the hole you're looking at is actually a void in the middle of the part. You must fill the hole so that AutoCAD can return the accurate properties of the part, and so that the 2D drawing won't include hidden lines that represent the void.

1. Type **cylinder**↵, and at the Specify center point for base of cylinder or [Elliptical] <0,0,0>: prompt, use the Osnap marker to find and pick

the center of either end of the cylinder that you want to fill. At the `Specify radius for base of cylinder or [Diameter]:` prompt, use the Quadrant Osnap to pick any of the four quadrants of the existing cylinder. If you remember the size of the existing cylinder, you can type in a diameter the same as, or slightly larger than, the diameter of the existing cylinder. The size of the cylinder must be larger but does not have to be exact to fill in the void.

2. At the `Specify height of cylinder or [Center of other end]:` prompt, type **c↵** to specify the location of the other end, rather than the height of the cylinder. You could give the location by coordinates, or use the geometry on the screen. At the `Center of other end:` prompt, use the Center Osnap to locate and pick the other end of the existing cylinder.

3. Union the base/post and the cylinder. Type **uni↵**. At the `Select objects:` prompt, pick the base/post, and at the next `Select objects:` prompt, type **l↵** to select the cylinder (which was the last object created). At the `Select objects:` prompt, press ↵ to complete the selection set and the command.

TIP   You could have saved a step if you had made a single box in the previous step large enough to fill both slots and the hole also.

Using the Fillet Tool with Solids

A fillet would greatly improve the strength of the base/post.

1. Type **f↵** and at the `Current settings: Mode = TRIM, Radius = 0.3800 Select first object or [Polyline/Radius/Trim]:` prompt, pick the line/edge between the front of the post and base.

2. At the `Enter fillet radius <0.3800>:` prompt, type **.38↵** and at the `Select an edge or [Chain/Radius]:` prompt, press ↵ to complete the selections and complete the command. Your drawing will look like Figure 12.3.

3. Save your work. If you opened this drawing from the CD-ROM, save your work as `Press.dwg`.

You can use the same Boolean operations to add and subtract complex models as well as primitives. Once the model is whole, the same editing commands that you used in Chapter 11, "Introducing 3D," can be used on more complex modes.

FIGURE 12.3:

The fillet between the front of the post and the base

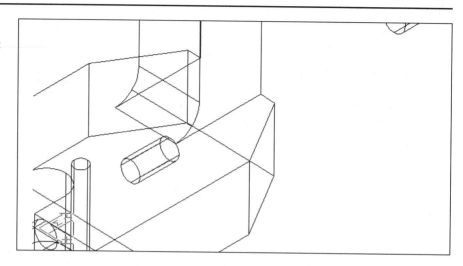

Drawing a Few More Parts

Next you'll draw a few new parts very quickly, using the techniques you learned in the last chapter. If you've forgotten how to use a command or tool, you can review the appropriate chapter, or use the AutoCAD Help feature by choosing the question mark (?) icon after you've begun a command.

We don't have space here to draw all of the parts of the press. The CD-ROM contains the missing parts' drawings that you'll need to complete the press assembly. The next part to construct is called a *table*. This part will sit on the base and will have four cutouts. See Figure 12.4 for the table's dimensions.

1. Start a new drawing using the Versions template. If AutoCAD is already started, type **new**↵. Although you could keep the press drawing open while starting a new drawing, you will find it advantageous to close it and free up some system resources. If you haven't saved your changes, at the Save changes to Press.dwg? prompt, type **yes**↵.↵.

2. Click the Save tool in the Standard toolbar to give your new drawing a name. When the Save Drawing As dialog box opens, type **table** in the File Name input box and click the Save button. A new file named Table has been generated.

3. Change the viewpoint to the Isometric view so that you can more easily see and edit the table. Click the SE Isometric tool in the View toolbar.

FIGURE 12.4:

The table with dimensions
for the press

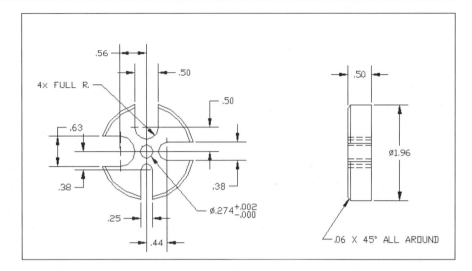

4. See Figure 12.4 for the dimensions of the table. Start by drawing a cylinder at coordinate 2.5,2.5. Make the cylinder 2 units in diameter and 0.5 units in height by typing **cylinder**↵, followed by **2.5,2**↵, **d**↵, **2**↵, and finally, **.5**↵.

5. To create the hole at the center of the table, draw another cylinder .274 units in diameter and 1 unit high. Type **cylinder**↵ and use the Center Osnap to find the bottom center of the existing cylinder. Type **d**↵, then **.274**↵, then **1**↵.

6. To subtract the second cylinder from the first cylinder, click the Subtract tool in the Solids Editing toolbar or type **su**↵. At the `Select solids and regions to subtract from...Select objects:` prompt, pick the first cylinder. At the `Select solids and regions to subtract...Select objects:` prompt, pick the large cylinder and then type **l**↵ to select the last object created. Finally, hit ↵ to finish selecting objects and execute the command.

You've just created the basic shape of the circular table for the press. If you want to see how this part will be used in the assembly of the press, you can look at the figure in the last section of this chapter, "If You Want to Experiment...."

Using the Chamfer Tool on a Solid

Because this is a hand tool and sharp edges tend to cut hands, it's a good idea to add a chamfer around the top of the table.

The Chamfer and Fillet tools are able to discern the difference between 2D and solid objects. Solids require the object's edge be selected. Notice the different prompts for the solid.

1. Type **cha↵**, click the Chamfer tool in the Modify toolbar, or choose Modify ➢ Chamfer to begin the Chamfer command.

2. At the (TRIM mode) Current chamfer Dist1 = 0.5000, Dist2 = 0.5000 Select first line or [Polyline/Distance/Angle/Trim/Method]: prompt, pick the circle that is the top of the table. The surface—not just the line—is highlighted. The Base surface selection.. Enter surface selection option [Next/OK (current)] <OK>: prompt confirms that the correct surface has been selected. Press ↵ for OK if the circle is highlighted.

3. At the Specify base surface chamfer distance <0.5000>: prompt (unless the prompt already is <0.0600>), type **.06↵** to chamfer the surface 0.06 units along the top surface.

4. At the Specify other surface chamfer distance <0.5000>: prompt, type **.06↵** to create a 0.06 × 0.06 chamfer.

5. At the Select an edge or [Loop]: prompt, pick the top circle again to define the edge that you want to chamfer. At the Select an edge or [Loop]: prompt, press ↵ to finish selecting edges and execute the command.

Now you need to make four cutouts. Use the Polyline tool to make the first shape.

1. To draw the small cutout, draw a closed polyline that is a little longer than it needs to be. Type **pl↵**, and at the Specify start point: prompt, type **from↵**, and at the Base point: prompt, Shift+right click, select the

Quadrant Osnap, and pick the front quadrant of the lower circle. See Figure 12.5 for this point. Next enter the distance from the quadrant to begin the polyline. At the <Offset>: prompt, type **@.125,.375**↵.

FIGURE 12.5:

The table with the polyline drawn under it

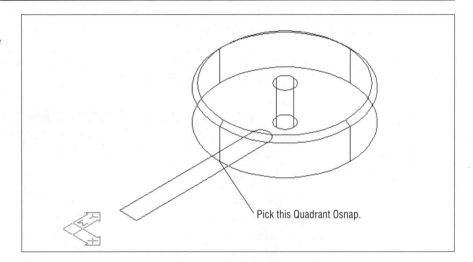

Pick this Quadrant Osnap.

2. Draw the polyline counterclockwise from this point. (However, you could draw it in either direction.) Type **@2<270**↵ at the Specify next point or [Arc/Close/Halfwidth/Length/Undo/Width]: prompt.

3. At the Specify next point or [Arc/Close... prompt, type **@.25<0**↵. At the next Specify next point or [Arc/Close... prompt, type **@2<90**↵.

4. To close the polyline and draw the arc, at the fourth Specify next point or [Arc/Close... prompt, type **a**↵ to draw an arc rather than another line, and at the Specify endpoint of arc or [Angle/CEnter/CLose/Direction/Halfwidth/Line/Radius/Second pt/Undo/Width]: prompt, type **cl**↵. You've drawn the arc and closed the polyline. Your drawing will look like Figure 12.5.

Next, make the polyline into a solid using the Extrude tool.

1. Type **ext**↵, and at the Select objects: prompt, pick the polyline.

2. To make the cut-out three dimensional, give a 1-unit height at the Specify height of extrusion or [Path]: prompt by typing **1**↵, and at the Specify angle of taper for extrusion <0>: prompt, press ↵ to make the sides of the cut-out perpendicular to the top and bottom of the table.

Using the Polar Array Tool with Solid Objects

The table drawing (Figure 12.4) shows four evenly spaced cut-outs around the edge of the table. You used a rectangular array in Chapter 5, "Editing for Productivity." Here is an opportunity to use the same technique to draw a polar array; this time you'll use solid objects. The Array command's polar option creates evenly spaced copies of objects in a circular pattern around a point which you specify.

1. To start the Array command type **ar.⏎**. At the `Select objects:` prompt, pick the extruded polyline and press ⏎.

2. At the `Enter the type of array [Rectangular/Polar] <R>:` prompt, type **p⏎** to begin a polar array, and at the `Specify center point of array:` prompt, use the Center Osnap to find and pick the center of the table.

3. You'll need four of the extruded polylines, so at the `Enter the number of items in the array:` prompt, type **4⏎**.

4. At the `Specify the angle to fill (+=ccw, -=cw) <360>:` prompt, your choice is to accept the default (360°) or enter an angle less than 360°. Choosing the default results in four extruded polylines equally spaced along the arc equal to the angle. Let's use the default: Press ⏎.

5. At the `Rotate arrayed objects? [Yes/No] <Y>:` prompt, you can accept the default or enter **n**. Type **n⏎**. Three copies of the original solid are created and are not rotated. Entering No (n) results in four copies equally spaced on the circumference of the circle. AutoCAD measures the distance and angle from the center of the circle to the beginning point of the polyline. The distance and angle are used to create and locate three more solids. Each copy will be oriented the same as the original. The result of such an array of objects is not intuitive. You should try this to better understand it and then undo it to continue the tutorial.

6. Undo the unrotated array. Go back through steps 1–4 and then press ⏎ to accept the default. Three copies of the original solid are created and rotated so that each one is in a good orientation.

7. Save your work. Your drawing should look like Figure 12.6.

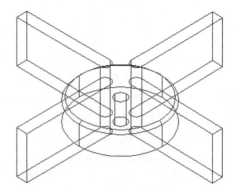

You now have four equally sized solids that you'll use to cut out the four solids on the table. Next, you need to change the size of the solids to agree with the four different-sized solids shown in the dimensioned drawing (Figure 12.4). The original solid is the correct size, so only the other three need adjustment. The first solid located counterclockwise from the original should be .375 units across, or 1.5 times larger, and 0.062 units farther from the center of the table. The next solid should be two times larger and 0.125 units farther from center. The third solid should be 2.5 times larger and 0.188 units farther from center. Let's make the changes.

Using the Scale Tool to Work on Solids

Let's use the tools in AutoCAD that allow us to make proportional changes.

1. Type **sc↵** to start the Scale command, click the Scale tool in the Modify toolbar, or select Modify ➤ Scale from the menu bar.

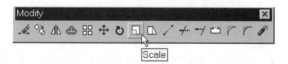

2. At the Select objects: prompt, select the first solid going counterclockwise from the original. Press ↵.

3. At the Specify base point: prompt, use the Center Osnap to locate and pick the center of the arc of the polyline. The scale value will be applied from this point in all directions, which will leave the center located where it

is now. If you were to pick any other point, the solid would scale from that point and move the center. (You might try choosing different points to see what happens, and then use Undo when you're finished.)

4. At the `Specify scale factor or [Reference]:` prompt, type **1.5.⏎** to get a shape 1.5 times larger than the original.

5. Type **m⏎** to begin the Move command. At the `Select objects:` prompt, pick the larger solid and press ⏎.

6. At the `Specify base point or displacement:` prompt, type **.0625<0⏎⏎** to move the solid 0.0625 units to the right to agree with the dimensioned drawing.

Make the same changes to the next solid (the one pointing in the Y axis away from the original solid), except this solid will be twice the width and .125 units farther from the center. Type **sc⏎**, and select the third solid.

1. At the `Specify base point:` prompt, use the Center Osnap to find and pick the center of the arc on the solid. At the `Specify scale factor or [Reference]:` prompt, type **2⏎**.

2. Type **m⏎** to move this solid into position.

3. At the `Select objects:` prompt, pick the new, larger solid and press ⏎. At the `Specify base point or displacement:` prompt, pick any point on the screen, and at the `Specify second point of displacement:` prompt, type **@.125<90⏎**.

For the last solid, the one that is 2.5 times the size of the original and .188 units farther from the center, follow these steps:

1. Type **sc⏎** to begin the Scale command. At the `Select objects:` prompt, pick the left-facing solid and press ⏎.

2. At the `Specify base point:` prompt, pick the center of the arc of the solid. At the `Specify scale factor or [Reference]:` prompt, type **2.5⏎**.

3. Type **m⏎** to start the Move command, and at the `Select objects:` prompt, pick the last solid that you scaled and press ⏎.

4. At the `Specify base point or displacement:` prompt, pick any point on the screen. At the `Specify second point of displacement:` prompt, type **@.188<180⏎**.

Now you need to subtract the solid shapes from the table.

1. Type **su**⏎ to begin the Subtract command. At the `Select solids and regions to subtract from` prompt, pick the table cylinder. At the `Select solids and regions to subtract` prompt, pick the four solids; press ⏎ to finish. Your drawing should look like Figure 12.7.

FIGURE 12.7:

The completed table

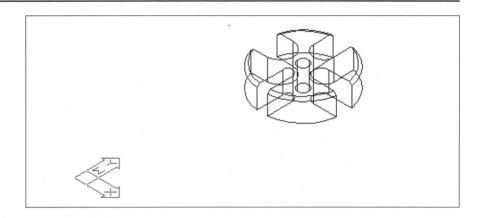

2. Save your drawing. You can take a breather now if you like.

Using the Revolve Tool to Make New Parts

Start a new drawing using the Versions template and name it `Revolutions` when you save it. You'll be drawing more of the press' parts: a sleeve, a button, an end cap, and a special screw. You'll draw the sleeve using a rectangle (a closed polyline). The button, end cap, and special screw will be constructed of lines, arcs, and circles gathered into closed polylines and revolved to make them solid.

Let's start by making the rectangle, which you'll revolve to become the sleeve.

1. Type **rec**⏎ to begin the Rectangle command. You could also click the Rectangle tool in the Draw toolbar or choose Draw ➢ Rectangle from the menu bar.

2. At the `Specify first corner point or [Chamfer/Elevation/Fillet/Thickness/Width]:` prompt, type **3,1**⏎ to begin the rectangle in the lower-left corner of the drawing space. At the `Other corner:` prompt, type **@.208,1.68**⏎ to draw the profile of the cross section of a cylinder that is 0.75 ID × 1.166 OD × 1.68 L. The arithmetic for the size of the rectangle is (OD–ID)/2.

3. Type **cha**↵ to chamfer the bottom-right corner by 0.06 × 45 degrees. At the `(TRIM mode) Current chamfer Dist1 = 0.5000, Dist2 = 0.5000 Select first line or [Polyline/Distance/Angle/Trim/Method]:` prompt, type **d**↵ because the distances are the same for 45°, and accepting the defaults in the distance version of the prompts will require less typing. At the `Specify first chamfer distance <0.5000>:` prompt, type **.06**↵, and at the `Specify second chamfer distance <0.0600>:` prompt, press ↵ to accept the default. You have now set the distance that will be used at the next Chamfer command.

4. Press ↵ to begin the Chamfer command again. After the `CHAMFER (TRIM mode) Current chamfer Dist1 = 0.0600, Dist2 = 0.0600 Select first line or [Polyline/Distance/Angle/Trim/Method]:` prompt, pick the bottom line. At the `Select second line:` prompt, pick the right-side line.

Using the Revolve Tool to Make Complex Round and Cylindrical Solid Objects

The Revolve command is similar to the Extrude command, but instead of extruding the shape in a straight line, you define an axis that is the center of the revolution; AutoCAD revolves the shape about the axis to form a solid. In this case, the axis is located .375 units from the left side of the polyline, and becomes the center line of the sleeve. The chamfered rectangle revolved around the axis defines a hollow cylinder.

1. Type **rev**↵, click the Revolve tool from the Solids toolbar, or choose Draw ➢ Solids ➢ Revolve to begin the Revolve command.

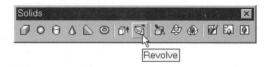

2. At the `Select objects:` prompt, pick the polyline and press ↵.

3. At the `Specify start point for axis of revolution or define axis by [Object/X (axis)/Y (axis)]:` prompt, use the From Osnap. At the `Base point:` prompt, use the Endpoint Osnap to pick the endpoint of the polyline at the lower-left corner. At the `<Offset>:` prompt, type **@.375<180**↵ to begin the axis definition .375 units to the left of the lower-left corner. At

the <End point of axis>: prompt, drag the mouse up (ensure that Ortho mode is on) and pick a point or type **@1<90**↵. This will define an axis parallel to the rectangle and .375 units to the left.

4. At the Angle of revolution <full circle>: prompt, press ↵ to accept the default. You could draw an arc less than a full circle by entering an angle less than 360°.

5. Click the SE Isometric tool in the View toolbar or type **vpoint**↵. At the Current view direction: VIEWDIR=0.0000,0.0000,1.0000 Specify a view point or [Rotate] <display compass and tripod>: prompt, type **1,–1,1**↵.

The sleeve has a radial hole through it that is 1.178 units from the far end. See Figure 12.8 to better understand this verbal picture.

FIGURE 12.8:

The completed sleeve model

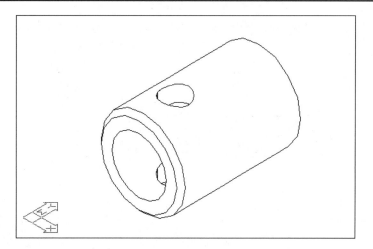

1. Type **cylinder**↵ to begin the Cylinder command, and at the Specify center point for base of cylinder or [Elliptical] <0,0,0>: prompt, use the Quadrant Osnap by typing **qua**↵ or Shift+right-click to use the pop-up menu. At the qua of: prompt, pick the lower quadrant of the circle defining the end of the cylinder farthest from you (opposite the chamfered end).

TIP   If you aren't sure which end is which, use the Hide command (**hi**↵). Hide is not a transparent command, so you'll have to escape the Cylinder command to use it.

2. At the `Specify radius for base of cylinder or [Diameter]:` prompt, type **d**↵ and at the `Specify diameter for base of cylinder:` prompt, type **3/8**↵ to set the interior dimensions of the hole.

3. At the `Specify height of cylinder or [Center of other end]:` prompt, type an arbitrary number, such as **2**↵.

4. Type **m**↵ to move the cylinder (hole) 1.178 units from the end of the sleeve. At the `Select objects:` prompt, type **l**↵ to use the last object created. At the `Select objects:` prompt, press ↵ to complete the selection set. At the `Specify base point or displacement:` prompt, pick any point on the screen, and at the `Specify second point of displacement:` prompt, type **@0,–1.178**↵.

5. Type **su**↵ to begin the Subtract command. At the `Select solids and regions to subtract from` prompt, pick the sleeve. At the `Select objects:` prompt, press ↵ to complete the command.

6. At the `Select solids and regions to subtract` prompt, pick the cylinder. At the `Select objects:` prompt, press ↵ to complete the selection set.

7. Change the *Dispsilh* system variable to display the silhouette only. Type **dispsilh**↵, and at the `Enter new value for DISPSILH <0>:` prompt, type **1**↵.

8. Issue the Hide command by typing **hi**↵ to see the finished sleeve. Your drawing should look like Figure 12.8.

9. Save your work.

Using the Torus Tool

The torus (donut) shape may be unfamiliar to you. This shape can be very useful in the creation of complex geometry. In this exercise, you're going to draw a button with the inverse of a fillet cut out of it. The dimensioned drawing in Figure 12.9 describes the shape.

NOTE   The following exercise is only one of many ways to create this shape. Another way would be to draw half of the outline as a closed polyline, as you did with the sleeve, and then revolve the shape.

FIGURE 12.9:

The dimensions for the button

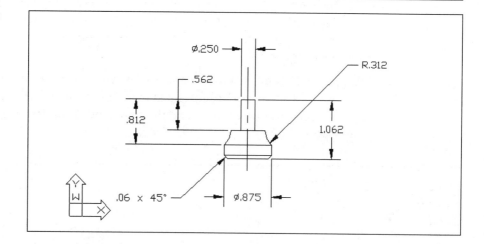

1. Type **cylinder.⏎** to begin the Cylinder command, and at the `Specify center point for base of cylinder or [Elliptical] <0,0,0>:` prompt, type **1,1⏎** to place the center of the cylinder.

2. At the `Specify radius for base of cylinder or [Diameter]:` prompt, type **7/16⏎**.

3. At the `Specify height of cylinder or [Center of other end]:` prompt, type **.5⏎**.

4. Click Torus in the Solids toolbar, select Draw ➤ Solids ➤ Torus, or type **tor⏎** to begin the Torus tool. Look at the icon on the toolbar button if this donut shape is unfamiliar to you.

5. At the `Center of torus <0,0,0>:` prompt, use the Center Osnap to find and pick the top center of the cylinder.

6. At the `Diameter/<Radius> of torus:` prompt, type **.625⏎**. This is the dimension to the center of the tube, not the outside diameter.

7. At the `Diameter/<Radius> of tube:` prompt, type **5/16⏎** (as on the dimensioned drawing).

TIP

You might have some trouble seeing this object. Take a look at it from some different angles using the View toolbar, and selecting the Front, Side, and Top tools. Return to the SE Isometric view at the end of your exploration.

8. Next you'll subtract the torus from the cylinder. Type **su↵**, press ↵, and at the SUBTRACT Select solids and regions to subtract from: prompt, pick the cylinder. At the Select objects: prompt, press ↵ to complete the selection set.

9. At the Select solids and regions to subtract: prompt, type **l↵** to use the last object created, and at the Select objects: prompt, press ↵ to complete the selection set and the command. Your drawing should look like Figure 12.10.

FIGURE 12.10:

The cylinder with the torus subtracted

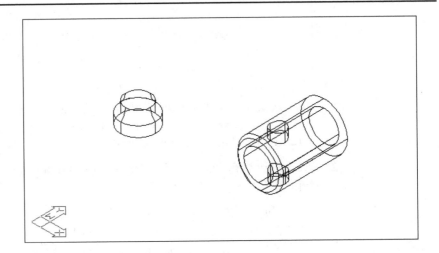

Next, you'll add the rest of the detail to the button. The bottom of the button gets a chamfer and the top has a pin projection.

1. Type **cha↵** to begin the Chamfer command.

2. At the (TRIM mode) Current chamfer Dist1 = 0.0600, Dist2 = 0.0600 Select first line or [Polyline/Distance/Angle/Trim/Method]: prompt, pick the circle at the bottom of the cylinder. At the Base surface selection Enter surface selection option [Next/OK (current)] <OK>: prompt, remember that the highlighted surface is the base surface and that this is how the distance dimensions are applied. In this case, the two distances are the same, so either surface is acceptable.

3. At the `Specify base surface chamfer distance <0.0600>:` prompt, press ↵ only if your default is 0.06. If it's not, type **.06**↵ to match Figure 12.9. At the next prompt, `Enter other surface distance <0.0600>:`, do the same thing: Press ↵ to accept a default of 0.06 or enter **.06**↵ if you have a different default distance.

4. At the `Select an edge or [Loop]:` prompt, pick the bottom edge again to tell AutoCAD which edge you want to chamfer. There is only one edge on the surface you selected, but there could have been many, and AutoCAD needs to know which one you want to use.

5. At the `Loop/<Select edge>:` prompt, press ↵ to tell AutoCAD that you've finished selecting edges.

This also completes the command, and you'll see the chamfered edge. Next you need to draw the pin projection.

1. Type **cylinder**↵ to begin the Cylinder command and draw the pin projection.

2. At the `Specify center point for base of cylinder or [Elliptical] <0,0,0>:` prompt, use the Center Osnap marker to find and pick the center of the top of the button. At the `Specify radius for base of cylinder or [Diameter]:` prompt, type **d**↵.

3. At the `Specify diameter for base of cylinder` prompt, type **.25**↵ for the dimension (from Figure 12.9) and at the `Specify height of cylinder or [Center of other end]:` prompt, type **9/16**↵ (also from Figure 12.9).

Let's put the chamfered button and the pin together.

1. Type **uni**↵ to start the Union tool, and at the `Select objects:` prompt, pick the button and the cylinder. Then, at the `Select objects:` prompt, press ↵ to complete the selection and complete the command. Your drawing should look like Figure 12.11.

2. Save your work now.

Next you'll use the sphere primitive to construct the end cap.

FIGURE 12.11:

The completed button

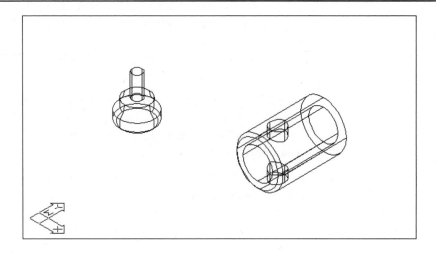

The Sphere Tool

Let's try a new tool to make a **9/16** diameter spherical end cap to be pressed onto the press' handle bar.

1. Click the Sphere tool in the Solids toolbar, or select Draw ➤ Solids ➤ Sphere from the menu bar, or type **sphere.⏎** to begin the Sphere tool. Look at the icon in the toolbar button if this ball shape is unfamiliar to you.

2. At the Specify center of sphere <0,0,0>: prompt, type **1,3.⏎** to locate the sphere.

3. At the Specify radius of sphere or [Diameter]: prompt, type **d.⏎** to enter the diameter of the sphere.

4. At the Specify diameter: prompt, type **9/16.⏎** to draw a 9/16-unit diameter sphere.

The sphere is truncated 0.21 units from the center and a 0.4-unit deep hole is drilled from the flat into the sphere. One way to create the flat is to draw a box using the Center Placement option of the Box command and then move the box into position to subtract it from the sphere.

1. Type **box.⏎** to begin the Box tool. At the Specify corner of box or [CEnter] <0,0,0>: prompt, type **c.⏎** to use the Center option to locate the box. At the Specify center of box <0,0,0>: prompt, use the Center Osnap to pick the

center of the sphere. You want the box located in a known position relative to the sphere so that it is easy to move the box in relationship to the sphere. If you had not placed the box using the sphere, you would have had to move it into a known, identifiable position; this would require a number of additional steps.

2. To use the Cube option of the Box tool, at the `Specify corner or [Cube/ Length]:` prompt, type **c**↵ and at the `Specify length:` prompt, type **.75**↵ to create a cube 0.75 units on a side. You could use any value to define the cube (but the math in step 4 would have to change to suit a different value).

TIP

When you created the cube in step 2, you created a tool to slice off one side of the sphere and make one surface flat. You did this because it is very hard to drill into a small, spherical object without it rolling.

---

3. The box completely covers the sphere. Type **m**↵ to start the Move tool, and at the `Select objects:` prompt, type **l**↵ to move the last object created. At the `Select objects:` prompt, press ↵ to complete the selection set.

4. At the `Specify base point or displacement:` prompt, pick any point on the screen, and at the `Specify second point of displacement:` prompt, type **@.585<0**↵ to move the box 0.375 (half the length of a side of the box) + 0.21 (the distance from the center of the sphere to the flat). The box is now positioned to slice off a portion of the sphere.

TIP

You could take a quick look from the top view (Top tool in the View toolbar). If you do, be sure to return to the SE Isometric view so that your drawing will look the same as the figures.

---

5. Type **subtract**↵ to subtract the box from the sphere and create the flat.

6. At the `Select solids and regions to subtract from...` `Select objects:` prompt, pick the sphere, and at the `Select objects:` prompt, press ↵.

7. At the `Select solids and regions to subtract...Select objects:` prompt, pick the box, and at the `Select objects:` prompt, press ↵. The box disappears and you have only the remaining flat on the sphere.

8. Use the Cylinder command to create the solid to subtract from the sphere that creates the hole. Type **cylinder**↵ and at the `Specify center point for base of cylinder or [Elliptical] <0,0,0>:` prompt, use the Center

Osnap to pick the center of the flat. At the `Specify radius for base of cylinder or [Diameter]:` prompt, type **d**↵ and at the `Specify diameter for base of cylinder:` prompt, type **.25**↵.

9. At the `Center of other end/<Height>:` prompt, type **c**↵ to enter the location of the center of the other end of the cylinder. At the `Center of other end:` prompt, type **@.4<180**↵ to define the center of the other end as a point .4 units in the –X axis.

TIP

If you were using the WCS (look at the icon if you aren't sure), the cylinder would have been drawn with the axis in the positive Z direction. However, the Center option of the Cylinder command allows you to define the center as any coordinate in any direction.

---

10. Subtract the cylinder from the sphere. Type **subtract**↵ and at the `Select solids and regions to subtract from...Select objects:` prompt, pick the sphere. At the `Select objects:` prompt, press ↵ and at the `Select solids and regions to subtract...Select objects:` prompt, pick the cylinder. At the `Select objects:` prompt, press ↵.

11. Save your work and close this drawing.

TIP

If you like, take a look at a Hidden Line view of your work. Type **hi**↵ to use the Hide command.

That's it. You've finished the sleeve, button, and handle end cap. Your drawing should look like Figure 12.12.

FIGURE 12.12:

The completed sleeve, button, and handle end cap

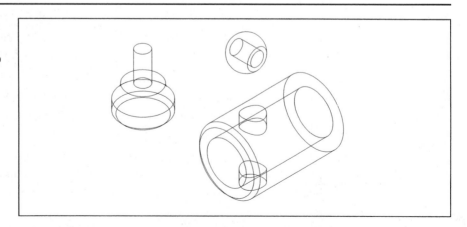

NOTE   We used the Cylinder, Array, Subtract, and Osnap commands that you've already learned to make the screw holes and the cover. If you wish to see how this was done, the images are included on the CD-ROM. Try it yourself and compare your results to the drawing called Ch15press.dwg on the CD-ROM.

Making Slides to Improve Communication

A *slide* is a presentation tool that allows you to share the work you've done. It can be viewed by anyone with a copy of AutoCAD Release 12 or later. A slide cannot be edited, measured, zoomed, or panned. However, even with these limitations, it is a very good tool for informing others of the current status of your work, or for providing a historical record of the status of a specific drawing at a specific time.

The Mslide and Vslide Tools

Mslide and Vslide are tools that make and view an externally saved file with the extension .sld, which is a snapshot (raster image) of your work. The slide file is independent of the current drawing file, and it can be viewed at any time from within any AutoCAD editing session with Vslide command. In this exercise, you'll zoom the extents of the press, change the UCS to View so that you can write an easy-to-read note, hide the view to get rid of any hidden lines, and make a slide. Then you'll zoom into the face of the press so that you are looking at a different zoom scale. Finally, you'll view the slide.

First, let's zoom the extents of the press and place a note.

1. Open Ch12press2.dwg from the companion CD-ROM.

2. Type z.⏎e.⏎, to zoom the extents of the drawing.

3. Type ucs.⏎ to change the UCS to a view in which the text will be easy to read. At the Enter an option [New/Move/orthoGraphic/Prev/Restore/ Save/Del/Apply/?/World] <World>: prompt, type v.⏎.

4. Type dt.⏎ to begin the Dtext tool.

5. At the Current text style: "Standard" Text height: 0.2000 Specify start point of text or [Justify/Style]: prompt, pick a point above

and to the left of the press. At the `Height <0.2000>:` prompt, type **.25**↵ (your prompt may have a different default value). At the `Specify rotation angle of text <0>:` prompt, press ↵ unless your default is something other than 0.

6. At the `Enter text:` prompt, type **This is the press before adding the rack, pinion, and the press parts.**↵↵.

7. The UCS icon is in the way of the drawing, so let's turn it off by changing its setting to No Origin. Type **ucsicon**↵ and at the `Enter an option [ON/OFF/All/Noorigin/ORigin] <ON>:` prompt, type **n**↵.

8. Type **hi**↵. The hidden lines are gone, but the text remains. Your drawing should look like Figure 12.13.

FIGURE 12.13:

The press with a note

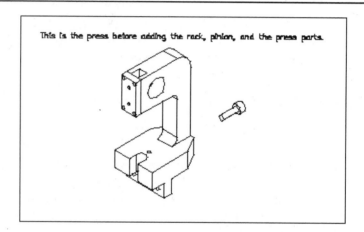

This is the press before adding the rack, pinion, and the press parts.

Making a Slide

Now let's make a slide and view it.

1. With the press and note drawing on your screen, type **mslide**↵. (There aren't any toolbar or menu shortcuts for this command.)

2. In the File dialog box, click Save to accept the default filename (the default slide name is the same name as the current drawing, with the extension .sld). The location of the slide file will be the same as the drawing file unless you change the location in the dialog box. You can always move the file later.

3. To view the slide, type **vslide⏎**. In the File dialog box, the default slide is the one that matches the name of the file that you're working in. Accept the default by clicking Open. The slide should look like Figure 12.13.

4. To return to your drawing, type **r⏎** or run any command (such as Redraw) that causes a regeneration.

5. Save your work.

For a slide presentation, you'll want to group your slides and run them automatically. You can do with with the Script command, which is described in Chapter 19, "Introduction to Customization."

More Primitives

You've probably noticed that there are two more solid primitives that we have not yet used or talked about. The Wedge and the Cone are tools that are easy to define now that you've used the other primitives. You'll find a use for them from time to time. The cone could be the point on a set screw or the nose cone on a rocket. The wedge can be used to add inclined surfaces to solids, and as the name implies, to create a wedge that takes up tolerance in a design.

The Wedge Tool

The Wedge tool can be started by clicking the tool in the Solids toolbar, by choosing Draw ➢ Solids ➢ Wedge from the menu bar, or by typing **wedge⏎**.

The `Specify first corner of wedge or [CEnter] <0,0,0>:` prompt gives you the same options as the Box tool to either locate a corner or the center of the wedge. The `Specify corner or [Cube/Length]:` options allow you to define the wedge as a cube, which really means the length, width, and height are the same. You do this either by defining lengths or by selecting points for the opposite corners and then giving the height. The length is defined by the size in the current X axis, the width is defined by the size in the current Y axis, and the height is defined by the size in the current Z axis. Figure 12.14 shows the wedge drawn squarely in a cube, and a variation with a length of 2, a width of 1, and a height of 0.5.

FIGURE 12.14:

A wedge drawn as a cube
in a 2-unit cube where
Length=2, Width=1, and
Height=.5

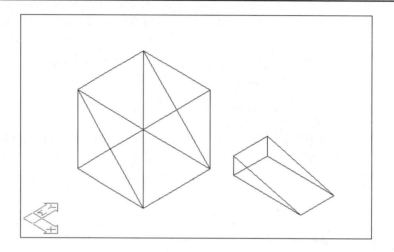

FIGURE 12.14:

A wedge drawn as a cube
in a 2-unit cube where
Length=2, Width=1, and
Height=.5

The Cone Tool

The Cone tool resembles the Cylinder tool in the same way that the Wedge tool resembles the Box tool. Many of the same options appear when you use this tool. The Cone tool is found in the Solids toolbar, by choosing Draw ➤ Solids ➤ Cone from the menu bar, or by typing **cone**↵.

The Specify center point for base of cone or [Elliptical] <0,0,0>: prompt allows you to create a cone based on an ellipse or a circle. Pick a point for the center of the default circle. At the Specify radius for base of cone or [Diameter]: prompt, type your choice. At the Specify height of cone or [Apex]: prompt, type your choice of height (where the height is a point along the Z axis in the current UCS), or enter a point that will be the apex (it can be any point in space). If you use the Apex option, the end of the cone will be perpendicular to the axis. This system works in the same way as the Cylinder tool when you select the Center option.

If you select the Elliptical option from the Specify center point for base of cone or [Elliptical] <0,0,0>: prompt, the rest of the prompts look a little different from the circular option because first you must define the ellipse. You'll see the following prompts:

- Specify axis endpoint of ellipse for base of cone or [Center]: (to pick a center point or axis endpoint and the default)

- Specify second axis endpoint of ellipse for base of cone:

- Specify length of other axis for base of cone: (to complete the definition of the ellipse)

- Specify height of cone or [Apex]: (to determine whether the height is along the Z axis or perpendicular to the axis)

Figure 12.15 shows a circular cone and elliptical cone.

FIGURE 12.15:

A circular cone drawn with a 1-unit radius and a 2-unit height, and an elliptical cone drawn from the center option with a 1-unit first axis endpoint, a .5-unit other axis endpoint, and a 1-unit height

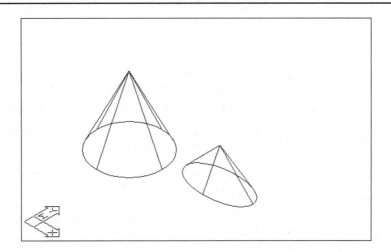

Editing Solids with Slice and Solidedit

Although you've been editing solids all along, the Slice tool may be one of the most valuable for editing. The Slice tool allows you to define a plane and cut a shape in two along the plane. You can choose whether to save the two parts or to discard one of them.

The post on the press needs to be lengthened .5 units. This would be very difficult without the Slice tool. You would have to delete the top or the bottom section and redraw it. So let's try lengthening the post using the Slice tool.

1. If the press file is still open, you can continue with this exercise. If not, or if you aren't sure that it's right, open the one on the CD-ROM called C12 slice.dwg.

2.   Click the Slice tool in the Solids toolbar, choose Draw ➤ Solids ➤ Slice from the menu bar, or type **slice**↵.

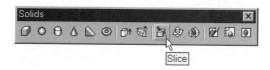

3.   At the Select objects: prompt, pick the post/base and at the next Select objects: prompt, press ↵.

4.   At the Specify first point on slicing plane by [Object/Zaxis/ View/XY/YZ/ZX/3points] <3points>: prompt, you have several options:

> **Object** allows you to define a plane by the object that you select. Any object that describes a plane is eligible. See Chapter 11 for some examples.
>
> **Zaxis** defines the cutting plane by a specified origin point on the Z axis (normal) of the X, Y plane.
>
> **View** defines a plane parallel to the current view, and the point that you pick is a point on that plane.
>
> **XY/YZ/ZX** define the cutting plane parallel to the plane of the axes indicated and the point that you pick is a point on that plane.
>
> **3points** allows you to identify three specific points to form your own plane. The first point is the origin, the second point is a point on the X axis, and the third point is a point on the Y axis.

5.   For this exercise, press ↵ to use the 3points (default) option. At the Specify first point on plane: prompt, pick the midpoint of one of the lines from the vertical section of the post. At the Specify second point on plane: prompt, with Ortho mode on, drag the mouse to the right and pick a point below and to the right of the original point. At the Specify third point on plane: prompt, drag the mouse and pick a point to the right and above the original point.

6.   At the Specify a point on desired side of the plane or [keep Both sides]: prompt, you have a choice of saving either side or both. Type **b**↵ to save both sides. (You would point to the side you wanted to preserve if you only needed half of your drawing.) A square appears at the point of the slice. Although it is not obvious, you now have two separate solids.

7. Type **m**↵ to start the Move tool and at the Select objects: prompt, pick the top half of the post. At the Select objects: prompt, press ↵. At the Specify base point or displacement: prompt, pick any point on the screen, and at the Specify second point of displacement: prompt, type **@0,0,.5**↵. Now your two solids are easy to identify. Your drawing should look like Figure 12.16.

FIGURE 12.16:

The post sliced in two and separated

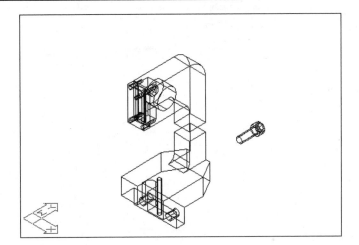

Now, let's add a section to extend the length of the post. You'll use the Solidedit command to create a solid to fill in the gap between the two parts. The Solidedit command has many options, which are also available as tools on the Solids Editing toolbar. In this exercise, you will use the Face and Extrude options. The other options are discussed in the sidebar at the end of this section.

1. Type **Solidedit**↵. At the Enter a solids editing option [Face/Edge/ Body/Undo/eXit] <eXit>: prompt, type **f** [cr].

2. At the Enter a face editing option [Extrude/Move/Rotate/Offset/Taper/Delete/Copy/coLor/Undo/eXit] <eXit>: prompt, type **e**[cr] to select the Extrude option.

3. The next prompt is Select faces or [Undo/Remove]:. Select the exposed portion of the press, where you made the slice. You'll see 1 face found and then Select faces or [Undo/Remove/ALL]:. Press [cr] to complete this portion of the command.

4. At the `Specify height of extrusion or [Path]:` prompt, type **1**[cr]. The face will be extruded 1 inch upward.

5. At the `Specify angle of taper for extrusion <0>:` prompt, press [cr] to accept the default of no taper. Press [cr] again to complete the command.

6. To union the two parts, type **uni**↵. At the `Select objects: Other corner:` prompt, pick all three parts. At the `Select objects:` prompt, press↵. Now the post/base is one piece again, but longer. Your drawing should look like Figure 12.17.

FIGURE 12.17:

The post/base united but longer

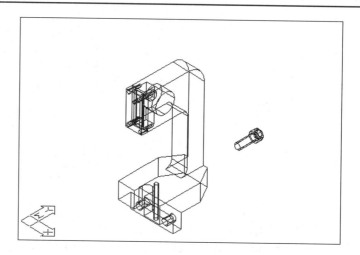

7. Save your work. If you opened this file from the CD-ROM, save your work to your hard disk as `Press.dwg`.

Editing Solids

AutoCAD 2000 provides the Solids Editing toolbar for easy access to tools for editing solids. For example, you could have used the Extrude Faces tool in the previous exercise.

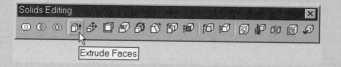

Continued on next page

The Solidedit command (and corresponding Solids Editing toolbar tools) allows you to perform a variety of editing operations, such as extruding, moving, rotating, offsetting, tapering, and deleting faces and edges. After you enter **Solidedit.⌐**, you see the prompt

```
Enter a solids editing option [Face/Edge/Body/Undo/eXit] <eXit>:
```

You then choose the Face, Edge, or Body option to edit the various parts of a solid.

If you select the Face option, you can edit the selected 3D faces. You're prompted

```
Enter a face editing option
[Extrude/Move/Rotate/Offset/Taper/Delete/Copy/coLor/Undo/eXit]
<eXit>:
```

If you select the Edge option, you can copy individual edges or change their color. You're prompted

```
Enter an edge editing option [Copy/coLor/Undo/eXit] <eXit>:
```

Selecting the Body option lets you modify the whole solid object. For example, you can change its geometry or separate it into individual objects.

The Section Tool

The Section tool creates a region object on the cutting plane that you define. The region created by the section cut will only have the shape of the sections it actually touches. The region can be hatched as a separate command. If you use Section to create cross-sectional views, you'll need to add missing geometry to complete the section.

You'll find that you can draw sections very well using other tools. However, the Section tool allows you to capture geometry that you can use to create additional solids. You also can isolate areas that may not be clear after numerous edits and Boolean operations.

To explore the Section tool, let's take a section of the press assembly. We won't go over the methods for selecting the plane because they are the same as those for Slice. The result, as mentioned above, will be an object called a *region*, which will be discussed in the next section.

Don't save any of the work that you do in this exercise.

1. Click the Section tool in the Solids toolbar, or choose Draw ➤ Solids ➤ Section from the menu bar.

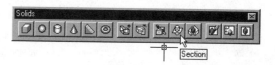

2. At the `Select objects:` prompt, use a crossing window to select the post/base and the cover.

3. At the `Select objects:` prompt, press ↵.

4. At the `Specify first point on Section plane by [Object/Zaxis/View/XY/YZ/ZX/3points] <3points>:` prompt, type **yz**↵ so you can pick a point and generate a plane on the Y, Z axes of the current UCS.

5. At the `Specify a point on YZ plane <0,0,0>:` prompt, use the Midpoint Osnap to pick the midpoint of one of the top lines that you can see on the cover. This is a point through which the Y, Z plane will pass and at which the section will be created. See Figure 12.18 to help you find the midpoint.

FIGURE 12.18:

The section of the post/base, cover, and gib

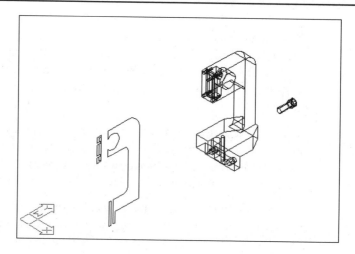

6. Type **z↵** and adjust Zoom so that you can see the three sections.

7. Move these new sections so you can see them. Type **m↵** to start the Move tool, and at the `Select objects:` prompt, pick the new sections with a crossing window, and then press and hold the Shift key while you select the press and cover objects. This will remove them from the selection. If you don't use the preceding method, they will be difficult to pick. You can also use object cycling to pick the sections. Once you have them, at the `Select objects:` prompt, press ↵.

TIP   Prior to making the section, you could have created a new layer named Section, set its color to something that stood out, and made it current. The section would be created on this layer, and would have been easier to see. Also, if you wished to use the region made by the section operation, you could have aligned the UCS to the location you wished to use, prior to using Section. This would make it very easy to capture the section as a block object. Any block object, when inserted into your drawing, will remember its relation to the current UCS when created and inserted. So, you could insert the section block object to any UCS setting, or you could insert it into Paper Space.

---

8. At the `Specify base point or displacement:` prompt, pick any point on the screen, and at the `Specify second point of displacement:` prompt, type **@10<270↵**. You've just used the displacement method to move the selected objects. AutoCAD adds what you've typed to the current coordinates, and moves the objects to the resulting new coordinate.

9. Zoom All to see the move. Your drawing will look like Figure 12.18.

Don't save these changes.

The Region—A 2D Object for Boolean Operations

The sections you created (and didn't save) are objects called *regions*. A region acts like a surface in that it is a closed area that will obscure objects behind it, but unlike a surface, a region can be trimmed or edited with other regions. A region can be extruded to become a 3D solid. When the region is exploded it becomes a collection of line and arc objects. All of the Boolean operations will work with regions.

Let's look at what would happen if we were to add a rectangle 1 unit wide and 2 units high to the post section of our press drawing.

We're just going to take a quick look at Boolean operations and regions, so don't save any of the work from this exercise.

1. Type **ucs.↵r↵s↵** to restore the UCS you saved in the press drawing in Chapter 11 called *s* for *side*. If you didn't save it, be sure that the UCS is set to World and use the UCS presets to change the UCS to *s*.

2. Draw a rectangle near the section of the post. Type **rectang↵**.

3. At the Specify first corner point: prompt, pick any point on the screen.

4. At the Specify other corner point: prompt, type **@1,2↵**.

5. Type **m↵** to move and align the rectangle, and at the Select objects: prompt, type l↵. At the next Select objects: prompt, press ↵ to complete the selection set.

6. At the Specify base point or displacement: prompt, pick the midpoint of the left side of the rectangle, and at the of Specify second point of displacement: prompt, pick the midpoint of the right side of the post.

7. You need to pick one more point to move the rectangle 0.125 past the edge of the post and create a notch. Press ↵ to reissue the Move command, and at the Select objects: prompt, type l↵ to use the last object. At the Select objects: prompt, press ↵.

8. At the Specify base point or displacement: prompt, type **.125<180 ↵↵**. Your drawing should look like Figure 12.19.

The section and rectangle

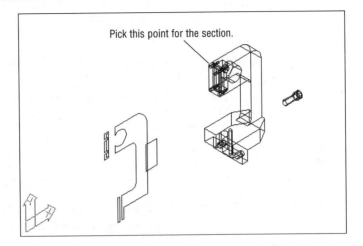

Pick this point for the section.

If Your UCS Doesn't Look the Way You Expected

If your WCS or UCS icon doesn't look like the one in Figure 12.19, take the following steps:

1. Type **ucs**⏎ to return to the World UCS from wherever you are.

2. Press ⏎ again to reissue the last command, then type **x**⏎ to rotate the UCS about the X axis.

3. Type **90**⏎ to rotate the UCS 90°.

4. Press ⏎ again to reissue the last command, and then type **s**⏎ to save the UCS.

5. Type **s**⏎ to save a UCS named *s*.

You'll need to make the polyline rectangle into a region before you can try out some of the Boolean operations.

1. Click the Region tool in the Draw toolbar, select Draw ➤ Region, or type **reg**⏎.

2. At the `Select objects:` prompt, pick the rectangle, and at the `Select objects:` prompt, press ↵. You'll see the validation `1 loop extracted. 1 Region created`.

3. Type **union**↵ to union the section region and the rectangle, and at the `Select objects:` prompt, pick both regions. At the next `Select objects:` prompt, press ↵. You've added the two shapes together. Your drawing will look like the top half of Figure 12.20.

4. Type **u**↵ to undo the last command, so you can try a subtraction instead of a union.

5. Type **subtract**↵ to subtract the rectangle from the section.

6. At the `Select solids and regions to subtract from...Select objects:` prompt, pick the section, and at the `Select objects:` prompt, press ↵.

7. At the `Select solids and regions to subtract...Select objects:` prompt, pick the rectangle. At the `Select objects:` prompt, press ↵. Now your drawing should look like the bottom half of Figure 12.20.

Don't save the changes you've made. You could also have tried the Intersection tool to save only the part of the two regions common to both of them.

FIGURE 12.20:

The section and the rectangle. The top is after a Boolean union, and the bottom is after a Boolean subtract.

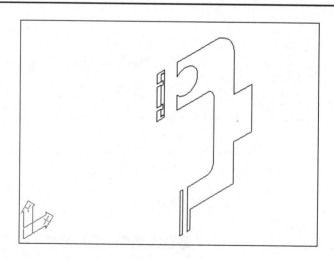

FIGURE 12.20 CONTINUED:

The section and the rectangle. The top is after a Boolean union, and the bottom is after a Boolean subtract.

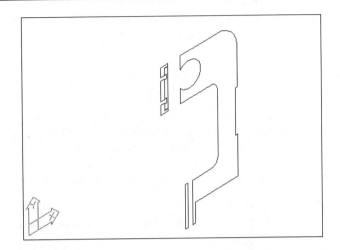

Interference Checking—Looking for the Not-So-Obvious

Interference checking is a buzz phrase today. In industry, the reference is to assembly tolerances and interference. In AutoCAD, the interference you can check for is caused by overlapping of the objects in ways you didn't anticipate. You may have noticed that you haven't applied any tolerances to the size of your solids. This is because you can't. AutoCAD has no capacity to do this. However, you can check the nominal fit of your parts.

So far, you've drawn parts that fit. Now you'll move the cover 0.125 units into the press to create an obvious interference between the cover and the post of the press, and let AutoCAD check for it. Restore the last saved version of the press, and let's see how interference checking works.

1. Type **open**⏎, and in the Select File dialog box, be sure that the filename is Press.dwg. If you can't use yours, you'll find one on the CD-ROM called ch12 Interfere.dwg.

2. Type **m**⏎ to start the Move command. At the Select objects: prompt, pick the cover. Press ⏎ at the next Select objects: prompt.

3. At the `Specify base point or displacement:` prompt, pick any point on the screen, and at the `Specify second point of displacement:` prompt, type **@.125<90**↵.

This move has created a definite interference between the post and the cover.

NOTE   If you had not moved the cover, there would have been no interference and Auto-CAD would have returned a null finding. This wouldn't have made a very interesting exercise. If you had made some error earlier in the tutorial, the Interfere tool would find it for you.

---

1. Start the Interfere tool by clicking Interfere in the Solids toolbar, by selecting Draw ➤ Solids ➤ Interfere, or by typing **interfere**↵.

2. At the `Select the first set of solids: Select objects:` prompt, pick the post, and at the `Select objects:` prompt, press ↵.

3. At the `Select the second set of solids: Select objects:` prompt, pick the cover, and at the `Select objects:` prompt, press ↵.

4. You'll see the following validation:

   ```
   Comparing 1 solid against 1 solid. Interfering solids (first set):
   1 (second set): 1
   Interfering pairs: 1
   ```

5. At the `Create interference solids? [Yes/No] <N>:` prompt, type **y**↵.

NOTE   You don't want to create the solid but simply to check for interference, but if you don't let AutoCAD make the solid, all you'll know is that the two objects interfere. You won't know *how* they interfere.

---

6. Type **m**↵ to move the new solid so that you can see it. At the `Select objects:` prompt, type **l**↵, and at the next `Select objects:` prompt, press ↵.

7. At the B `Specify base point or displacement:` prompt, pick any point on the screen, and at the `Specify second point of displacement:` prompt, type **@3<180**↵. You can now see the solid, which should look like Figure 12.21.

FIGURE 12.21:

The interference solid generated by the Interfere tool

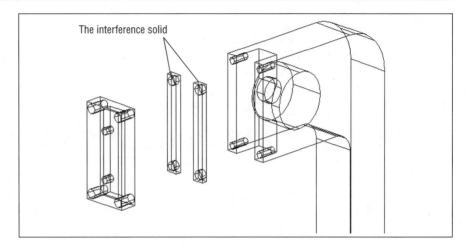

The interference solid

The block created by the Interfere tool could be used to remove material from one of the parts in order to make them fit, or it could just be used to create new solids. Move the cover back to the original position, erase the interference solid, and check for interference again.

1. Type **m.⏎** to move the cover 0.125 units in the Y axis. At the `Select objects:` prompt, pick the cover, and at the `Select objects:` prompt, press ⏎. At the `Specify base point or displacement:` prompt, pick any point on the screen, and at the `Specify second point of displacement:` prompt, type **@.125<270.⏎**.

2. Type **e⏎**, and at the `Select objects:` prompt, pick the interference solid. At the `Select objects:` prompt, press ⏎.

3. Type **interfere.⏎** or click the Interfere tool in the Solids toolbar.

4. At the `Select the first set of solids: Select objects:` prompt, pick the post and at the `Select objects:` prompt, press ⏎.

5. At the `Select the second set of solids: Select objects:` prompt, pick the cover, and at the `Select objects:` prompt, press ⏎.

6. You should see the following validation:

    `Comparing 1 solid against 1 solid. Solids do not interfere.`

7. Save your work. If you opened this file from the CD-ROM, save your work as `Press.dwg`.

NOTE
If you want to see what your work looks like now, skip ahead to the rendering section in Chapter 15, "3D Rendering in AutoCAD." Rendering consumes huge amounts of memory, so if you don't have a lot of time, save this slow task for later.

Enhancing the 2D Drawing Process

You've already created a solid from a 2D drawing in Chapter 11, "Introducing 3D." The technique in the next section is a good way to reuse any existing 2D drawing work you might have already drawn, rather than trying to redraw it using 3D techniques.

AutoCAD has provided a way to make 2D multiple-view working drawings from the solid models that you've created. Fabrication shops and inspection departments often require these drawings. Two-dimensional drawings are often used to estimate the cost of making a part, even if the part will be made from the data in the model. Many fabrication shops are only equipped to work with 2D drawings.

The next few exercises will show you one way to let your computer convert your 3D work into 2D drawings. AutoCAD will take care of all of the projections, view creation, layer creation, and layer management. The only bad news is that these drawings appear in Paper Space, which you might find difficult to manage at first.

You'll probably want to take a quick review of the discussion of Paper Space in Chapter 9, "Advanced Productivity Tools," before going on to make the 2D drawing view. Be sure to pay particular attention to viewport scale.

TIP
How do you know if you're in Paper Space or Model Space? If you're in Model Space, the cursor and UCS icon are visible in the viewport.

Copying Solid Models Using the Clipboard

You could create the 2D drawings of each part without the next step, but you'll find this process difficult to manage as the number of Paper Space viewports and layers increases. We recommend that you create a separate file for each part, which will keep the number of layers and viewports to a minimum. You could use the Wblock tool that you learned about in Chapter 4, "Organizing Your Work," to export the part to a new file and then open the new file to begin the 2D conversion. However, if you do this, you'll have to reset the system variables in the new file to be the way that you want them. Remember that the new file used the default system variables. Or, you could begin a new file using the template of your choice, and then insert the file created with the Wblock.

There is another alternative that requires fewer steps and won't clutter up the hard disk with unnecessary files. You can use the Copy to Clipboard and Paste from Clipboard tools.

There are several ways to use the Copy to Clipboard command: Click the Copy to Clipboard tool in the Standard toolbar (not the AutoCAD copy tool in the Modify toolbar), choose Edit ➤ Copy from the menu bar, press Ctrl+c, or type **copyclip**↵. Select the item you wish to duplicate, and the copy is stored temporarily as a block on the Clipboard.

NOTE   The *Clipboard* is a feature of the Windows 95/98/NT operating system and is available in most programs designed to run in Windows. The Clipboard is a temporary space and can only hold one item or selection set at a time. New items copied to the Clipboard will overwrite the existing items.

The next step is to open a file that is set up the way that you like—in our case this is the Versions template—and paste the contents from the Clipboard. Click the Paste tool in the Standard toolbar, choose Edit ➤ Paste from the menu bar, press Ctrl+v, or type **pasteclip**↵.

Let's try it.

1. Open the assembly of the press. If you don't think your version is correct or if you're just starting here, you can open the file named Ch12post.dwg on the companion CD-ROM.

2. Click the Copy to Clipboard tool in the Standard toolbar, choose Edit ➤ Copy, press Ctrl+c, or type **copyclip**↵.

3. Pick the post, then press ↵ to complete the selection set and execute the command.

NOTE   If you're a DOS holdout, you may be wondering why you can't use Ctrl+c to exit or escape from a command. The Windows operating system uses that particular combination to copy text and graphics to the Clipboard. AutoCAD is Windows-compliant, so if you try Ctrl+c, the Copyclip command will be issued.

4. Start a new drawing using the Versions template. Save the new drawing as Base.dwg.

5. To insert or paste the contents of the Clipboard into the base drawing, click the Paste tool in the Standard toolbar, choose Edit ➣ Paste from the menu bar, press Ctrl+v, or type **pasteclip**⏎.

6. At the Specify insertion point: prompt, type **4,4**⏎ to place the base point of the part at the coordinate 4,4. This is an arbitrary location that is some distance from 0,0. At the X scale factor <1>/Corner/XYZ: prompt, press ?⏎. At the Y scale factor (default=X): prompt, press ?⏎, and at the Rotation angle <0>: prompt, press ?⏎.

7. You have inserted a block that will not act like a solid until it is exploded. Use the Explode tool from the Modify toolbar and pick the base.

NOTE   A block is a special group of AutoCAD objects that cannot be edited (see Chapter 4). The Solprof, Solview, and Soldraw commands that you are about to use will not recognize solids grouped in a block. For these commands to function, a block must be exploded so that the solids can be edited and the block identity of the solids removed. Warning: The solid or mass identity will be lost if you explode the objects again.

NOTE   Edit ➣ Pasteclip does not ask for scaling, and does not insert blocks. If you use Pasteclip to bring in the post, and then test it with the List command, it will say that it is a 3D solid.

It's time to look at the post from an isometric viewpoint.

1. Click the SE Isometric View tool to enter this command. Your drawing should look like Figure 12.22.

2. Save this drawing as Base.dwg.

FIGURE 12.22:

The post/base in the new drawing

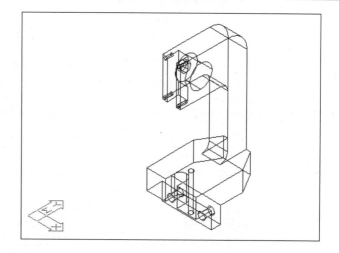

The Setup View Tools

The next step is to create and arrange the drawing views that you'll need for your multiple-view 2D drawing. You'll use the Solview tool for this purpose. Then you'll use the Soldraw tool to project the view onto a single plane, determine which lines are hidden lines and place them on a hidden-lines layer, and manage the viewport-layer visibility. Finally, you'll create a few dimensions for the part.

The Setup View Tool—Solview

Solview is used to create the views on your drawing. Next you'll change to Paper Space and create floating viewports easily with the aid of Solview.

1. Click the Setup View tool in the Solids toolbar or choose Draw ➤ Solids ➤ Setup ➤ View. If no Layouts pages have been set up yet, you'll open the Page Setup Layout dialog box. (A page setup allows you to prepare the parameters for the drawing layout, and ultimately the plot. This will be covered in detail in Chapter 14.) Select OK.

TIP   If you type **Solview.↵**, you'll bypass this step and go directly to the Solview command.

2. At the next prompt, `Enter an option, select. Use MVIEW to insert Model space viewports. Regenerating paperspace. Ucs/Ortho/Auxiliary/Section <eXit>:`, type **u↵** to use the UCS option.

TIP
Use a UCS to define the plane from which your first view will be generated. Throughout this book, we've used a convention of drawing the base with the bottom parallel to the WCS. If you wish to make this first view the top or plan view, it's a good idea to use the World UCS plane option.

---

3. You could choose either a named UCS or the current UCS (if the current UCS is different from the WCS). All of these options are designed to allow you to place your 3D object in this first view with the X and Y axes in a good relationship to the part. You'll be placing dimensions on the 2D drawing that will be oriented on the drawing relative to the UCS.

4. At the `Enter view scale<1.0000>:` prompt (you could enter any scale here—your drawing will work at any size, including full size), type **.5↵** to set it to half size.

5. At the `Specify view center:` prompt, pick a point approximately in the upper-right corner of the screen.

6. You're allowed to move the view around by moving its center. At the `Specify view center<specify viewport>:` prompt, press ↵ to complete the centering.

7. At the `Specify first corner of viewport:` prompt, pick a point near the middle of the screen.

8. At the `Specify opposite corner of viewport:` prompt, pick a point near the upper-right corner.

9. At the `Enter view name:` prompt, type **top↵**.

10. At the `UCSVIEW = 1 UCS will be saved with view. Enter an option [Ucs/Ortho/Auxiliary/Section]:` prompt, press ↵ to exit. Your drawing should look like Figure 12.23.

FIGURE 12.23:

The first view created by
Setup View

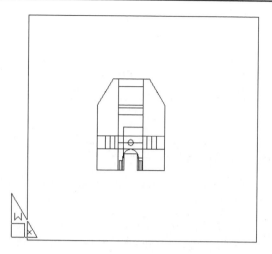

If you find that you don't have sufficient room for your view, right-click the Layout
tab at the bottom of your drawing viewport. In the Page Setup Layout dialog box,
you'll see a tab that allows you to choose the plot device. If you have other more
suitable choices for paper size, select one of them. You can also adjust the plot
scale setting in this dialog box.

When you exited the Solview command, AutoCAD placed you in Model Space,
within the newly created viewport, ready to work on the model. The first view
that you defined (a plan or top view) was placed on the screen arbitrarily. Click
Model in the status bar or type **ps.⏎** to enter Paper Space. Then type **m.⏎** to move
the viewport. Pick the viewport—not the objects in the viewport—and move the
view to a more appropriate place toward the upper-left corner of the screen.

Another item to clean up before continuing is an extra viewport. By default, the
Create Viewport in New Layouts option in the Display tab of the Options dialog
box is checked. Unless you have deselected this option, a new viewport with no
scale was automatically created when you moved into Paper Space. Erase this
viewport before proceeding with the next exercise. Type **e.⏎** and select the viewport that you didn't create by picking its corners. Press ⏎ to remove it.

You need to start Solview again to create a front view, a side view, and what will be a scaled isometric view.

1. Type **solview.**↵ and at the Enter an option Ucs/Ortho/Auxiliary/ Section]: prompt, type **o**↵ to create a new orthogonal front view.

2. At the Specify side of viewport to project: prompt, pick the bottom of the top view.

3. At the Specify view center: prompt, drag the mouse down about 5.5 units and pick this point.

4. At the next Specify view center: prompt, press ↵ to set the center of the view.

5. At the Specify first corner of viewport: prompt, pick a point approximately in line with the left side of the top viewport and very near the bottom of the screen.

6. At the Specify opposite corner of viewport: prompt, pick a point approximately in line with the right side of the top viewport and very near the bottom of the top viewport.

7. At the Enter view name: prompt, type **front**↵ to name this view. Auto-CAD will respond with UCSVIEW = 1 UCS will be saved with view.

8. At the Enter an option [Ucs/Ortho/Auxiliary/Section]: prompt, type **o**↵ to create the orthogonal side view.

9. Repeat the steps above, choosing the right edge of the front view for the side to project: and dragging the center of the new viewport to the right. Name the viewport **side**.

Creating an isometric view is a similar process. You'll need to make one adjustment to the UCS first. Click the Model tab to return to Tiled Model Space. If your Model Space view doesn't look like the SE isometric in Figure 12.22, select View ➢ 3D View ➢ SE Iso. Now set the UCS to align to the viewport by typing **ucs**↵**v**↵.

1. Start the Solview command. At the Enter an option [Ucs/Ortho/Auxiliary/Section]: prompt, type **u**↵. Use the current UCS.

2. At the Enter view scale <1.0000>: prompt, type **.5**↵ to set the view scale to half size.

3. At the Specify view center: prompt, drag the mouse and pick a point that looks suitable.

4. At the Specify view center: prompt, press ⏎ to set the center of the view.

5. At the Specify first corner of viewport: prompt, pick a point very near the top-right corner of the side viewport.

6. At the Specify opposite corner of viewport: prompt, pick a point very near the upper-right corner of the top viewport.

7. At the Enter view name: prompt, type **iso**⏎ to name this view.

8. That's it, your drawing should look like Figure 12.24.

9. Save your work.

FIGURE 12.24:

The drawing layout with all four views

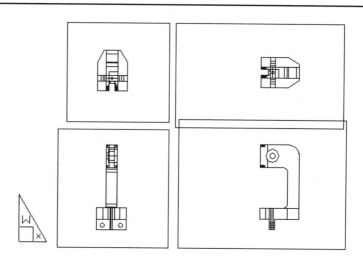

The Setup Drawing Tool—Soldraw

The arrangement of the views was a neat trick. Did you notice that all of the objects were aligned vertically and horizontally? The floating viewports are still showing the model. Next you'll want AutoCAD to create the 2D drawings. This is the job of the Setup Drawing tool. Let's see what it does.

1. Click the Setup Drawing tool in the Solids toolbar, choose Draw ➤ Solids ➤ Setup ➤ Drawing from the menu bar, or type **soldraw**⏎.

2. At the Select viewports to draw: prompt, pick the top, front, iso, and side views.

3. At the `Select objects: 1 found, 4 total Select objects:` prompt, press ↵. The hidden line views have been created.

TIP

You don't want your scaled and aligned views to change in scale and alignment unless you're controlling such changes. Remember not to move the viewports while working in Model Space on any of the viewports because this will change the alignment of the view. Once you realize you've changed the alignment of some viewports, use the Undo tool until you're back to the original arrangement, or realign the viewports by using the Move command. You can also lock the Zoom scale of any or all viewports. Simply start Mview, type **l** to select the Viewport Lock option. You'll see the `Viewport View Locking [ON/OFF]:` prompt. Type **on**↵ and then select the borders of the viewports you wish to (zoom) lock.

The Setup Drawing tool only works in viewports created with the Setup View tool.

The Setup Profile Tool—Solprof

The Setup Profile tool is a subset of the Setup View tool and you can create hidden line views with it. The Setup Profile tool works in any floating viewport, not just those created by the Setup View tool.

In the previous exercise, we finished the isometric viewport, but it is scaled too large for the sheet. We could have used this tool to create the isometric view also. Create a new floating viewport with the Mview tool. Locate it near the first isometric view. Erase the first isometric view. Look at the status bar. If the box next to LWT contains the word *Paper*, click to toggle to Model Space.

1. This tool doesn't use the UCS setting like Solview, so set the UCS back to World. Type **ucs**↵ to begin the UCS command. Pick in the Iso window, and at the prompt, press ↵ to use the default WCS.

2. Click the SE Isometric tool in the View toolbar to be sure the view is isometric.

3. The orthographic views are half scale. To make the isometric view half the scale of the other views, use the Zoom tool to set the scale of the viewport to 0.25 Paper Space. Type **z**↵, and at the prompt, type **.25xp**↵.

4. Click the Setup Profile tool in the Solids toolbar, choose Draw ➤ Solids ➤ Setup ➤ Profile, or type **solprof**↵.

5. At the `Select objects:` prompt, pick the solid (not the viewport).

6. At the `Select objects:` prompt, press ↵.

7. At the `Display hidden profile lines on separate layer? <Y>:` prompt, press ↵ to accept the default. If you answer No, you won't be able to manage the hidden lines with the Layer tool.

8. At the `Project profile lines onto a plane? <Y>:` prompt, press ↵ to accept the default. (If you answered No, AutoCAD will create a 3D wireframe copy of the solid model.) Projecting them onto a plane makes them an easier to manage, 2d "flat" projection.

9. At the `Delete tangential edges? <Y>:` prompt, press ↵ to accept the default, or AutoCAD can project each edge as many times as it finds one. Choosing Yes here simplifies the drawing should it be necessary to edit it. However, note that you may also loose the small lines that represent fillets and small chamfers.

10. You'll see the validation `One solid selected`, and the Setup Profile command is finished. Your drawing will look like Figure 12.25.

FIGURE 12.25:

The four views with an Isometric view

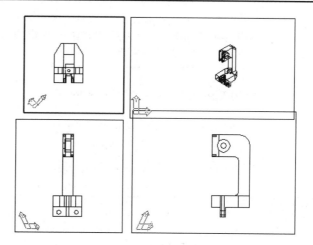

11. Save your work.

Managing the New Layers

These Setup tools have made a few not-so-obvious changes to your drawing. A number of new layers have been created and special properties have been assigned to them.

Click the Layer tool in the Object Properties toolbar or choose Format ➤ Layers from the menu bar to display the Layer Properties Manager dialog box. Each of the new views that you created using Setup View has three layers assigned to it, beginning with the view name. The layer called **hid** is where the hidden lines are, the **vis** layer is where the object lines are, and the **dim** layer is where you place the dimensions for that view. The **ph** (Profile Hidden) and **pv** (Profile Visible) layers were created by the Setup Profile tool. You may need to use the scroll bar to see all of the available layers

To see the isometric view correctly, you have to freeze the ph layer and also the 0 layer, which is now the current layer. You can't freeze the current layer, so you need to set another layer to be current. You also need to create some dimensions, so make the front-dim layer current by clicking it.

Adding Dimensions

Before placing a dimension in a viewport, make sure that you are in Model Space. If not, type **ms.↵** or click in the Model/Paper toggle box in the status bar.

TIP   You might find it easier to dimension the various views if, while in Paper Space, you use the Zoom tool to fill the screen with the view and then change to Model Space to pick and place the dimensions.

Next, prepare and place a few dimensions.

1. Decide which view you're going to dimension first and then use the layer drop-down list on the Object Properties toolbar to set the dim layer to current.

2. Set the UCS to the view of the viewport.

WARNING   If you attempt to pan or change the zoom scale while in Model Space, you'll receive a surprise. While a floating viewport is zoom locked, if you attempt to pan or zoom, all of the viewports in Paper Space will move instead. You may zoom or pan to your heart's content while in Paper Space with no damage.

3. Look at Figure 12.26 to place the dimensions as shown. If you aren't sure how to pick and place these dimensions or which commands to use, refer to Chapter 8.

4. Save your work.

FIGURE 12.26:

The part with some dimensions

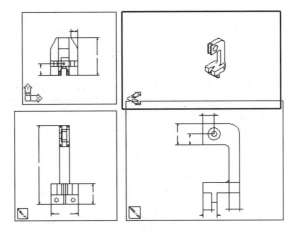

Finding the Mass Properties

You'll need to know some of the *mass properties* of your solid objects. Calculating mass can be done for 2D regions as well as 3D objects. The regions that you created and edited earlier in this chapter are candidates for mass properties.

To find the mass properties of a solid or region, use the Mass Properties tool. Click Mass Properties in the Standard toolbar (a flyout usually located under the Distance tool), choose Tools ➢ Inquiry ➢ Setup ➢ Mass Properties, or type **massprop.⌐**. Pick a solid (in this case, we chose the base).

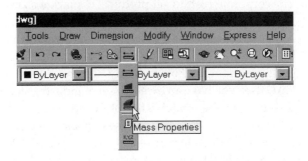

AutoCAD will automatically bring up the text screen and list all of the information shown in Table 12.1. At the end of the list, you'll see the prompt `Write analysis to a file? [Yes/No] <N>:`. If you answer Yes, you'll be able to save this information to a file for use in another document or spreadsheet.

TABLE 12.1: Mass Property Data Returned by the Mass Properties Tool

Property	Data
Mass	17.7384
Volume	17.7384
Bounding box	X: −0.3125–2.3125Y: 1.5000–5.0000Z: −1.3750–6.0000
Centroid	X: 0.9909Y: 3.5108Z: 1.8372
Moments of inertia	X: 387.8420Y: 175.5605Z: 258.7756
Products of inertia	XY: 61.7523YZ: 121.8875ZX: 31.8773
Radii of gyration	X: 4.6760Y: 3.1460Z: 3.8195
Principal moments and X-Y-Z directions about centroid	J: 99.0065 along [0.0081 0.9952–0.0975]K: 21.9833 along [−0.0047 0.0975 0.9952]

Taking Advantage of Rapid Prototyping

The corporate rush to get products to market as fast as possible has generated a demand to create prototypes from solid models just as quickly. *Rapid prototyping* has been used to describe a process of creating parts as solid models that are directly interpreted and fabricated by exotic and humble methods such as stereolithography (STL) and Computer Numerically Controlled (CNC) mill, respectively. As an Auto-CAD user, you have an excellent set of industry-standard tools to transfer your data directly from the model to the fabrication shop. This is euphemistically known as *art-to-part*. Many corporations use various forms of art-to-part to make prototypes—and even production parts—that have no conventional drawings to support them. Art-to-part is not yet considered a standard method of operation.

Here are some suggestions for solid modelers who wish to use rapid prototyping, or even use their models for production the same day.

- Draw all of your models to perfect nominal scale. Your model won't be evaluated by a human eye, and whatever discrepancies exist in the model will also exist in the product. If you're unable to draw your design in complete, accurate detail because of a limitation in the software, you'll have to note the problem area so that it can get special attention in the shop.

- Add as much information to the file as you think is necessary to build the part—and then add a little more. You can use leaders to point to surfaces requiring special finishes, such as with paint, without paint, ground, or ground to a specific finish. (You'll use leaders with an attribute in Chapter 17, "Storing and Linking Data with Graphics.") You can point to welds or even highlight surfaces and use GD&T symbology to define contour, texture, and tolerance.

- Remember that the solid model generates much less ambiguity in definition than conventional orthogonal-view drawings.

You'll want to export your solid and surface models to computer aided manufacturing (CAM) software that is the front end for rapid prototyping machines. A few of these systems can accept AutoCAD drawing files directly. These are rare because the drawing-file format of the AutoCAD drawing file is binary and very difficult to read. If the receiving software can't read the AutoCAD binary file, the file must be translated into a readable format.

Data Translation and AutoCAD

AutoCAD has three primary data translation tools: DXF, SAT, and stereolithography (STL). Each of these translators has a significantly different purpose.

> **NOTE** You may have heard of the International Graphics Exchange Standard (IGES) translation, or a vendor may have requested an IGES translation. Autodesk offers an add-on package for IGES translation.

This section is not intended to be an exercise in file-export options; rather, it is an exploration of the concept of importing files as it relates to rapid prototyping.

All of the File Export options are combined in a discussion of tasks in Chapter 18, "Getting and Exchanging Data from Drawings." Go there for specific information on the mechanics of importing and exporting models, text, and graphics.

Read on here to discover how file translation is applied to rapid prototyping.

The DXF Tool

By far, the most commonly used of the translators included with AutoCAD is the *DXF translator*. DXF stands for Drawing Interchange Format. This is an ASCII or binary format of an AutoCAD drawing file for exporting to other applications or for importing drawings from other applications. Such a file carries the file extension .dxf. This translator is capable of exporting 2D data, 3D surface data, and solid-model data in a simple-to-read ASCII format. The ASCII format can be viewed and edited in a text editor such as Write and Notepad. The DXF standard is published in the documentation that came with your copy of AutoCAD, so we won't go into it here. You need to know quite a lot about AutoCAD to edit a DXF file.

The SAT and ACIS Tools

At the core of AutoCAD solid modeling is the *ACIS solid-modeling engine*. ACIS, a solid modeler produced by Spatial Technology, Inc. and licensed by Autodesk, provides a solid modeling file format that AutoCAD users can use. AutoCAD reads the model stored in the ACIS file format and creates a region or solid body object in the AutoCAD drawing. This tool makes all of the wonderful solid models possible. ACIS provides a solid modeling file format that AutoCAD can use to store solid objects. An AutoCAD solid, a body, or a region can be stored as an SAT (ASCII) file. A translation of a drawing file into an SAT file is not strictly a translation; rather, it is an export of the ACIS data. Some users consider this lack of translation to be a great advantage because no errors induced by translation can affect the model. You could say that no translation occurred at all, but that the drawing data was stripped from the ACIS model. An SAT-translated file carries the file extension .sat, and it is not an editable or user-readable file. Like the AutoCAD drawing file, the SAT file is binary.

Unlike the DXF file, no other drawing data is exported. If you have placed annotation in this file, you'll have to export that data using another tool, and then have users merge the SAT file and the data-export files when they import these files into their systems.

One of the advantages of ACIS is that it is a standard engine used by a good number of CAD and CAM system manufactures.

Stereolithography (STL)

AutoCAD also allows for the export of AutoCAD drawing data in .stl format. The earliest version of the STL process used a vat of photosensitive plastic that hardened in the presence of a laser. When the machine read the file, it played a laser across the surface, creating a thin layer of solid plastic. The machine then lowered the solid by one-slice increments, and the laser hardened another layer. The process continued until the entire model was described, and the part was lifted from the vat and cleaned to provide a reasonable simulation of the finished part. Today, many machines use similar processes to create parts from materials ranging from paper to sintered metals.

Complex models can be created from a number of materials using processes similar to stereolithography, and many tools can accept the .stl file. The translation of a solid model into the .stl format creates a number of slices through your model at a pitch that you define. .stl files don't carry any annotation, so you'll have to provide supporting documentation.

> **NOTE**   Move your model to the positive octant prior to attempting to export an .stl file. And just what is the *positive octant*? Coordinates in which the X, Y, and Z values are all positive are in the positive quadrant.

Talking to the Shop

A lot of 2D data goes to fabrication shops in the form of 2D multiple-view drawings saved to disk and sent by e-mail, bulletin board, or floppy disk. If the shop can use your accurate drawing's geometry in their process of creating the program for the CNC, you'll save them time and reduce the possibility of making an error in transposing the information.

If you have drafting experience, you know that there can be a lot of ambiguity in an orthographically projected view using hidden lines and the silhouettes of parts to describe the shape of the part. These problems don't exist in 3D modeling because the CAM software creates the evaluated shape. However, you'll still need a creative approach to indicate thread size, pitch, and type. Tolerances and surface-contour specifications are not covered in any ANSI specification, so it's a good idea to keep a running dialogue going with the fabrication shop personnel, whether they are in-house or a vendor. If you and the shop are both proactive, you'll soon arrive at the appropriate methods for defining this information. For instance, you might use an inspection document, which can easily be created using the Solview and Soldraw commands to help you communicate with the fabrication shop. The inspection document would carry special instructions for the shop and assist your inspection department in interpreting the part.

3D and Solid Modeling Design Tips

The joy and the curse of solid modeling is the accuracy of the model. When you design a part, keep in mind that some kind of machine will have to make it. For instance, the cover that fits on the face of the post vplayer has a pocket for the gib. The pocket has square inside corners, and the outside has square corners all around. If the part were to be machined, the pocket could not be made as drawn. If the cover were fabricated by a mill, it would have either rounded corners or corner reliefs. The shop would certainly call, asking you which way to proceed.

As you'll recall from chamfering the edges of the table, you also have to keep in mind how the piece will be used once it is manufactured. Our press' cover on the face of the post might need chamfered outside corners because this is a hand tool, and you must consider how hands will be placed when someone is using the tool.

A machine shop is only one of many places where AutoCAD solid models can be found. The sheet metal shop, the tool shop, and the welding shop often see the work of AutoCAD solid modelers. Each of these specialties has certain standards and process limitations. For example, a precision sheet metal shop has a certain minimum-bend radius for the thickness and type of material; a need for corner relief to keep the material from tearing; and special features that can be made, such as half shears and extruded tapped holes. Each of these must be drawn and defined with the help of the shop.

If You Want to Experiment...

You've covered a lot of territory in this chapter, so it would be a good idea to play with these commands to help you remember what you've learned. Try finishing the dimensions in the 2D base drawing that you made.

Other options for creating views in setup view are a section view and an auxiliary view. Open the base drawing and use setup view and setup drawing to create two additional views. You'll have to change the limits to 44,34 (E size) to accommodate the new views.

You will want to add the rack, pinion, shaft collar for the pinion, handle and table pin, jack screws, nuts, and cover-mounting screws to make your drawing look like Figure 12.28.

FIGURE 12.28:

The complete press assembly

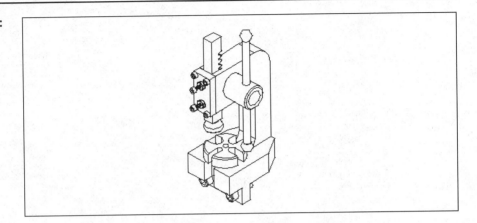

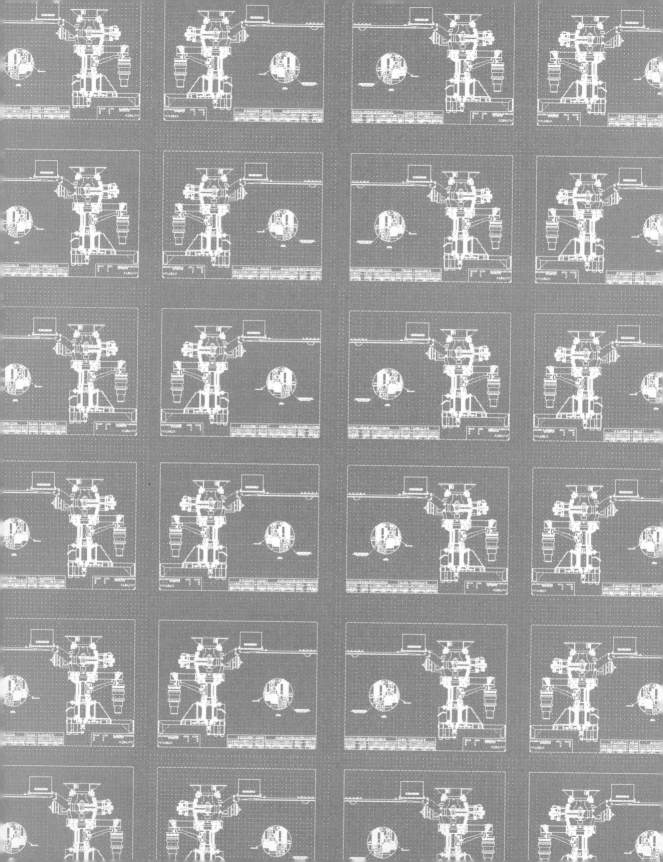

CHAPTER
THIRTEEN

13

Using 3D Surfaces

- Creating a Surface Model

- Getting the 3D Results You Want

- Drawing 3D Surfaces

- Creating Complex 3D Surfaces

- Other Surface Drawing Tools

- Editing a Mesh

- Moving Objects in 3D Space

- Viewing Your Model in Perspective

A *3D surface* is an object that can take any shape—from a simple plane to the convoluted undulations of a contour map. A surface has no thickness, only area, thus a surface has no volume. You can create surfaces that cannot be described by Auto-CAD solid modeling because they aren't solids—they're several combined surfaces.

For example, the elliptical shape of an airplane fuselage and the blend of the wing to the body require complex shape descriptions. The complex contours of many consumer products that have molded components, such as a computer keyboard, are good candidates for surface models. The hull of a boat or the body of a car are very difficult solid models because they aren't confined to a single plane, but these represent much less of a challenge to surface modeling than something with many parts. AutoCAD surface models, like their solid-model counterparts, are basic tools.

Creating a Surface Model

There are at least two ways to create surfaces using plain AutoCAD: by giving objects extrusion thickness (see Figure 13.1) and by using geometry-generated surface commands (shown later in this chapter).

FIGURE 13.1:

A cube created using thickness

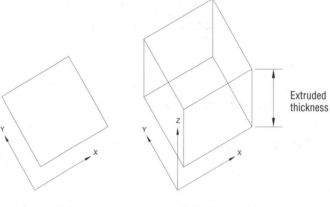

Square drawn with lines   Lines extruded to form cube

Extruded thickness

Another object property related to 3D is elevation. You can set AutoCAD so that everything you draw has an elevation. By default, objects have a 0 elevation. This means that objects are drawn on the imagined 2D plane of Model Space, but you can set the Z coordinate for your objects so that whatever you draw is above or below that surface. An object with an elevation value other than 0 doesn't rest on the imagined drawing surface but above it (or *below* it if the Z coordinate is a negative value). Figure 13.2 illustrates this concept.

FIGURE 13.2:

Two identical objects at different Z coordinates

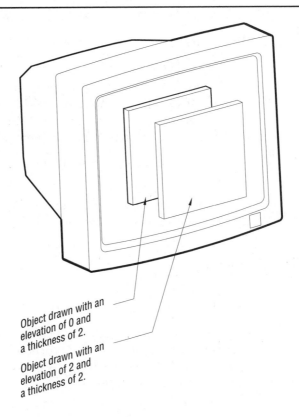

Object drawn with an elevation of 0 and a thickness of 2.

Object drawn with an elevation of 2 and a thickness of 2.

TIP

Thickness is measured from the base of an object, while elevation is the drafting term for the height of an object above the X, Y plane. If an object's thickness property is set to a negative value, the thickness will extend downward below its base.

Changing a 2D Polyline into a 3D Model

In this exercise, you'll change the thickness of a 2D polyline to create 3D surfaces by changing the polyline properties. You'll also create a new polyline by offsetting the first one and change the elevation of the new polyline.

1. Start AutoCAD, open a new file using the Versions template, and save it as surfaces. Turn off the grid if it's on.

2. Select View ➤ 3D Views ➤ SE Isometric from the menu bar.

3. Draw a rectangle 4 × 7 from point 2,2.

4. Click the Fillet tool in the Modify toolbar. At the Select first object or Polyline/Radius/Trim]: prompt, type r↵. At the Specify fillet radius <0.5000>: prompt, enter .56↵↵. At the Select first object or [Polyline/Radius/Trim]: prompt, type p↵ to use the Polyline option and pick the polyline. You have now filleted all four corners.

5. Click the Properties button in the Standard toolbar. Then click the polyline. Its grips appear.

6. In the Properties dialog box, double-click the Thickness input box, highlight the current thickness (which should be 0.000), enter **5.5**, and click OK.

7. You'll see that the polyline has tall sides. Press the Esc key twice to turn off the grips.

8. Type **hide**↵ to use the Hide command. You'll see that the polyline is a surface. It can obscure all that is behind it from your point of view.

9. To offset this polyline, click the Offset tool in the Modify toolbar. At the Specify offset distance or Through <0.000>: prompt, type .75↵. At the Select object to offset: prompt, pick the polyline. At the Specify point on side to offset: prompt, pick anywhere outside the polyline. At the Select object to offset: prompt, press ↵ to finish the command.

10. Type **hide**↵ to see the two polylines with the hidden lines removed.

11. Click the Properties button in the Standard toolbar. Then click the inside polyline.

12. In the Geometry section of the Properties dialog box, double-click the Elevation input box, highlight 0.0000, and type **2**↵. The polyline's elevation changes.

Figure 13.3 shows the polylines. You can see through the "surfaces" because this is a wire-frame view. Similar to solids, a wire-frame view displays the volume of a 3D object by showing the lines representing the intersections of surfaces.

FIGURE 13.3:

The two polylines after they have been given thickness, and one has had the elevation changed

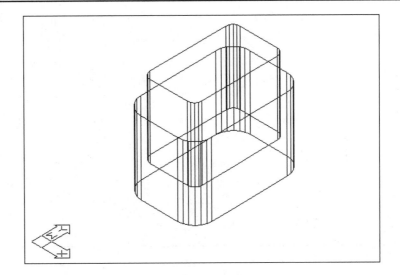

Creating a 3D Object

Although you may visualize a design in 3D, you'll often start sketching it in 2D and later generate the 3D views. When you know the thickness and elevation of an object from the start, you can set these values so that you don't have to extrude the object later. The following exercise shows you how to set elevation and thickness before you start drawing.

1. Choose Format ➤ Thickness.

2. Enter **1.2**⏎ at the Enter new value for THICKNESS <0.0000>: prompt. Now as you draw objects, they will appear 1.2 units thick.

3. Draw a circle next to the polylines at coordinate 9,2. Make your circle 3 units in diameter. The circle appears as a 3D object with the current thickness and elevation settings, as in Figure 13.4.

The circle and polylines

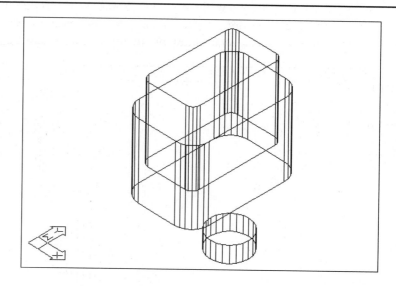

If you use the same thickness and elevation often, you can create a template file with these settings so that they are readily available when you start your drawings. The command for setting thickness and elevation is Elev. You can also use the Elevation and Thickness system variables to set the default elevation and thickness of new objects. Last but not least, the Properties dialog box will show general settings until an object is selected.

After you've set the thickness setting to 1.2, everything you draw will have a thickness of 1.2 units until you change it back to 0 or some other setting.

Giving objects thickness and modifying their elevation is a very simple process. These two properties let create quite a few three-dimensional forms.

Getting the 3D Results You Want

Working in 3D is tricky because you can't see exactly what you're drawing. Experienced AutoCAD users learned to alternately draw and then hide or shade their drawings from time to time to see exactly what is going on. That's changed in this version of AutoCAD. Here are a few tips on how to keep control of your 3D drawings.

Making Horizontal Surfaces Opaque

To make a horizontal surface appear opaque, you must draw it with a wide polyline, a solid hatch, a 3D Face, or a region. For example, consider the polylines you drew in the previous exercise for the changing elevation process; the shape formed by the two lines is open at both the top and the bottom. Only the sides of the polyline have become opaque. To make the entire top "surface" opaque, you can use a solid hatch or a region. The Hide command will make the top formed by the region and sides formed by the extruded polylines appear to be opaque (see Figure 13.5).

FIGURE 13.5:

The polyline before and after adding a region to the top

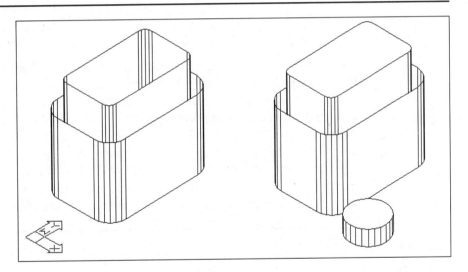

When a circle is extruded, the top surface appears opaque when you use the Hide command. Where you want to show an opening at the top of a circular volume, as in a circular chimney, you can use two 180° arcs, or the Donut command (see Figure 13.6).

To create more complex horizontal surfaces, you can use a combination of wide polylines, solids, and 3D faces.

FIGURE 13.6:

A circle and two joined arcs

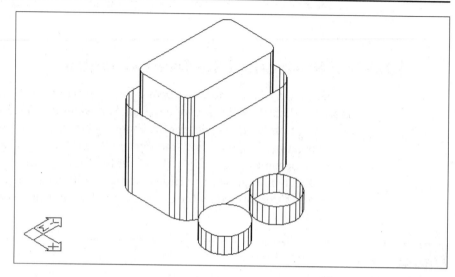

Careful layer control must be exercised. Bear in mind that the Hide command hides objects that are obscured by other objects on layers that are turned off. For example, if the inside polyline is on a layer that is either on or off when you use Hide, the lines behind the polyline are hidden, even if the polyline doesn't appear in the view (see Figure 13.7). However, you can freeze any layer containing objects that you don't want to affect the hidden-line removal process. You can also use solid model objects with the Hide command. In other words, create a 3D model with a mixture of solids and surfaces to draw complex 3D objects with holes and other features and they will all respond to the Hide command.

FIGURE 13.7:

A polygon hiding a line when the polygon's layer is turned off

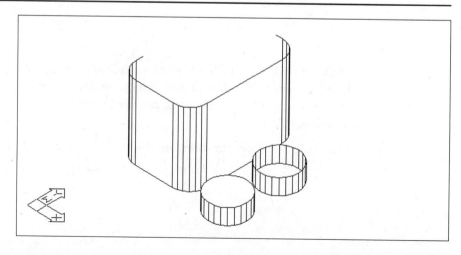

Drawing 3D Surfaces

In your work with 3D so far in this chapter, you've simply extruded existing forms, or you've set AutoCAD to draw extruded objects. However, extruded forms have their limitations. Using extruded forms only, it's hard to draw diagonal surfaces in the Z axis. AutoCAD provides the 3D face object to give you more flexibility in drawing surfaces in three-dimensional space. The 3D face produces a 3D surface where each corner can be given a different X, Y, and Z value. By using 3D faces in conjunction with extruded objects, you can create a 3D model of just about anything. When you view these 3D objects in a 2D plan view, you'll see them as 2D objects showing only the X and Y positions of their corners or endpoints.

> **TIP**   If you plan to view a 3D drawing in 2D at some later stage, you can use Dview (Dynamic View) or 3D Viewpoint presets.

Using Point Filters

Before you start working with 3D surfaces, you should have a good idea of what the Z coordinate values are for your model. The simplest way to construct surfaces in 3D space is to first create some layout lines to help you place the endpoints of 3D faces.

AutoCAD offers a method for 3D point selection called *point filtering*, which simplifies the selection of Z coordinates. Point filtering allows you to enter an X, Y, or Z value by picking a point on the screen and telling AutoCAD to use only the X, Y, or Z value of that point, or any combination of those values. If you don't specify a Z coordinate, the current elevation setting is assumed.

> **NOTE**   If you use filters, you'll be asked to supply any value that is not already set.

Laying Out a 3D Form Object

In the following exercises, let's imagine that you've been asked to design a decorative addition to a new product. The marketing head wants a five-pointed star projected to a point as a logo on the injection-molded front panel of the Mark Five widget. In creating the star, you'll practice using 3D faces and filters. You'll start by doing some setup so you can work on a like-new drawing.

Spherical and Cylindrical Coordinate Formats

In many of these exercises, you used relative Cartesian coordinates to locate the second point for the Move command. For commands that accept 3D input, you can also specify displacements by using the *spherical* and *cylindrical coordinate* formats.

The spherical coordinate format lets you specify a distance in 3D space while specifying the angle in terms of degrees from the X axis of the current UCS, and degrees from the X, Y plane of the current UCS (see the top image of Figure 13.8). In other words, you create an object formed by defined distances from the X and Y axes rather than spherically exploding a two-dimensional object along three planes. You don't need to create a sphere to use this coordinate formatting; you just need to be able to think of the shape as a piece of a sphere.

The cylindrical-coordinate format, on the other hand, lets you specify a location in terms of a distance on the plane of the current UCS and a distance on the Z axis. You also specify an angle from the X axis of the current UCS (see the bottom image of Figure 13.8). In other words, you create a cylindrical object at a defined distance from the Z axis along the planes formed by the UCS and the X axis, rather than expanding a two-dimensional X, Y-oriented rectangle along the Z axis. Again, you don't need to be drawing a cylinder to use this tool; you just need to be able to think of your object as a segment from a cylinder.

Now you're ready to lay out your star.

1. Choose Format ➤ Thickness and set the thickness to 0. Erase the objects from the previous exercise.

2. Draw a circle 4 units in diameter at coordinates 2,2. Click the Circle tool in the Draw toolbar. At the `Specify center point for circle or [ 3P/2P/TTR (tan tan radius)]:` prompt, type **2,2**↵ and at the `Specify radius of circle or [Diameter] <0.000>:` prompt, type **d**↵. Then, at the `Specify diameter of circle <0.000>:` prompt, type **4**↵.

3. Place a five-sided polygon about the circle. Click the Polygon tool in the Draw toolbar and at the `Enter number of sides <4>:` prompt, type **5**↵. At the `Specify center of polygon or [Edge]:` prompt, use the Center Osnap marker to pick the center of the circle, and at the `Enter an option [Inscribed in circle/Circumscribed about circle] <C>:` prompt, press ↵ to accept the default. At the `Specify radius of circle:` prompt, use the Quadrant Osnap marker to pick the bottom quadrant.

FIGURE 13.8:

The spherical and cylindrical coordinate formats

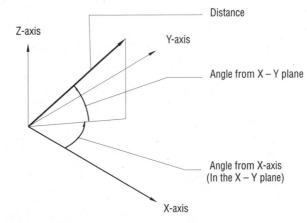

[Distance] < [Angle from X-axis] < [Angle from X – Y plane]

The Spherical Coordinate Format

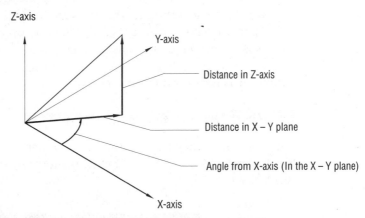

[Distance in X – Y plane] < [Angle from X-axis] , [Distance in Z-axis]

The Cylindrical Coordinate Format

Next you'll draw the lines that make up a five-pointed star. If you aren't sure what this shape looks like, see Figure 13.9.

4. Use the Line tool in the Draw toolbar and the Endpoint Osnap marker to pick the endpoints of the lines of the pentagon.

FIGURE 13.9:

The five-pointed star drawn
from the points of the pentagon circumscribed about
the circle

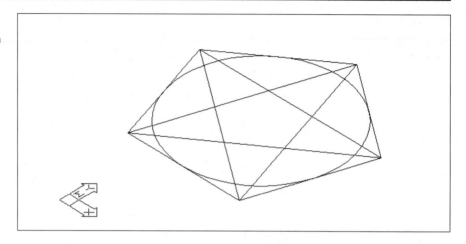

5. Draw another line from the center of the circle 2 units in the positive Z axis. Press ↵ to start the Line command and at the `Specify first point:` prompt, use the Center Osnap marker to locate and pick the center of the circle. At the `Specify next point or [Close/Undo]:` prompt, type **@0,0,2**↵ to draw the vertical line. Press ↵ to finish the command.

Now we need to use the Surfaces tools to begin giving the star three dimensions.

Loading the Surfaces Toolbar

The 3D Face command and AutoCAD's 3D shapes are located in the Surfaces toolbar. To display this toolbar, right-click any toolbar, and then click the Surfaces check box.

Adding a 3D Face

Now that you've opened the Surfaces toolbar, you can begin to draw 3D faces on the star.

1. Zoom into the lines you just created, so you have a view similar to Figure 13.10.

FIGURE 13.10:

Zooming in on the star lines so you can pick the points to make the 3D Face

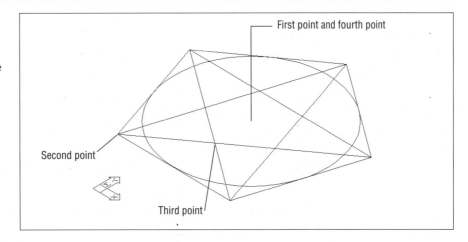

First point and fourth point

Second point

Third point

2. Click the 3D Face tool in the Surfaces toolbar or type **3f**↵. You can also choose Draw ➤ Surfaces ➤ 3D Face from the menu bar.

3. At the `Specify first point or [Invisible]:` prompt, use the Osnap overrides to pick the first of the endpoints of the 3D lines you drew. (You could pick any endpoint for a starting point.) Be sure Ortho mode is off.

TIP

The Running Osnap mode can help you select endpoints quickly in this exercise. See Chapter 4, "Organizing Your Work," if you need to review how to set up the Running Osnap mode.

---

4. As you continue to pick the endpoints, you'll be prompted for the second, third, and fourth points.

NOTE

With the 3D Face tool, you pick three points per surface of the star in a circular fashion, as shown in Figure 13.10. When prompted for the fourth point, press ↵. Once you've drawn one 3D face, you can continue to add more by selecting more points.

---

5. When the `Specify third point:` prompt appears again, press ↵ to end the 3D Face command. A 3D face appears between the two 3D lines. It's difficult to tell if the 3D points are actually there until you use the Hide command, but you should see vertical lines connecting the endpoints of the 3D lines. These vertical lines are the edges of the 3D face (see Figure 13.11).

NOTE   When the `Specify third point:` prompt reappears, you can draw more 3D faces if you like. The next 3D face will use the last two points selected as the first two of its four corners—hence, the prompt for a third point.

---

6. Use the 3D Face command to put another surface on the star, as demonstrated in the top image of Figure 13.12.

FIGURE 13.11:

The 3D face

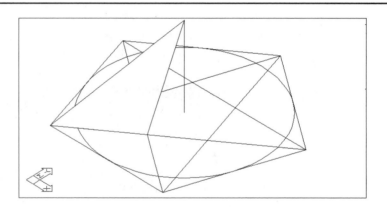

FIGURE 13.12:

The two faces of the star and the star with hidden lines removed

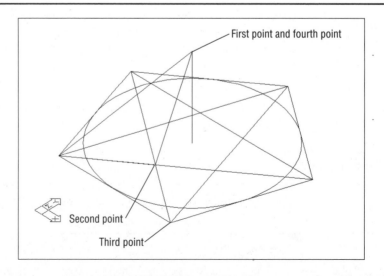

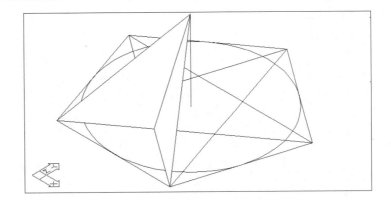

FIGURE 13.12 CONTINUED:

The two faces of the star and the star with hidden lines removed

7. Use the Intersection Endpoint Osnap override to snap to the corners of the 3D faces.

8. Use the Hide command in the Render toolbar to get a view that looks like the bottom image of Figure 13.12.

Use the Array tool in the Modify toolbar to complete the star. The Polar Array option will create five sets of the surfaces around the center of the circle.

1. Select the Array tool, and at the `Select objects:` prompt, pick the two surfaces that you just created. At the next `Select objects:` prompt, press ↵. At the `Enter the type of array [Rectangular/Polar]<R>:` prompt, type **p**↵ to create a polar array, and at the `Specify center point of array:` prompt, pick the center of the circle using the Center Osnap marker. At the `Enter the number of items in the array:` prompt, type **5**↵ to make five sets.

2. At the `Specify the angle to fill (+=ccw, -=cw) <360>:` prompt, press ↵ to accept the default of 360°.

3. At the `Rotate arrayed objects? [Yes/No] <Y>:` prompt, press ↵ to accept the default of Yes.

4. Use the Hide tool again and your star will look like the one in Figure 13.13.

5. Save the `Surfaces.dwg` file.

You might be pretty happy about the way the star looks, but if you want to remove some of the lines, read the next section. Otherwise, skip to the section "Creating Complex 3D Surfaces," and we'll play with our star shape a little more.

FIGURE 13.13:

The complete star with hidden lines removed

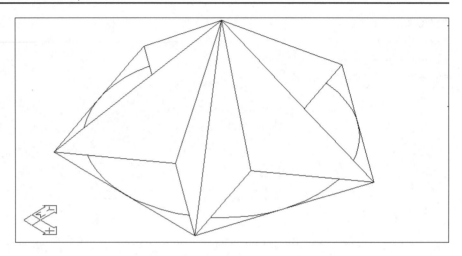

Hiding Unwanted Surface Edges

When using the 3D Face command, you're limited to drawing surfaces with three or four straight sides. However, you can create more complex shapes by simply joining several 3D faces. Figure 13.14 shows an odd shape constructed of three joined 3D faces. Unfortunately, you're left with extra lines that cross the surface, as shown in the top image of Figure 13.14. You can hide those lines by using the Invisible option under the 3D Face command, in conjunction with the Splframe variable.

To make an edge of a 3D face invisible, start the 3D Face command as usual. While selecting points, just before you pick the first point of the edge to be hidden, enter **i**↵, as shown in the bottom image of Figure 13.14. When you're drawing two 3D faces sequentially, only one edge needs to be invisible to hide their joining edge.

You can make invisible edges visible for editing by setting the Splframe system variable to 1. Setting Splframe to 0 will cause AutoCAD to hide the invisible edges. Bear in mind that the Splframe system variable can be useful in both 3D and 2D drawings.

FIGURE 13.14:

Hiding the joined edge of
multiple 3D faces

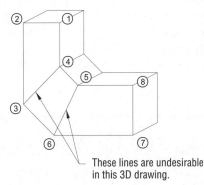

Drawing an odd-shaped surface
using 3D Face generates extra lines.
The numbers in the drawing to the
left indicate the sequence of points
selected to create the surface.

These lines are undesirable
in this 3D drawing.

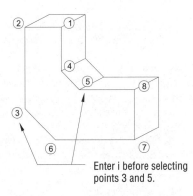

If you draw the same surface using
the i option before selecting the
appropriate points, the unwanted
lines will be hidden. This drawing
indicates where the i option is issued
in the point-selection process.

Enter i before selecting
points 3 and 5.

The Edge tool in the Surfaces toolbar lets you change an existing, visible 3D face edge to an invisible one. Click the Edge tool and then select the 3D face edge to be hidden.

Using Predefined 3D Surface Shapes

You may have noticed that the Surfaces toolbar offers several 3D surface objects, such as cones, spheres, and tori (which are donut-shaped). All are made up of

3D faces, which may not be apparent until after you've tried exploding one of these primitive surfaces. To use them, click the appropriate button in the Surfaces toolbar. When you select an object, AutoCAD will prompt you for the points and dimensions that define that 3D object; then AutoCAD will draw the object. This provides quick access to shapes that would otherwise take substantial time to create.

Things to Watch Out for When Editing 3D Objects

You can use the Move, Copy, and Stretch commands on 3D lines, 3D faces, and 3D shapes, but you have to be careful with these commands when editing in 3D. Here are a few tips to keep in mind.

- The Scale command will scale an object's Z coordinate value, as well as the standard X and Y coordinates. (Click Scale in the Modify toolbar.) Suppose you have

  an object with an elevation of 2 units. If you use the Scale command to enlarge that object by a factor of 4, the object will have a new elevation of 2 units times 4, or 8 units. On the other hand, if that object has an elevation of 0, its elevation will not change, because 0 times 4 is still 0.

- Array, Mirror, and Rotate (in the Modify toolbar) can also be used on 3D lines, 3D faces, and 3D objects, but these commands won't affect their Z coordinate values. Z coordinates can be specified for base and insertion points, so you may be surprised by the results when you use these commands with 3D models.

- Using the Move, Stretch, and Copy commands (in the Modify toolbar) with Osnaps can produce some unpredictable and unwanted results. As a rule, it is best to use point filters when selecting points with Osnap overrides. For example, to move an object from the endpoint of one object to the endpoint of another on the same Z coordinate, invoke the .XY point filter at the **Base Point:** and **Second Point:** prompts before issuing the endpoint override. Pick the endpoint of the object you want. Then enter the Z coordinate, or just pick any point to use the current default Z coordinate.

- When you create a block, the block will use the UCS that is active at the time the block is created. When that block is later inserted, it will orient itself to the coordinate system currently in use. (UCS is discussed in more detail in Chapter 11, "Introducing 3D.")

Creating Complex 3D Surfaces

In the previous examples, you drew objects that were mostly straight lines or curves with a thickness. All of the forms in your star were defined in planes perpendicular to each other. However, at times you'll want to draw objects that don't fit so easily into perpendicular or parallel planes. The following exercise demonstrates how you can create more complex forms using some of AutoCAD's other 3D commands.

Creating Curved 3D Surfaces

Next imagine that the people in marketing have changed their minds about the shape of the star. They would like to soften and contour some of the edges of the star while keeping its basic shape. They'd like to clip the point of the star, but they're not sure whether they'd like it clipped a lot or if it should just get a trim.

To soften the contour, first you need to define the perimeter of one of the sides of the star using arcs, and then use the Edge Surface tool in the Surfaces toolbar to form the shape of the draped surface. Edge Surface creates a surface based on four objects defining the edges. In this example, you'll use arcs to define the edges of the star.

Before you draw the arcs defining the star edge, you must erase the 10 surfaces that you drew in the last exercise, then create the top and bottom arcs and create a UCS in the plane of one of the edges.

1. Click the Erase tool in the Modify toolbar and pick the surfaces one at a time until you have them all, then press ↵. Or, you can pick and erase each surface one at a time to be sure that you're not erasing geometry that you want to keep. After you've erased all of the surfaces that need to be erased, erase the lines that you drew inside the polygon. Don't erase the line that you drew in the Z axis.

TIP A quick way to erase all the 3D faces is to use the Filter tool (discussed in Chapter 9). Type **e**↵ to start the Erase tool. At the `Select objects:` prompt, type **'filter**. In the Object Selection Filters dialog box, choose 3dface from the Select Filter drop-down list, click the Add to List button, and then click the Apply button. You'll see the message `Applying filter to selection`, followed by the Erase tool's `Select objects:` prompt. Type **All**↵ to have AutoCAD find and select all of the 3D faces. Press ↵ to complete the command and erase all of the 3D faces in the drawing.

---

2. Choose View ➢ Viewports ➢ New Viewports ➢ Three:Right. Click in the upper-left viewport to make it active. (You can do this while you're still in the dialog box, before or after you've started a new command.) Select 3D from the setup drop-down list. Then click OK.

WARNING You can't change the viewpoint focus after you've entered the Pan or Zoom commands.

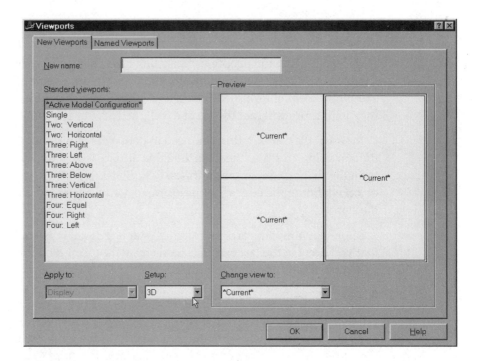

3. Click the Arc tool in the Draw toolbar. You're going to create an arc at the bottom of the star that begins at one point, ends at the adjacent point, and has a radius of 2.00 units.

4 At the ARC Specify start point of arc or [CEnter]: prompt, pick the lower-right corner of the polygon. At the Specify second point of arc or [CEnter/ENd]: prompt, type **en**↵ to use the End option to specify the end of the arc instead of the midpoint (Midpoint is the Second point default option). At the Specify end point of arc: prompt, pick the lower-left

corner of the polygon, and at the `Specify center point of arc or [Angle/Direction/Radius]:` prompt, type **r↵** to specify the radius value. At the `Specify radius of arc:` prompt, type **2↵**. Your drawing should look like Figure 13.15.

FIGURE 13.15:

The polygon with an arc across the corners

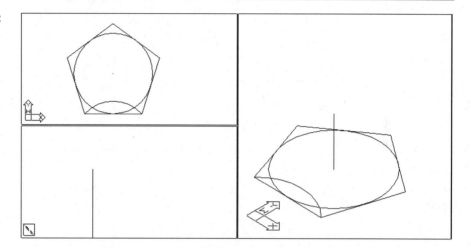

The next thing that you'll do is draw an arc from the corner of the polygon to the top endpoint of the Z-axis line. Before you can draw the arc, you must align a UCS with this plane, with which the arc will also align. Remember that all 2D geometry is drawn on the current UCS plane.

1. To create a UCS that is aligned with the center and one point of the polygon, click the 3 Point UCS tool in the UCS toolbar, or choose Tools ➢ New UCS ➢ 3 Point, then create a UCS using the three points shown in the Isometric view in Figure 13.16.

2. Choose Draw ➢ Arc ➢ Start, End, Radius.

3. Draw the arc defining another edge of the star (see Figure 13.17). Use the Endpoint Osnap marker to pick the endpoints of the polygon and the Z-axis line. Pick the corner of the polygon first, then pick the top end of the Z-axis line and enter the radius (3 units will do it) in response to the prompts. (If you need help with the Arc command, refer to Chapter 3.)

You still need to create the arc on the other side of the surface. Click the WCS tool in the UCS toolbar to return the UCS to the World plane.

FIGURE 13.16:

The points to pick to create the UCS

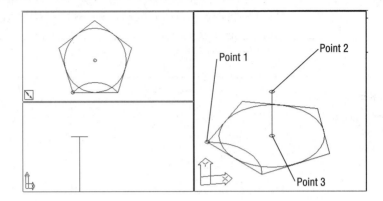

FIGURE 13.17:

Drawing the arc in the new UCS

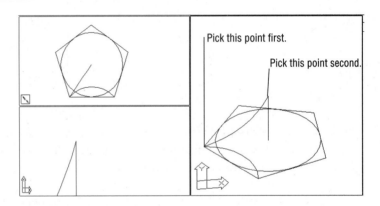

NOTE Did you notice that each viewport can hold its own UCS setting? Top view was still in WCS, even after setting 3point UCS in Isometric view.

Next you'll mirror the side-edge arc to the opposite side. This will save you from having to define a UCS for that side.

1. Click the Mirror tool in the Modify toolbar. Click in the upper-left viewport to change the focus to that viewport. The arc that you're going to mirror looks like a straight line going from the center of the circle to the lower-left corner of the polygon. At the Select objects: prompt, pick the arc that you just drew and press ↵ to complete the selection set. At the Specify first point of

`mirror line:` prompt, you can pick the center of the circle or the midpoint of the bottom arc. At the `Specify second point:` prompt, type **@1<90**↵.

2. At the `Delete source objects? [Yes/No] <N>:` prompt, press ↵ to accept the default and not delete your original arc. Your drawing should look like Figure 13.18.

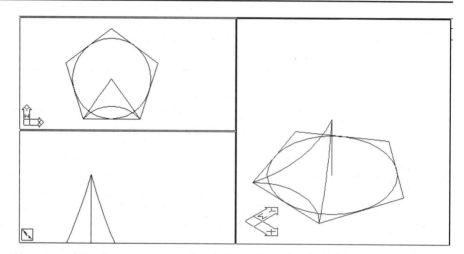

The top of the two-sided arc needs to be trimmed 0.2 units below the point and connected with a line to clip the point of the star. Study the Isometric viewport carefully. Notice that the arcs curve away from the vertical line that supports them. This could make the creation of the next object tricky. However, we still have a few tricks up our sleeves.

1. Click in the lower-left viewport to change viewports. Choose Tools ➤ New UCS ➤ View to set a UCS parallel to this view.

2. Draw a horizontal line from the top of the arcs 0.5 units in the X direction. Click the Line tool in the Draw toolbar, and at the `Specify first point:` prompt, pick the top of the arcs. At the `Specify second point:` prompt, type **@.5<0**↵.

TIP   Another method (with Ortho mode on) is to drag the mouse directly to the right and type **.5**↵ to use the direct distance entry (DDE) feature.

---

3. Move the line 0.25 units in the –X direction and 0.2 units in the –Y direction. To do this, click the line to highlight the grips and pick any of them to turn on the Move tool. Right-click to pop up the Grips menu, and select Move. At the <Move to point>/Base point/Copy/Undo/eXit: prompt, type @–.25,–.2↵.

NOTE   Although the new horizontal line doesn't touch the arcs, it appears to do so when viewed in the lower-right viewport. Here we can take advantage of the Apparent Intersection Osnap to actually draw a line that will align to the new line, and still touch the arcs.

---

4. Start the Line command, and at the Specify first point: prompt, apply the Override Appint Osnap. Be very careful to select one of the arcs first, and when prompted by the and prompt, select the line created in steps 2 and 3.

NOTE   If you don't remember how to start an Override Osnap, review Chapter 2, "Creating Your First Drawing."

WARNING   You must select the arc first if you want your new line to touch it, otherwise, it will touch the line instead.

---

5. When the Line command prompts you to Specify next point or [Undo]:, apply the Override Appint Osnap again and select the other arc. You'll now see the and prompt once more. Select the line created in steps 2 and 3. Your new line should look like Figure 13.19.

NOTE   The object created in steps 2 and 3 (whose grips are turned on in Figure 13.19) can be erased now.

---

6. Use the Zoom Window tool in the Standard toolbar to enlarge the view of the point on the star in the lower-left viewport.

NOTE   After you completed each command to draw, edit, or delete objects in one viewport, all of the other viewports were updated.

FIGURE 13.19:

Creating a new line using
the Override Appint Osnap

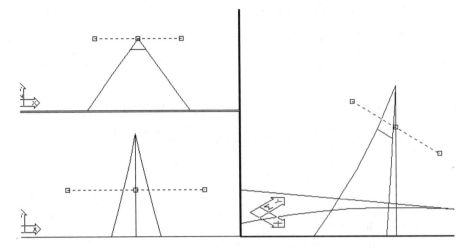

7. Click the Trim tool in the Modify toolbar. At the `Current settings: Projec-
    tion=UCS Edge=None Select cutting edges: Select objects:` prompt,
    press ↵ to use the Automatic Trim feature. At the `Select object to trim
    or [Project/ Edge/Undo]:` prompt, type **p** to use the Project option. At
    the `Enter a projection option [None/Ucs/View] <Ucs>:` prompt, type ↵.
    You'll now be prompted to `Select object to trim`. Type **f**↵ to use the
    Fence Selection Set option. Then draw a fence across all three objects (the two
    arcs and the line that we've been calling the Z-axis line), above the last line
    that you drew. Press ↵ when you're done. All three objects are trimmed.
    Figure 13.20 shows several views of the trimmed arcs.

8. The Edgesurf command that you'll use to apply surfaces to your arcs requires
    four edges, and you've just drawn the fourth one.

Finally, let's finish off this star by adding the surface representing the star's side,
and then arraying it around the star. Marketing has mentioned that they like a
mesh texture.

1. Change the viewport focus to the Isometric viewport (the one on the right).
    Click the Edge Surface tool in the Surfaces toolbar, or enter **edgesurf**↵ at the
    command prompt.

FIGURE 13.20:

The line has been drawn and the lower-left viewport is highlighted after trimming the arcs at the new line.

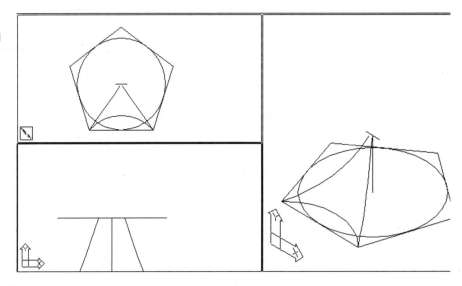

You'll be notified of two very important settings:

```
Current wire frame density: SURFTAB1=6 SURFTAB2=6
```

The variables Surftab1 and Surftab2 can be entered at the command prompt and set to higher values to increase the smoothness of the Edgesurf. They'll be discussed in the next section.

2. At the `Select object 1 for surfaceedge:` prompt, pick the arc on the WCS.

    At the `Select object 2 for surfaceedge:` prompt, pick either side arc.

    At the `Select object 3 for surfaceedge:` prompt, pick the line.

    At the `Select object 4 for surface edge:` prompt, pick either side arc.

In order for the Edgesurf command to work properly, the arcs—or any set of objects—used to define the boundary of a mesh with the Edge Surface option must be connected exactly, end to end.

3. Change the viewpoint focus to the lower-left viewport (click once in the viewport to highlight it) for the most detailed view of the surface, and pick the last line that you drew. Change the viewport focus once more back to the Isometric view and pick the last unselected arc. (The arcs should be picked in a circular fashion, not crosswise.) A mesh will appear, filling the space between the four arcs. Your star is now almost complete.

TIP   Before starting the Array command, be sure that the UCS has been set to World.

---

4. Click the Array tool in the Modify toolbar. At the `Select objects:` prompt, type l↵ to select the last object drawn, then at the `Enter the type of array [Rectangular/Polar]<R>:` prompt, type p↵.

5. At the `Specify center point of array:` prompt, pick the center of the circle, and at the `Enter the number of items in the array:` prompt, type **5**↵. At the `Specify the angle to fill (+=ccw, -=cw) <360>:` prompt, press ↵, and at the `Rotate arrayed objects? [Yes/No] <Y>:` prompt, press ↵ to accept the default.

6. Choose View ➤ Hide to get a better view of the star. Be sure that the viewport is the Isometric view. You should have a view similar to Figure 13.21.

7. Save this file.

FIGURE 13.21:

The completed star with meshes, in the Isometric viewport

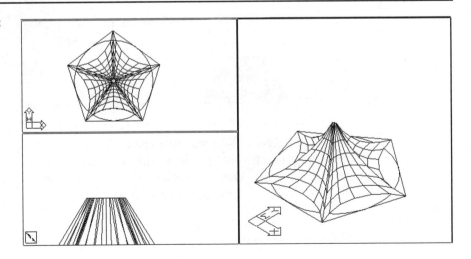

At this point, you've tried a few of the tools in the Surfaces toolbar. You'll get a chance to use more of them throughout the rest of this chapter. Next, you'll learn how to edit mesh objects like the star.

Adjusting the Settings That Control Meshes

As you can see from Figure 13.21, the draping mesh of our star is made up of rectangular segments. If you want to increase the number of segments in the mesh (see Figure 13.22), you can change the Surftab1 and Surftab2 system variables. Surftab1 controls the number of segments along edge 1, the first edge you pick in the sequence. Surftab2 controls the number of segments along edge 2. AutoCAD refers to the direction of edge 1 as m and the direction of edge 2 as n. These two directions can be loosely described as the X and Y axes of the mesh, with m being the X axis and n being the Y axis.

FIGURE 13.22:

The star with different *Surftab* settings

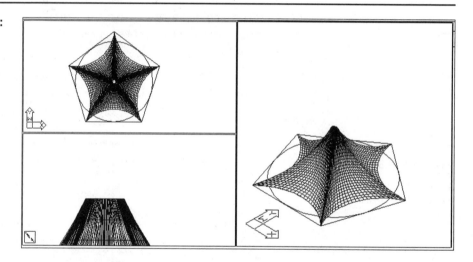

In Figure 13.22, Surftab1 and Surftab2 are both set to 25. The default value for both settings is 6. If you'd like to try different Surftab settings on the star mesh, you must erase the existing mesh, change the Surftab settings, and then use the Edge Surface tool again to define the mesh.

Creating a 3D Mesh by Specifying Coordinates

If you need to draw a mesh like the one in the previous example, but you want to give exact coordinates for each vertex in the mesh grid, you can use the 3D Mesh command. Suppose you have data from a survey of a part from a coordinate-measuring machine; you can use 3D Mesh to convert your data into a graphic representation of its topography. Another use for 3D Mesh is to plot mathematical data to get a graphic representation of a formula.

Because you must enter the coordinate for each vertex in the mesh, 3D Mesh is better suited to scripts or AutoLISP programs, where a list of coordinates can be applied automatically to the 3D Mesh command in a sequential order. See Chapter 19, "Introduction to Customization," and the companion CD-ROM for more on AutoLISP.

Other Surface Drawing Tools

In the last exercise, you used the Edge Surface tool to create a 3D surface. There are several other 3D surface commands available that allow you to generate complex surface shapes easily.

TIP   All of the surface drawing tools described in this section, along with the meshes you're already familiar with, are actually composites of 3D faces. This means these 3D objects can be exploded into their component 3D faces, which in turn can be edited individually.

Using Two Objects to Define a Surface

The Ruled Surface tool in the Surfaces toolbar draws a surface between two 2D objects, such as a line and an arc or a polyline and an arc. This command is useful for creating extruded forms that transform from one shape to another along a straight path. Let's see how this command works.

1. Open the file called `Rulesurf.dwg` from the companion CD-ROM. It looks like the top image of Figure 13.23. This drawing is of a simple half circle drawn using a line and an arc. Ignore the diagonal blue line for now.

2. Move the connecting line that is between the arc endpoints 10" in the Z axis.

3. Now you're ready to use a 3D surface to connect the half circle with the line that used to connect the arc's ends. Click the Ruled Surface tool in the Surfaces toolbar, or choose Draw ➤ Surfaces ➤ Ruled Surface.

4. At the Select first defining curve: prompt, place the cursor toward the right end of the arc and click it.

5. At the Select second defining curve: prompt, move the cursor toward the right end of the line and click it, as shown in the bottom image of Figure 13.23. The surface will appear as shown in Figure 13.24.

NOTE   The position you use to pick the second object will determine how the surface is generated.

FIGURE 13.23:

Drawing two edges for the Ruled Surface option

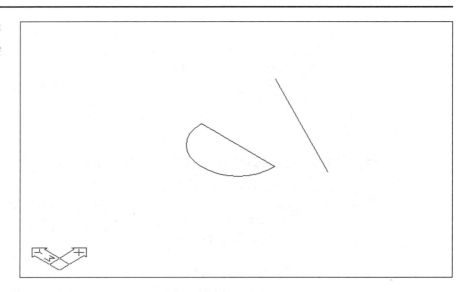

**FIGURE 13.23
CONTINUED:**

Drawing two edges for the
Ruled Surface option

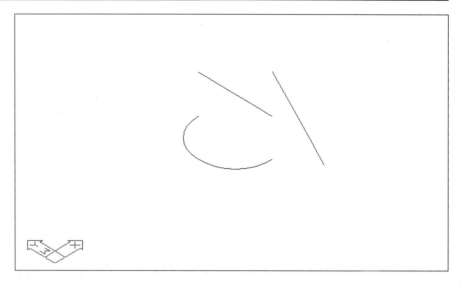

The location you use to select the two objects for the ruled surface is important. We asked you to select specific locations on the arc and line so that the ruled surface would be generated properly. Had you selected the opposite end of the line, for example, your result would look more like Figure 13.25. Notice that the segments defining the surface cross each other. This crossing effect is caused by picking the defining objects near opposite endpoints. The arc was picked near its lower end, and the line was picked toward the top end. If you needed to produce a pinwheel or a propeller, you might actually want this twisted effect

FIGURE 13.24:

The Rulesurf surface

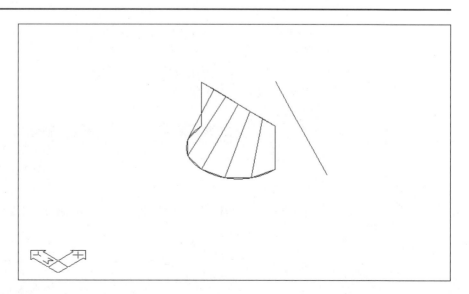

FIGURE 13.25:

The ruled surface redrawn by using different points to select the objects

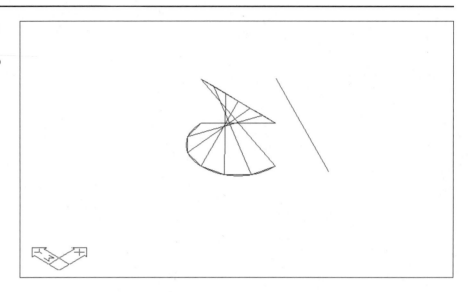

Extruding an Object along a Straight Line

The Tabulated Surface tool also uses two objects to draw a 3D surface, but instead of drawing the surface between the objects, Tabulated Surface extrudes one object in a direction defined by a direction vector. The result is an extruded shape the length and direction of the direction vector. To see what this means, try the following exercise:

1. While still in the Rulesurf drawing, click the Undo tool in the Standard toolbar to undo the ruled surface from the previous exercise.

2. Click the Tabulated Surface tool in the Surface toolbar, or choose Draw ➣ Surfaces ➣ Tabulated Surface.

3. At the Select object for path curve: prompt, click the arc.

4. At the Select object for direction Vector: prompt, click the lower end of the blue line farthest to the right. The arc is extruded in the direction of the blue line, as shown in Figure 13.26.

FIGURE 13.26:

Extruding an arc using a direction vector

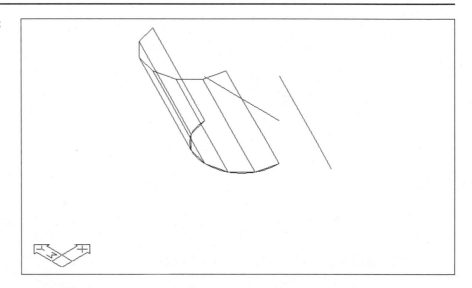

The direction vector can be any object, but AutoCAD will only consider the object's two endpoints when extruding the path curve. Similar to the ruled surface, the point at which you select the direction-vector object affects the outcome of the extrusion. If you had selected a location near the top of the blue line, the extrusion would have gone in the opposite direction.

Because the direction vector can point in any direction, the Tabulated Surface tool allows you to create an extruded shape that is not restricted to a direction perpendicular to the object being extruded.

The path curve defining the shape of the extrusion can be an arc, circle, line, or polyline. You can use a curve-fitted polyline or a spline polyline to create more complex shapes, as shown in Figure 13.27.

TIP   If you want to increase the number of facets in Ruled Surface or in Tabulated Surface, set the Surftab1 system variable to the number of facets you desire.

Some samples of other shapes created using Ruled Surface and Tabulated Surface

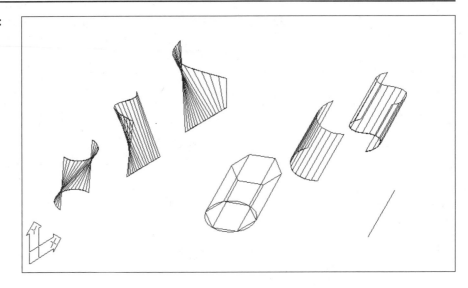

Creating a Circular Surface

The Revolved Surface tool allows you to quickly generate circular extrusions. Typical examples are vases or teacups. The following exercise shows how the Revolved Surface tool is used to draw a pitcher. You'll use an existing drawing that has a profile of the pitcher already drawn.

1. Open the Pitcher.dwg file on the companion CD-ROM. This file contains a polyline profile of a pitcher and a single line representing the center of the pitcher. (See the top image of Figure 13.28, which we'll come back to in a moment.) The profile and line have already been rotated to a position perpendicular to the WCS. The grid is turned on so you can better visualize the plane of the WCS.

2. Click the Revolved Surface tool in the Surfaces toolbar.

FIGURE 13.28:

Drawing a pitcher using the Revolved Surface tool

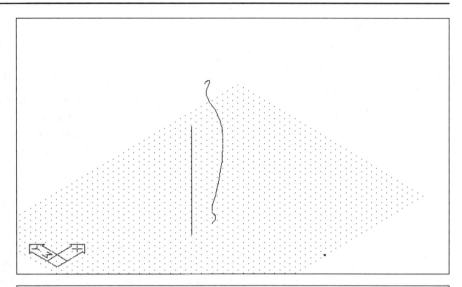

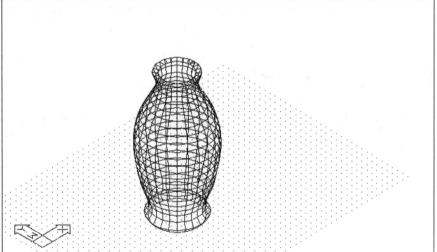

3. At the Current wire frame density: SURFTAB1=24 SURFTAB2=24 Select object to revolve: prompt, click the polyline profile, as shown in the top image of Figure 13.28.

4. At the Select object that defines the axis of Revolution: prompt, click near the bottom of the vertical line representing the center of the vase, as shown in the top image of Figure 13.28.

5. At the Specify start angle <0>: prompt, press ↵ to accept the 0 start angle.

6. At the Specify included angle (+=ccw, -=cw) <360>: prompt, press ↵ to accept the 360° default. The pitcher appears, as shown in the bottom image of Figure 13.28.

Notice that the pitcher is made up of a faceted mesh, like the mesh that is created by the Edge Surface tool. Similar to the Edge Surface tool, you can set the number of facets in each direction using the Surftab1 and Surftab2 system variable settings. Both Surftab1 and Surftab2 were already set to 24 in the Pitcher.dwg file, so the pitcher shape appears fairly smooth.

You may have noticed that in steps 5 and 6 of the previous exercise you had a few options. In step 5, you could specify a start angle. In this case, you accepted the 0 default. Had you entered a different value, 90 for example, the extrusion would have started in the 90° position relative to the current WCS. In step 6, you had the option of specifying the angle of the extrusion. Had you entered 180, for example, your result would have been half of the pitcher. You could also indicate the direction of the extrusion by specifying a negative or positive angle.

Editing a Mesh

Once you've created a mesh surface with either the Edge Surface or Revolved Surface tool, you can make modifications to it. For example, suppose you wanted to add a spout to the pitcher you created in the previous exercise. You could use grips to adjust the individual points on the mesh to reshape the object. Here, you must take care how you select points. The UCS will become useful for editing meshes, as shown in the following exercise:

1. Zoom into the area shown in the top image of Figure 13.29.

2. Click the pitcher mesh to expose its grips.

3. Shift+click the grips shown in the middle image of Figure 13.29.

4. Click the grip shown in the bottom image of Figure 13.29 and slowly drag the cursor to the left. As you move the cursor, notice how the lip of the pitcher deforms.

5. When you've dragged the cursor to the approximate location indicated in the bottom image of Figure 13.29, select that point. The spout will be fixed in the new position.

FIGURE 13.29:

Adding a spout to the
pitcher mesh

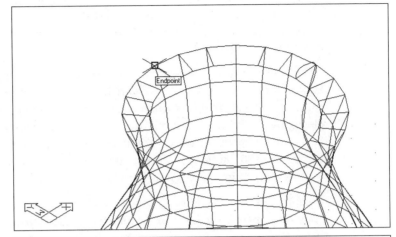

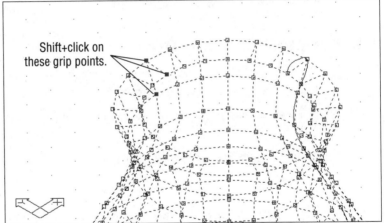

Shift+click on
these grip points.

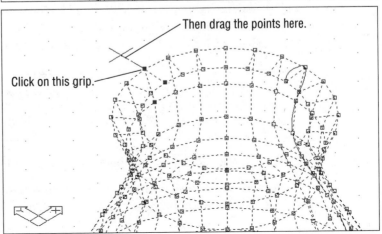

Then drag the points here.

Click on this grip.

You can refine the shape of the spout by carefully adjusting the position of other grip points around the edge of the pitcher. Later, when you render the pitcher, you can apply a smooth shading value so that the sharp edges of the spout are smoothed out.

This exercise shows how easy it is to make changes to a mesh by moving individual grip locations. However, be aware that when you move mesh grips manually (rather than by entering coordinates), the grips' motion is restricted to a plane parallel to the current UCS. You can use this restriction to your advantage. For example, if you want to move the spout downward at a 30° angle, you can rotate the UCS so it is tipped at a 30° angle in relation to the top of the pitcher, and then edit the mesh grips as you did in the previous exercise.

Another option would be to specify a *relative* coordinate as opposed to selecting a point. By specifying a coordinate, such as @.5<50, you would not have to move the UCS.

Other Mesh-Editing Options

You can use Modify ➤ Polyline to edit meshes in a way similar to editing polylines. When you choose this option and pick a mesh, you get this prompt:

```
Enter an option [Edit vertex/Smooth surface/Desmooth/Mclose/
Nclose/Undo]:
```

Here are descriptions of these options.

Edit Vertex   Allows you to relocate individual vertices in the mesh.

Smooth Surface   Is similar to the Spline option for polylines. Rather than having the mesh's shape determined by the vertex points, Smooth Surface adjusts the mesh so that mesh vertices act as control points that pull the mesh—much as a spline frame pulls a spline curve.

TIP   You can adjust the amount of pull the vertex points exert on a mesh by using Smooth Surface in conjunction with the Surftype system variable. If you'd like to know more about these settings, see Appendix D, "System and Dimension Variables."

Desmooth   Reverses the effects of Smooth Surface.

Mclose and **Nclose**   Allow you to close the mesh in either the *m* or *n* direction. When either of these options is used, the prompt line will change, replacing Mclose or Nclose with Mopen or Nopen, allowing you to open a closed mesh.

Undo   Undoes the last operation without leaving the command.

NOTE   The Edit Polyline tool in the Modify II toolbar performs the same function as Modify ➤ Polyline.

Moving Objects in 3D Space

AutoCAD provides two utilities for moving objects in 3D space: Align and 3D Rotate. All of these commands are found on the 3D Operation section of the Modify pull-down menu. They help you perform some of the more common moves associated with 3D editing.

Aligning Objects in 3D Space

In mechanical design you often create the parts in 3D and then show an assembly of the parts. The Align option can greatly simplify the assembly process. The following exercise describes how Align works:

1. Choose Modify ➤ 3D Operation ➤ Align, or type **al↵**.

2. At the `Select objects:` prompt, select the 3D source object you want to align to another part. (The *source object* is the object you want to move.)

3. At the `Specify first source point:` prompt, pick a point on the source object that is the first point of an alignment axis, such as the center of a hole or the corner of a surface.

4. At the `Specify first destination point:` prompt, pick a point on the destination object to which you want the first source point to move. (The *destination object* is the object with which you want the source object to align.)

5. At the `Specify second source point:` prompt, pick a point on the source object that is the second point of an alignment axis—such as another center point or other corner of a surface.

6. At the `Specify second destination point:` prompt, pick a point on the destination object indicating how the first and second source points are to align in relation to the destination object.

7. At the `Specify third source point:` prompt, you can press ↵ if two points are adequate to describe the alignment. Otherwise, pick a third point on the source object that, along with the first two points, best describes the surface plane you want aligned with the destination object.

8. At the `Specify third destination point:` prompt, pick a point on the destination object that, along with the previous two destination points, describes the plane with which you want the source object to be aligned. The source object will move into alignment with the destination object.

NOTE	The first source point will match the first destination point perfectly, assuming that Osnaps are used; the second and third points will be as close as possible. This technique can be used to make objects rotate in 3D space as they move to the desired alignment.

Figure 13.30 gives some examples of how the Align utility works.

FIGURE 13.30:

Aligning two 3D objects

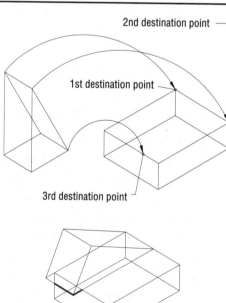

2nd destination point

1st destination point

3rd destination point

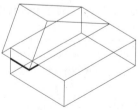

Result

Rotating an Object in 3D

If you just want to rotate an object in 3D space, the Modify ➤ 3D Operation ➤ Rotate 3D option on the menu bar can simplify the operation. Once you've selected this option and selected the objects you want to rotate, you get the following prompt:

```
Specify first point on axis or define axis by [Entity/Last/View/Xaxis/
Yaxis/Zaxis/2points]:
```

This prompt is asking you to describe the axis of rotation. Here are descriptions of the options presented in the prompt.

> **Entity**   Allows you to indicate an axis by clicking an object. When you select this option, you're prompted to pick a line, circle, arc, or 2D polyline segment. If you click a line or polyline segment, the line is used as the axis of rotation. If you click a circle, arc, or polyline arc segment, Auto-CAD uses the line passing through the center of the circle or arc and perpendicular to its plane.

> **Last**   Uses the last axis that was used for a 3D rotation. If no previous axis exists, you're returned to the Axis prompt.

> **View**   Uses the current view direction as the direction of the rotation axis. You're then prompted to select a point on the view direction axis to specify the exact location of the rotation axis.

> **Xaxis/Yaxis/Zaxis**   Use the standard X, Y, or Z axis as the direction for the rotation axis. You're then prompted to select points on the X, Y, or Z axis to locate the rotation axis.

> **2points**   Uses two points that you provide as the endpoints of the rotation axis.

This completes your look at creating and editing 3D objects. You've had a chance to use nearly every type of object available in AutoCAD. You might want to experiment on your own with the predefined 3D shapes offered in the Surfaces toolbar. In the next section, you'll discover how you can generate perspective views.

Viewing Your Model in Perspective

So far, your views of 3D drawings have been in *parallel projection*. This means parallel lines appear parallel on your screen. Although this type of view is helpful while constructing your drawing, you'll want to view your drawing in true perspective from time to time, to get a better feel for what your 3D model actually looks like.

AutoCAD 2000 has furnished two commands for 3D viewing. To view your model in perspective, you'll use the Dview (Dynamic View) command. This is an older command that is no longer found on the menus, so we'll type it in at the command prompt. We'll also try out a new tool named 3D Orbit that will allow us to rotate our model while it is shaded.

1. Open the Surfaces.dwg file you created earlier in this chapter in the section "Creating Complex 3D Surfaces," or use the Ch13surfaces.dwg file on the companion CD-ROM. Be sure you're in the WCS. Choose View ➢ 3D Views ➢ Plan View ➢ World UCS or type **plan**↵↵. This will display a plan view of the star (see Figure 13.31).

FIGURE 13.31:

The plan view of the star

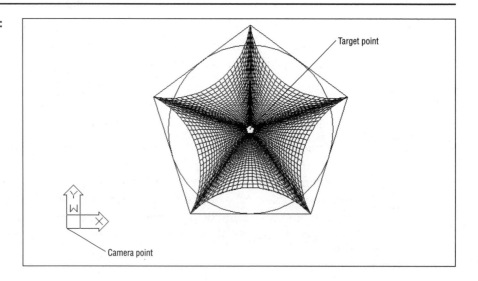

2. Choose View ➢ Zoom ➢ All to get an overall view of the drawing.

3. Enter **dv**↵ at the command prompt. At the `Select objects or <use DVIEWBLOCK>:` prompt, you have two options—to merely type ↵ and allow the use `Dviewblock` to be displayed, or to select one or more objects to use while you adjust the view of your model. The advantage is quite obvious on a large model. Commands of this type will use up all of the computer's resources quickly, and slow it down dramatically.

4. Because this model is quite small, at the `Select objects:` prompt, pick the bottom surface, the circle, and the polygon. You'll use these objects as references while using the Dview command.

5. Next you'll see a fairly lengthy prompt at which you can select the appropriate option.

    ```
    Enter option
    [CAmera/TArget/Distance/POints/PAn/Zoom/TWist/CLip/Hide/Off/Undo]:
    ```

    These options are discussed in the next section. The screen will change to show only the objects you selected.

Dview allows you to adjust your perspective view in real time. For this reason, you're asked to select objects that will allow you to get a good idea of your view without slowing the real-time display of your model. If you had selected the whole star, view selection would be slower because the whole star would have to be dragged during the selection process. For a file as small as this star, view selection isn't a big problem, but for larger files, you'll find that selecting too many objects can bog down your system.

Setting Up Your Perspective View

Similar to a camera, your perspective view in AutoCAD is determined by the distance from the object, camera position, view target, and camera lens type.

Follow these steps to determine the camera and target positions.

1. At the `Enter option…` prompt, enter **po**↵ for the Points option.

2. At the `Specify target point <current point>:` prompt, pick the center of the circle. This will allow you to adjust the camera target point (the point at which the camera is aimed).

3. At the Specify camera point <current point>: prompt, pick the lower-left corner of the screen. This places the camera location (the position from which you're looking) below and to the left of the star, on the plane of the WCS (see Figure 13.32).

WARNING
When selecting views in this set of exercises using Dview and its options, be sure you click the mouse/pick button as indicated in the text. If you press the ↵ key (or click the left button on the mouse), your view will return to the default orientation, which is usually the last view selected.

Your view now changes to reflect your target and camera locations, as shown in Figure 13.32. The Enter option prompt returns, allowing you to further adjust your view.

FIGURE 13.32:

The view with the camera and target positioned

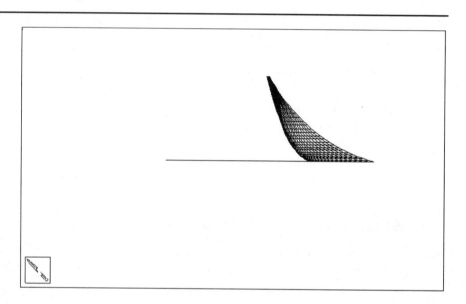

If you like, you can press ↵ at the object-selection prompt without picking any object. You'll get the default image—a house—to help you set up your view (see Figure 13.33). Or, you can define a block and name it Dviewblock. Dviewblock should be defined in a one-unit cubed space. AutoCAD will search the current drawing database, and if it finds `Dviewblock`, AutoCAD will use it as a sample image to help you determine your perspective view.

> **NOTE**   Make `Dviewblock` as simple as possible, but without giving up the detail necessary to distinguish its orientation.

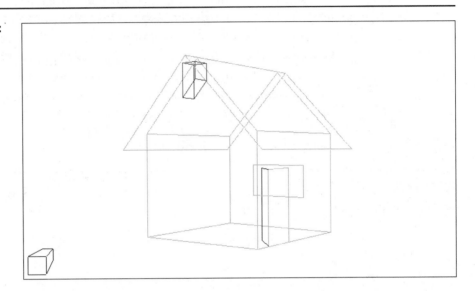

Adjusting Distances

Next you'll adjust the distance between the camera and target.

1. At the `Enter option` prompt, enter **d.⏎** for the Distance option. A slide bar appears at the top of the screen.

> **NOTE**   The Distance option actually serves two functions. Aside from allowing you to adjust your camera to target distance, it also turns on the Perspective View mode. The Off option of Dview changes the view back to a parallel projection.

2. At the `Specify new camera-target distance <current distance>:` prompt, move your cursor from left to right. The star appears to enlarge and shrink. You can also see that the position of the diamond in the slide bar moves. The slide bar gives you an idea of the distance between the camera and the target point in relation to the current distance.

3. As you move the diamond, you see lines from the diamond to the 1× value (1× being the current view distance). As you move the cursor toward the 4× mark on the slide bar, the star appears to move away from you. Move the cursor toward 0×, and the star appears to move closer.

4. Move the cursor farther to the left; as you get to the extreme left, the star appears to fly off the screen. This is because your camera location has moved so close to the star that the star disappears beyond the view of the camera as if you were sliding the camera along the floor toward the target point. The closer you are to the star, the larger and farther above you the star appears to be (see Figure 13.34).

FIGURE 13.34:

The camera's field of vision

5. Adjust your view so it looks like Figure 13.35. To do this, move the diamond to a spot between 1× and 4× in the slide bar.

6. When you have the view that you want, click the mouse button. The slide bar disappears and your view is fixed in place. Notice that you're now viewing the star in perspective.

FIGURE 13.35:

The perspective view of the star using the Distance option

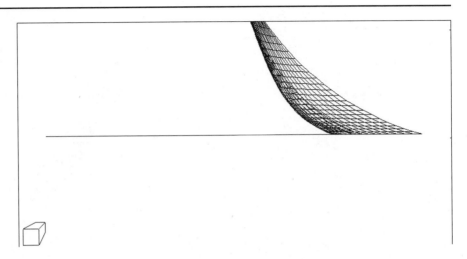

Adjusting the Camera and Target Positions

Next, you'll adjust your view so that you can see the whole star. You're still in the Dview command.

NOTE   Using Dview's Target option is like standing in one location while moving the camera around.

---

1. At the Enter option prompt, enter **ta.↵** for the Target option. The star will temporarily disappear from view.

2. When you see the Specify camera location, or enter angle from XY plane,or [Toggle (angle in)] <0.0000>: prompt, move your cursor very slowly in a side-to-side motion. Keep the cursor centered vertically or you may not be able to find the star. The star moves in an exaggerated manner in the direction of the cursor. The sideways motion of the cursor simulates panning a camera from side to side across a scene (see Figure 13.36).

3. Center the star in your view, and then move the cursor slowly up and down. The star moves in the opposite direction to the cursor. The up-and-down motion of the cursor simulates panning a camera up and down.

Adjusting the Target option is like panning your camera across a scene.

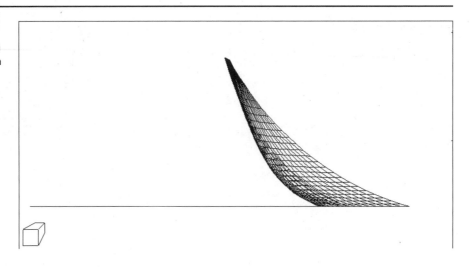

TIP

If you lose your view of the star but remember your camera angle, you can enter the angle at the **Enter angle in X-Y plane:** prompt to help relocate your view.

---

4. Enter **ta**⏎. You'll see the prompt

   Specify camera location, or enter angle from XY plane, or [Toggle (angle in)] <0.0000>:

   At this prompt, you can enter a value representing the vertical angle between the camera and the target point.

5. Press ⏎ to accept the default. This fixes the vertical motion of the target location to its current default location. However, you can still move the target horizontally. Here you can enter an angle value representing the angle between the camera and target in a horizontal plane measured from Auto-CAD's angle 0.

6. Position the view of the star so that it looks like Figure 13.37, and click the mouse button. You've now fixed the target position.

In steps 4 and 5, you could enter an angle value indicating either the vertical or horizontal angle to the target. Once you enter that value (or just press ⏎), the angle becomes fixed in either the vertical or horizontal direction. Then, as you move your cursor, the view's motion will be restricted to the remaining unfixed direction. In the exercise, you fixed the vertical angle and then visually selected a horizontal angle. If you would prefer to "fix" the horizontal angle and visually adjust the vertical, you can enter **ta.⏎t.⏎⏎** at the Dview prompt.

FIGURE 13.37:

While in the Dview Target option, set up your view to look like this.

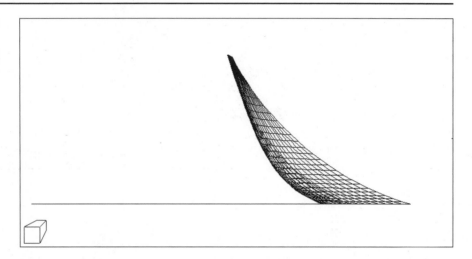

Changing Your Point of View

Next you'll adjust the camera location to one that is higher in elevation.

1. At the Enter option prompt, enter **ca⏎** to select the Camera option.

> **NOTE** Using Dview's Camera option is like changing your view elevation while constantly looking at the star.

2. At the following prompt:

    Specify camera location, or enter angle from XY plane, or [Toggle (angle in)] <0.0000>:

    move your cursor slowly up and down. As you move the cursor up, your view changes as if you were rising above the star (see Figure 13.38). If you know the vertical angle you want, you can enter it now.

In the Camera option, moving the cursor up and down is like moving your camera location up and down in an arc.

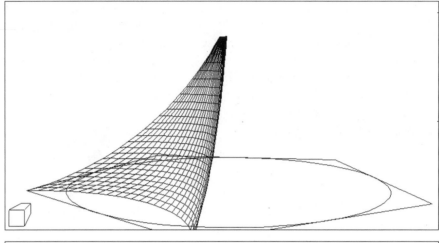

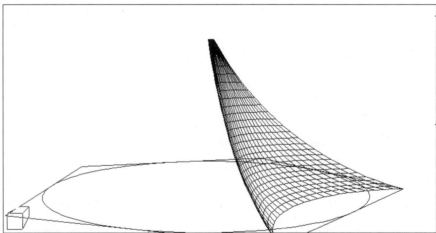

3. Move the cursor down so that you have a view roughly level with the star, and then move the cursor from side to side. Your view changes as if you were walking around the star, viewing it from different sides (see Figure 13.39).

4. When you're ready, enter t↵ to choose the [T]oggle angle in/Enter angle from XY plane option. The prompt changes to

   Specify camera location, or enter angle in XY plane from X axis, or [Toggle (angle from)]<-144>:

   If you know the horizontal angle you want, you can enter it now.

Moving the cursor from side to side is like walking around the target position.

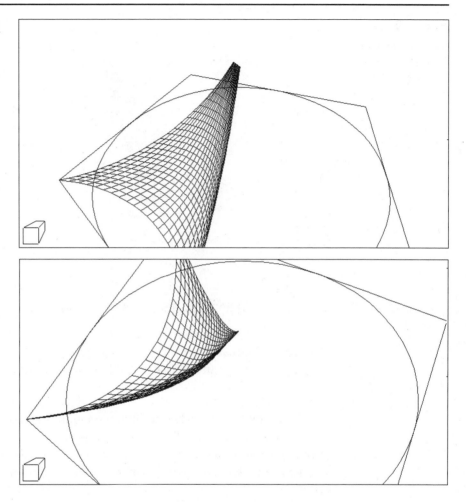

5. Position your view of the star so that it is similar to the one in Figure 13.40, and click the mouse/pick button.

In steps 2 and 4, you could enter an angle value indicating either the vertical or horizontal angle to the camera. Once you indicate a value, either by entering a new one or by pressing ↵, the angle becomes fixed in either the vertical or horizontal direction. Then, as you move your cursor, the view's motion will be restricted to the remaining unfixed direction.

FIGURE 13.40:

Set up your camera location so that your view resembles this one.

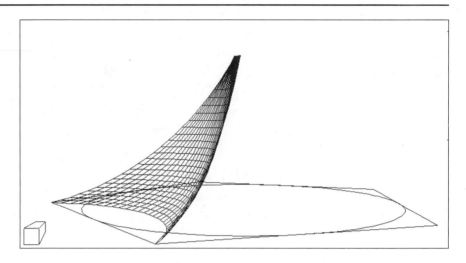

Using the Zoom Option as a Telephoto Lens

The Zoom option of Dview allows you to adjust your view's cone of vision, much like a telephoto lens in a camera. You can expand your view to include more of your drawing or narrow the field of vision to focus on a particular object.

1. At the Enter option prompt, enter z↵ for the Zoom option. Move your cursor from side to side, and notice that the star appears to shrink or expand. You also see a slide bar at the top of the screen, which lets you see your view in relation to the last Zoom setting, indicated by a diamond. You can enter a value for a different focal length, or you can visually select a view using the slide bar.

2. At the Specify Lens Length <50.000mm>: prompt, press ↵ to accept the 50.000mm default value.

NOTE
When you don't have a perspective view (obtained by using the Distance option) and you use the Zoom option, you'll get the prompt Specify zoom scale factor <1>: instead of the Specify Lens Length: prompt. The Specify zoom prompt acts just like the standard Zoom command.

Twisting the Camera

The Twist option lets you adjust the angle of your view in the viewport frame—like twisting the camera to make your picture fit diagonally across the frame.

1. At the Enter option prompt, enter **tw**↵ for the Twist option. Move your cursor, and notice that a rubber-banding line emanates from the view center; the star also appears to rotate, and the coordinate readout changes to reflect the twist angle.

2. At the Specify view twist <0>: prompt, press ↵ to keep the current 0° twist angle.

3. Press ↵ again. The drawing regenerates, showing the star in perspective (see Figure 13.41).

FIGURE 13.41:

A perspective view of the star

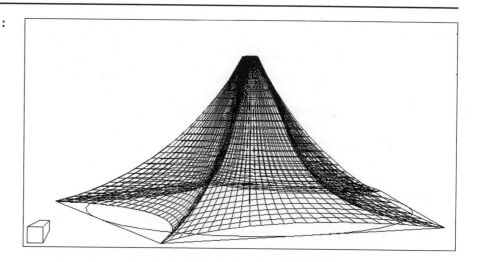

4. Choose View ➢ Hide to see a hidden-line perspective view.

In the next section, you'll look at one special Dview option—Clip—that lets you control what is included in your 3D views.

Using Clip Planes to Hide Parts of Your View

At times, you may want a view that would be normally obscured by objects in the foreground. For example, if you try to view the interior of an enclosure design, the objects closest to the camera obscure your view. The Dview command's Clip option allows you to eliminate objects in either the foreground or the background, so that you can control your views more easily. In the case of the star, set the Clip/Front option to delete any surfaces in the foreground that might obscure your view of its interior (see Figure 13.42).

FIGURE 13.42:

A view of the star interior using the Clip/Front plane

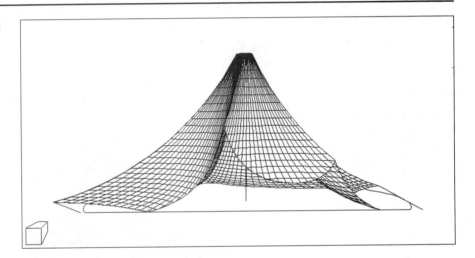

1. While still in the Dview command, enter **cl** ↵ for the Clip option.

2. At the Back/front <off>: prompt, enter **f** ↵ for the Front option. A slide bar appears at the top of the screen.

3. As you move the diamond on the slide bar from left to right, the star point's surfaces in the foreground begin to disappear, starting at the position closest to you. Moving the diamond the other way (right to left) brings the point's surfaces back into view. You can select a view either by using the slide bar or by entering a distance from the target to the Clip plane.

4. At the Specify Eye/<distance from target> <current distance>: prompt, move the slide-bar diamond until your view looks similar to Figure 13.42. Then click the mouse button to fix the view.

5. To make sure the Clip plane is in the correct location, preview your perspective view with hidden lines removed. Enter **h** ↵ at the Enter option prompt. The drawing regenerates, with hidden lines removed.

There are several other Dview Clip options that let you control the location of the Clip plane.

Eye   Places the Clip plane at the position of the camera itself.

Back   Operates in the same way as the Front option, but it clips the view behind the view target instead of in front (see Figure 13.43).

Off   Turns off any Clip planes you may have set up.

FIGURE 13.43:

Effects of the Back Clip plane option

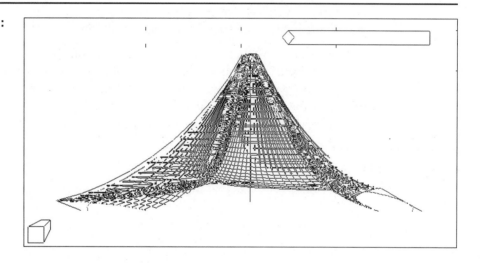

You've now completed the Dview-command exercises, and you should have a better understanding of how Dview can be used to get exactly the image you want.

If you like, you can use View ➤ Named Views to save your perspective views. This is helpful when you want to construct several views of a drawing to play back later as part of a presentation. You can also use the Hide and Shade options on the Tools pull-down menu to help you visualize your model.

TIP

The quickest and easiest way to establish a perspective view is to use the Viewpoint Presets dialog box to set up your 3D view orientation. Then start Dview and use the Distance option right off the bat to set your camera-to-target distance. Once this is done, you can use the other Dview options easily as needed, or exit Dview to see your perspective view.

Using the 3D Orbit Tool

New in AutoCAD 2000 is a tool named 3D Orbit. This tool offers many options for setting up perspective views of your three-dimensional drawings. The 3D Orbit tool has the same functions as the Dview command, described in the previous section, but it's easier to use. In the interactive 3D Orbit view, you can look at your entire model or at specific objects from different points.

With the 3D Orbit tool, you control your view with your mouse and an *arcball*, which is a circle with four smaller circles at its cardinal points. In 3D Orbit view, the target point you're viewing (the center of the arcball) remains stationary while you move the camera location around the target. Figure 13.44 shows an example of the press base drawing we've used in previous examples in 3D Orbit view.

FIGURE 13.44:

Viewing the press base with 3D Orbit

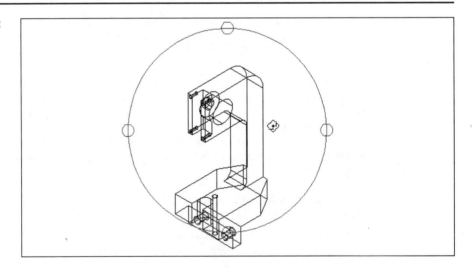

To try out the 3D Orbit tool, follow these steps:

1. Select the object or objects that you want to view. An easy way to set up your view is with the Camera tool in the View toolbar.

If you don't select any objects, you can view your whole drawing. However, the 3D Orbit tool works better when you view only selected objects.

2. Click the 3D Orbit tool in the 3D Orbit toolbar or in the Standard toolbar. You can also choose View ➤ 3D Orbit.

3. The arcball appears in the active viewport. Place the cursor on the small circle at the top of the arcball. The cursor changes shape to a vertically elongated ellipse. Click and drag the cursor downward from the top. The view follows your cursor, with the motion restrained vertically. Release the mouse where you want the view. Clicking and dragging from the small circle on the top or bottom of the arcball rotates your view around the X axis through the center of the arcball.

4. Place the cursor on the circle on the left side of the arc ball. The cursor now changes shape to an ellipse that is elongated horizontally. Click and drag the cursor to the left. The view follows your cursor, rotating around the target point from left to right. Clicking and dragging from the small circle on the left or right side of the arcball rotates the view around the Y axis through the center of the arcball.

5. Move the cursor outside the arcball. The cursor changes to a circle. Click and drag downward, and the view rotates in the direction that you move

the cursor. When the cursor is in this shape, you can rotate the view in the view plane.

6. Move the cursor inside the arcball. The cursor now looks like two superimposed ellipses. Click and drag the cursor in a circular motion. Notice how the view pivots in all directions around the target point.

7. Note that you cannot edit objects while 3D Orbit is active. To exit 3D Orbit, press Enter or Esc, or right-click and choose Exit from the pop-up menu.

TIP

You can change the target point, and therefore the point around which your 3D Orbit view rotates, by using the Camera tool in the View toolbar. Click the Camera tool, then press ↵ when you're prompted for a camera location. Then select the new target location. You can use an object in your drawing as a selection point.

When you're in 3D Orbit mode, right-click to see a pop-up menu of viewing options.

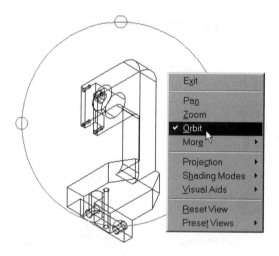

These options, many of which are also available as tools on the 3D Orbit toolbar, work as follows:

Exit Leaves 3D Orbit mode.

Pan Allows you to pan to adjust your 3D Orbit view (the arcball disappears, and the familiar panning cursor appears).

Zoom   Allows you to zoom in and out in 3D Orbit view. As you zoom, you change the field of view, or the focal length of the camera (similar to a telephoto lens on a camera).

More   Displays a submenu of options, which are described next.

Projection   Offers the choices Parallel and Perspective. Click Perspective, and AutoCAD will create a perspective view for you. Choose Parallel if you want to edit the drawing.

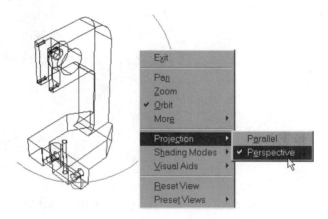

Shading Modes   Offers several shading methods. (AutoCAD will automatically represent the model as a simplified polygon if it's too complex to show normally.)

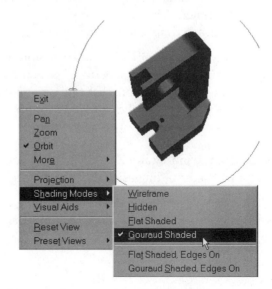

Visual Aids   Offers several visual aids for 3D Orbit view, including a compass and a grid.

Reset View and Preset Views   Allow you to return to the previous view or select from a list of saved views.

Click the More option on the pop-up menu to access more 3D Orbit choices.

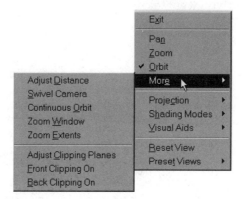

Adjust Distance   Allows you to adjust the camera distance from the target. The cursor turns into a double-headed arrow. Click and drag upward to bring the camera closer to the target. Click and drag downward to move the camera farther away.

Swivel Camera   Allows you to move the target of your view by rotating the camera. The cursor changes to a camera icon with a curved arrow. Click and drag to adjust the view.

Continuous Orbit   Provides a simple animated view by rotating your view about the target point. Click and drag to the left to spin your view in a clockwise direction; click and drag to the right to spin counterclockwise. The distance you click and drag controls the speed of the spin. Click again to stop spinning.

The Zoom Window and Zoom Extents options work as they do in other views, except with a three-dimensional perspective.

The Adjust Clipping Planes, Front Clipping On, and Back Clipping On options work as described in the previous section, "Using Clip Planes to Hide Parts of Your View."

NOTE   You cannot enter commands on the command line while 3D Orbit is active. However, if 3D Orbit is not active, you can enter a command that starts 3D Orbit and activates one of the options at the same time. For example, 3D Zoom starts 3D Orbit view and activates the Zoom option. Or, you can right-click, choose More from the pop-up menu, and then select Zoom Extents or Zoom Window. Both of these commands can be executed while 3D Orbit is still active.

If You Want to Experiment...

You've worked with a lot of new commands in this chapter, so now is a good time to play with these commands to help you remember what you've learned. Draw a propeller beanie using curved surfaces and the Arc tool, the Hide tool, the twisting effect that you learned in the section "Using Two Objects to Define a Surface," and the Mirroring tool. Remember to allow room for a head. Then try out Dview and 3D Orbit and their options as you develop your model.

PART IV

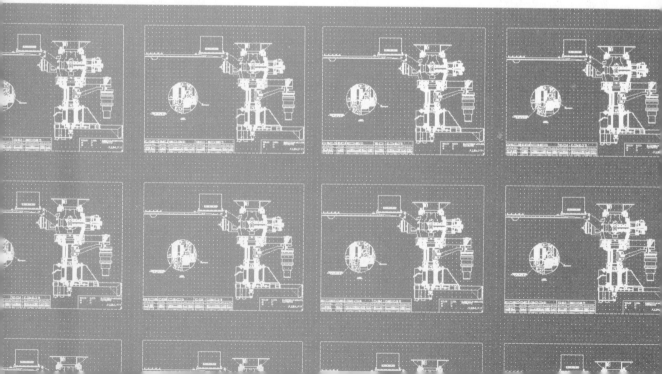

Printing and Plotting as an Expert

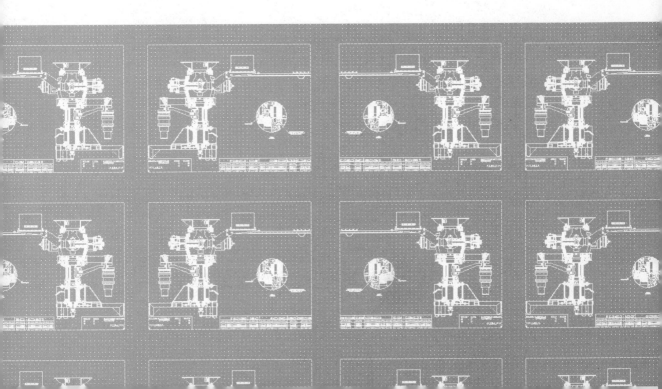

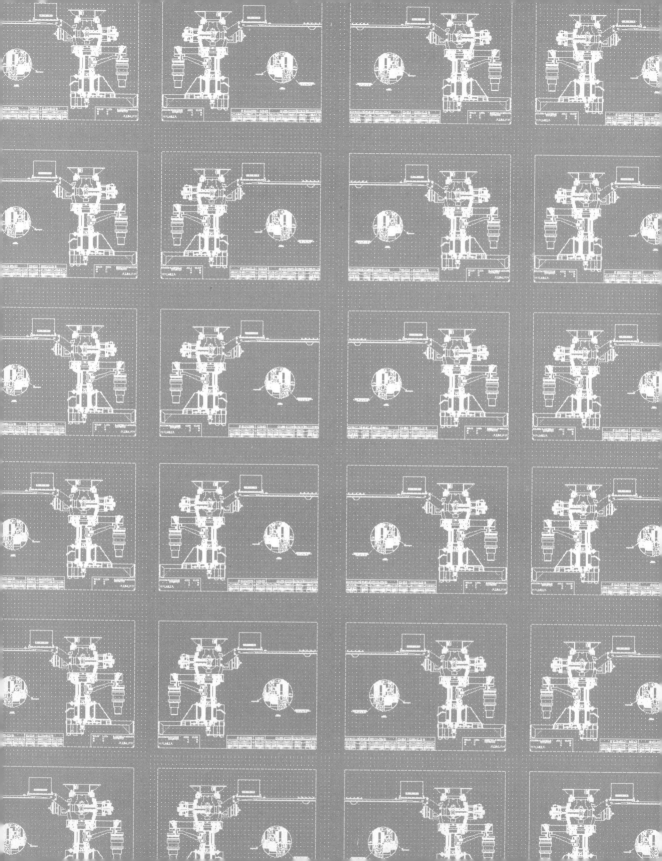

CHAPTER

FOURTEEN

14

Printing and Plotting

- Plotting the Base

- Understanding the Plotter Settings

- WYSIWYG Plotting Using Layout Tabs

- Setting Color, Line Corner Styles, and Shading Patterns with Plot Styles

- Assigning Plot Styles Directly to Layers and Objects

- Plotting Multiple Layout Tabs

- Adding an Output Device

- Storing a Page Setup

- Plotter and Printer Hardware Considerations

- Batch Plotting

- Sending Your Drawings to a Service Bureau

Getting hard-copy output from AutoCAD is something of an art. You'll need to be intimately familiar with both your output device and the settings available in AutoCAD. You'll probably spend a good deal of time experimenting with AutoCAD's plotter settings and with your printer or plotter to get your equipment set up just the way you want.

With the huge array of output options available, this chapter can provide only a general discussion of plotting. It's up to you to work out the details and fine-tune the way you and AutoCAD together work with your plotter. This chapter describes the features available in AutoCAD and discusses some general rules and guidelines to follow when setting up your plots.

You'll start out by getting an overview of the plotting features in AutoCAD. Then you'll delve into the finer details of setting up your drawing and controlling your plotter.

NOTE   If you've used an earlier version of AutoCAD, be sure you read through this chapter. Virtually every aspect of printing and plotting has changed in AutoCAD 2000.

Your First Plot—Plotting the Base

To see firsthand how the Plot command works, you'll plot the press base drawing using the default settings on your system. First, you'll look at a preview of your plot before you commit to actually printing your drawing.

1. Make sure your printer or plotter is connected to your computer and turned on.

2. Start AutoCAD and open the Ch14base drawing.

3. Choose View ➤ Zoom ➤ All to display the entire drawing.

4. Choose File ➤ Plot. You'll see the Plotting Help dialog box.

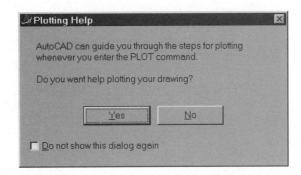

5. Here, you can choose to seek help from AutoCAD 2000 with the many plot options. Select No, and the Plot dialog box appears.

TIP   If you select Yes in the Plotting Help dialog box, you'll be in the new Fast Track to Plotting in AutoCAD help utility. It contains useful tips and audio/video tools.

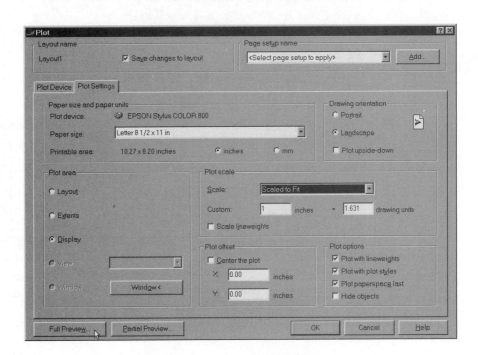

6.   Make sure that the Plot Settings tab is selected. Make the following selections:

- Click the Extents radio button in the Plot Area button group. This tells AutoCAD to plot the extents of the drawing much the same as Zoom Extents displays only the objects you've created.

- In the Plot Offset group, select Center the Plot.

- In the Plot Options group, select Plot with Lineweights, Plot with Plot Styles, and Hide Objects.

- In the Plot Scale group, open the Scale drop-down list and select Scaled to Fit.

- In the Drawing Orientation group, select the Portrait radio button.

7.   Click the Full Preview button in the lower-left corner of the dialog box. Auto-CAD works for a moment then displays a sample view showing you how your drawing will appear on your printer output.

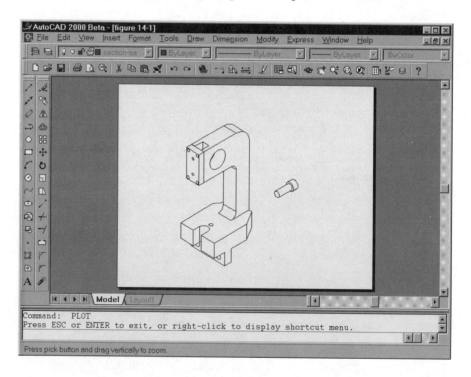

The print preview depends on the type of output device you chose when you installed AutoCAD or when you last selected a Plotter Device option.

NOTE   The print preview is also affected by other settings in the Plot dialog box, such as those in the Drawing Orientation, Plot Offset, and Plot Area groups. The previous example showed a typical preview using the Windows default system printer in landscape mode.

Notice that the view also shows the Zoom Realtime cursor. You can use the Zoom/Pan Realtime tool to get a close-up look at your print preview. Now go ahead and plot the file.

8. Right-click, then select Plot from the pop-up menu. AutoCAD sends the drawing to your printer.

9. Because you chose to plot to fit, your plotter or printer prints out the plan to no particular scale.

You've just done your first plot to see how the drawing looks on paper. You used the minimal settings to ensure that the complete drawing appeared on the paper. Next try plotting your drawing to an exact scale.

1. Choose File ➤ Plot again.

2. Select the Portrait option in the Drawing Orientation group.

3. Open the Scale drop-down list and select 1:4. In the Plot Offset group, select Center the Plot.

4. In the Paper Size and Paper Units group, select the Inches radio button then select Letter (8.5 × 11"). The options available in this drop-down list will depend on your Windows system printer or the output device you've configured for AutoCAD.

5. In the Plot Area button group, click the Layout option. This tells AutoCAD to use the layout of your drawing to determine which part of your drawing to plot.

6. Click Full Preview again to get a preview of your plot.

7. Right-click and select Plot. This time, your printout is to quarter scale.

Here, you were asked to make a few more settings in the Plot dialog box. A number of settings work together to produce a drawing to scale that fits properly on your paper. This is where it pays to understand the relationship between your drawing scale and your paper's size. This was discussed in Chapter 3, "Learning the Tools of the Trade," when we covered setting up the drawing limits.

NOTE
The next section is lengthy, but doesn't contain any exercises. If you prefer to continue with the exercises in this chapter, skip to the section "WYSIWYG Plotting Using Layout Tabs." Be sure to come back and read the following section while the previous exercises are still fresh in your mind.

Understanding the Plotter Settings

In this section, you'll explore all of the settings that are available in the Plot Settings tab of the Plot dialog box. These settings give you control over the size and orientation of the plotted image on the paper. They also let you control what part of your drawing gets printed. All of these settings work together to give you control over how your drawing will fit on your printed output.

WARNING
If you're a veteran AutoCAD user, be aware that AutoCAD 2000 now relies mainly on the Windows system printer configuration instead of its own plotter drivers. This gives you more flexibility and control over your output, but it also may create some confusion if you're used to the previous method that AutoCAD used to set up plots. Just be aware that you'll need to understand the Windows system printer settings in addition to those offered by AutoCAD.

Paper Size and Paper Units

The options in this group allow you to select the paper size and the measurement system you're using. You can select a paper size from the Paper Size drop-down list.

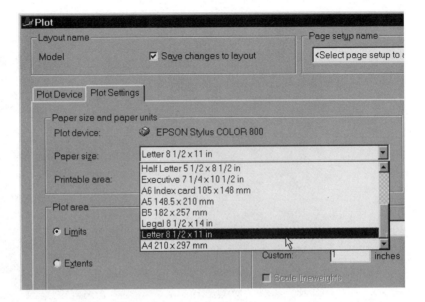

These sizes are derived from the sizes available from your currently selected system printer. You'll find out how to select a different printer in the section "Adding an Output Device" later in this chapter.

The Printable Area radio buttons offer inches and millimeters (mm). When you select one of these options, the printable area listing changes to show you the area in the units you selected. Options in the Plot Scale group also change to reflect the measurement system choice.

Drawing Orientation

When you used the Full Preview option in the first exercise of this chapter, you saw your drawing as it would be placed on the paper. In the illustration shown in our example, it was placed in what is called a *portrait orientation*. You can rotate the image on the paper into what is called a *landscape orientation* by selecting the Landscape radio button in the Drawing Orientation group (see the result in Figure 14.1). A third option, Plot Upside-Down, lets you change the orientation further by turning the landscape or portrait orientation upside down. These three settings let you change the image to any one of four orientations on the sheet.

FIGURE 14.1:

An example of a plot in the landscape orientation on an 8½" × 11" plot

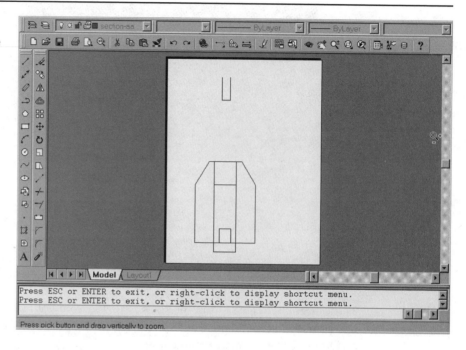

Note that in AutoCAD the preview displays the paper in the orientation that it leaves the printer. So for most small-format printers, if you're printing in the portrait orientation, the image appears in the same orientation as you see it when you're editing the drawing. If you're using the landscape orientation, the preview image is turned sideways. For large-format plotters, the preview may be oriented in the opposite direction.

Plot Area—Choosing What to Plot

The radio buttons on the lower-left side of the Plot dialog box let you specify which part of your drawing you wish to plot.

The Limits Option

The Limits printing option uses the limits of the drawing to determine what to print (see Figure 14.2). If you let AutoCAD fit the drawing onto the sheet (by selecting Scaled to Fit from the Scale drop-down list), the plot displays exactly the same thing that you would see on the screen had you selected View ➤ Zoom ➤ All.

FIGURE 14.2:

The screen display and the printed output when Limits is chosen

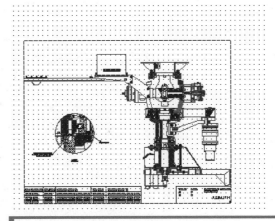

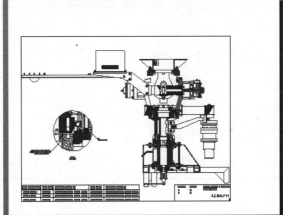

The Extents Option

The Extents option creates the entire drawing, eliminating any space that may border the drawing (see Figure 14.3). If you let AutoCAD fit the drawing onto the sheet (that is, you select Scaled to Fit from the Scale drop-down list), the plot displays exactly the same thing that you would see on the screen if you had selected View ➤ Zoom ➤ Extents.

FIGURE 14.3:

The printed output when Extents is chosen

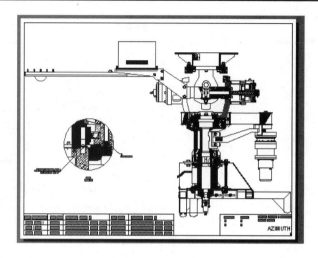

The Display Option

Display is the default option; it tells AutoCAD to plot what is currently displayed on the screen (see the top image in Figure 14.4). If you let AutoCAD fit the drawing onto the sheet (that is, you select the Scaled to Fit option from the Scale drop-down list), the plot is exactly the same as what you see on your screen (bottom image in Figure 14.4).

FIGURE 14.4:

The screen display and the printed output when Display is chosen and no Scale is used (the drawing is scaled to fit the sheet)

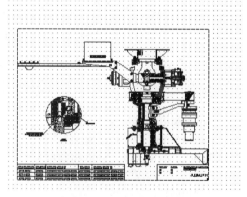

**FIGURE 14.4
CONTINUED:**

The screen display and the
printed output when Display
is chosen and no Scale is
used (the drawing is scaled
to fit the sheet)

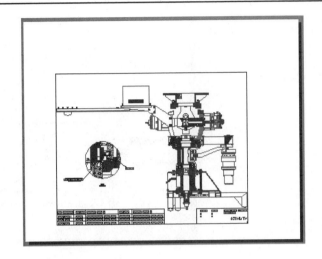

The View Option

The View printing option uses a previously saved view to determine what to
print (see Figure 14.5). To use this option, you must first create a view. Next click
the View radio button. You can then select a view from the drop-down list to the
right of the View radio button.

FIGURE 14.5:

A comparison of the saved
view and the printed output

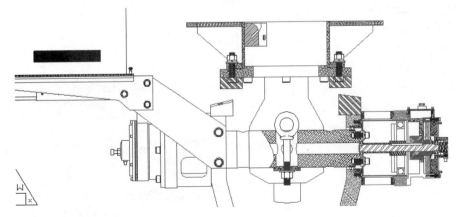

FIGURE 14.5 CONTINUED:

A comparison of the saved view and the printed output

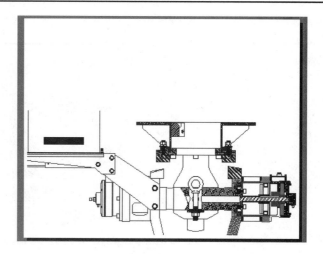

If you let AutoCAD fit the drawing onto the sheet (by selecting Scaled to Fit from the Scale drop-down list), the plot displays exactly the same thing that you would see on the screen if you had recalled the view you're plotting.

The Window Option

The Window option allows you to use a window to indicate the area you wish to plot (see Figure 14.6). Nothing outside the window prints.

FIGURE 14.6:

A selected window and the resulting printout

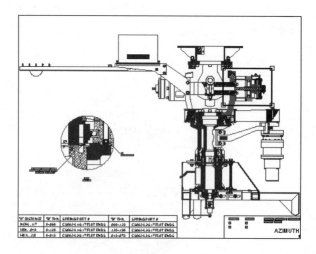

FIGURE 14.6
CONTINUED:

A selected window and the
resulting printout

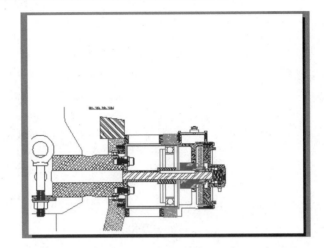

To use this option, click the Window button, then indicate a window in the drawing editor. The dialog box temporarily closes to allow you to select points. When you're done, click OK.

If you let AutoCAD fit the drawing onto the sheet using the Scale to Fit option in the Scale drop-down list, the plot displays exactly the same thing that you enclose within the window.

TIP   Do you get a blank print, even though you selected Extents or Display? Chances are the Scale to Fit option is not selected, or the Plotted Inches = Drawing Units setting is inappropriate for the sheet size and scale of your drawing. If you don't care about the scale of the drawing, make sure the Scale to Fit check box is checked. Otherwise, make sure the Plot Scale settings are set correctly. The next section describes how to set the scale for your plots.

Plot Scale—Choosing the Size of the Objects Being Plotted

In the previous section, the descriptions of several plot-area options indicated that the Scale to Fit check box must be checked. Bear in mind that when you apply a scale factor to your plot, it changes the results of the plot-area settings, and some problems can arise. This is usually where most new users have difficulty.

For example, an automobile drawn full size fits nicely on the paper when you use Scale to Fit. But if you tried to plot the drawing at full scale, you would probably get a blank piece of paper because, at that scale, hardly any of the automobile would fit on your paper. AutoCAD would tell you that it was plotting and then tell you that the plot was finished. You wouldn't have a clue as to why your sheet was blank.

If an image is too large to fit on a sheet of paper because of improper scaling, the plot image will be placed on the paper differently, depending on whether the plotter uses the center of the image or the lower-left corner for its origin. Keep this in mind as you specify scale factors in this area of the dialog box.

Selecting a Scale

You can select a drawing scale from a set of predefined scales. These are offered through the Scale drop-down list, and the options cover the most common scales you'll need to use.

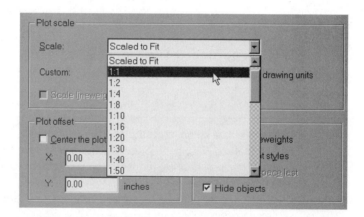

You've already seen how one option from this drop-down list, Scale to Fit, allows you to avoid giving a scale altogether and forces the drawing to fit on the sheet. This works fine if you're doing illustrations that aren't to scale. If you select another option like 1:4, you'll see the Custom input boxes change to reflect this scale. The Inches input box changes to 1 and the Drawing Units input box changes to 4, the scale factor for a 1/2 scale drawing.

The Custom Input Boxes

If you can't find the scale you want in the Scale drop-down list, you can use the two Custom input boxes that are provided in the Plot Scale group: Inches (or mm if you select mm in the Paper Size and Paper Units group) and Drawing Units.

Through these input boxes, you can indicate how the drawing units within your drawing relate to the final plotted distance in inches or millimeters. For example, if your drawing is of a scale factor of 1:2, follow these steps.

1. Double-click the Inches input box, enter **1,** and press the Tab key.

2. Double-click the Drawing Units input box, enter **2,** and press the Tab key.

Metric users who want to plot to a scale of 1:10 should enter **1** in the mm input box and **10** in the Drawing Units input box.

If you're more used to the English measurement system, you can enter a scale as a fraction. For example, for a 1/8" scale drawing, follow these steps:

1. Double-click the Inches input box, enter **1/8**, and press the Tab key.

2. Double-click the Drawing Units input box, enter **12**, and press the Tab key

You can specify a different scale from the one you chose while setting up your drawing, and AutoCAD will plot your drawing to that scale. You're not restricted in any way as to scale, but entering the correct scale is important: If it is too large, AutoCAD will think your drawing is too large to fit on the sheet, although it will attempt to plot your drawing anyway. (See Chapter 3, "Learning the Tools of the Trade," for a discussion on unit styles and scale factors.)

TIP   If you plot to a scale that is different from the scale you originally intended, objects and text will appear smaller or larger than would be appropriate for your plot. You'll need to edit your text size to match the new scale. Review text style in Chapter 7, "Adding Text to Your Drawings," and dimension style in Chapter 8, "Using Dimensions." It is a good idea to set up your drawing with a title block early in the process to avoid these types of problems.

The Scale Lineweights Option

New in AutoCAD 2000 is the option to assign lineweights to objects either by their layer assignments or by directly assigning a lineweight to individual objects. However, the lineweight option doesn't have any meaning until you specify a scale for your drawing. Once you do specify a scale, the Scale Lineweights option is available. Check this box if you want the lineweight assigned to layers and objects to appear correctly in your plots.

Plot Offset—Positioning the Plot on the Plot Media

Frequently, your first plot of a drawing shows the drawing positioned incorrectly on the paper. You can fine-tune the location of the drawing on the paper by using the Plot Offset settings.

To adjust the position of your drawing on the paper, enter the location of the view origin in relation to the plotter origin in X and Y coordinates (see Figure 14.7).

FIGURE 14.7:

Adjusting the image location on a sheet

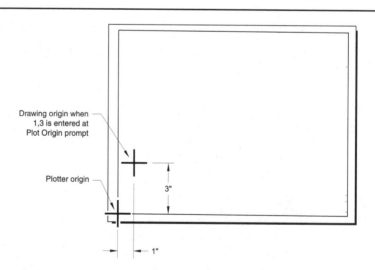

For example, suppose you plot a drawing, then realize that it needs to be moved 1" to the right and 3" up on the sheet. You can replot the drawing by making the following changes:

1. In the Plot Offset group, double-click the X input box, type **1**, and press the Tab key.

2. Double-click the Y input box, type **3**, and press the Tab key.

Now proceed with the rest of the plot configuration. With the above settings, when the plot is done, the image is shifted on the paper exactly 1" to the right and 3" up.

Plot Options

The options in the Plot Options group, with the exception of Hide Objects, are new in AutoCAD 2000. These new features offer a greater amount of control over your output and require some detailed instruction. Here is a brief description of these options. You'll learn more about them in the next section.

The Plot with Lineweights Option

You may recall that AutoCAD 2000 lets you assign lineweights to objects either through their layer assignment or by directly assigning a lineweight to the object itself. If you use this feature in your drawing, this option lets you turn lineweights on or off in your output.

The Plot with Plot Styles Option

Plot styles are new in AutoCAD 2000 and they give you a high degree of control over your drawing output. You can control whether your output is in color or black and white, and you can control whether filled areas are drawn in a solid color or a pattern. You can even control the way lines are joined at corners. You'll learn more about these options and how they affect your work in the next section.

The Plot Paperspace Last Option

When you're using Paper Space, this option determines whether objects in Paper Space are drawn before or after objects in Model Space. You'll learn more about Model Space and Paper Space printing in the section "WYSIWYG Plotting Using Layout Tabs," later in this chapter.

The Hide Objects Option

This option pertains to 3D models in AutoCAD. When you draw in 3D, you'll see your drawing as a *wire-frame view*. In a wire-frame view, your drawing looks like it's transparent even though it is made up of "solid" surfaces. You can view and plot your 3D drawings so that solid surfaces are opaque using hidden line removal. To view a 3D drawing in the editor with hidden lines removed, use the Hide command. To plot a 3D drawing with hidden lines removed, use the Hide Objects option.

WYSIWYG Plotting Using Layout Tabs

We talked about the Layout tabs and Paper Space in Chapter 9. You can have as many Layout tabs as you like, each set up for a different type of output. For example, you can have two or three different Layout tabs each set up for a different scale drawing or with various layer configurations for differing drawing requirements. Because the layout borders act like "picture frames," you can even hide parts of the drawing outside the layout that you don't wish to plot, such as sketches. You can even set up multiple views of your drawing at different scales within a single Layout tab.

NOTE   When you create a new file using AutoCAD 2000, you see two Layout tabs. If you open a pre-AutoCAD 2000 file, you only see one Layout tab.

To get familiar with the Layout tabs, try the following exercise.

1. With the `Figure 14-1.dwg` file open, click the tab labeled Layout1. You see the Page Setup dialog box.

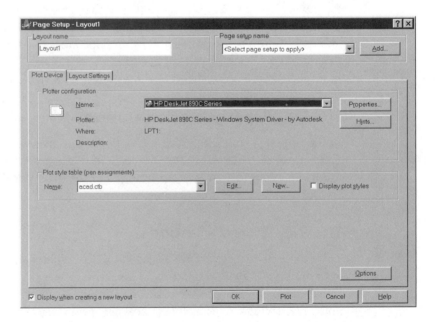

NOTE   When creating a new layout, if you remove the check from the display check box in the lower-left corner of the Page Setup dialog box, AutoCAD does not display the Page Setup dialog box the first time you select a Layout tab.

Make sure the Layout Settings tab is selected, then select Letter (8.5 × 11") from the Paper Size drop-down list. Notice that the Page Layout dialog box is identical to the Plot dialog box.

2. Click OK. A view of your drawing appears on a gray background as shown in Figure 14.8. This is a view of your drawing as it will appear when plotted on your current default printer or plotter. The white area represents the printer or plotter paper.

FIGURE 14.8:

A view of the layout

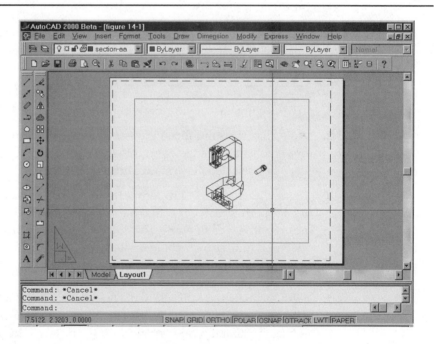

3. Try zooming in and out using the Realtime Zoom tool. Notice that the entire image zooms in and out, including the area representing the paper.

Let's take a moment to look at the different elements of the Layout1 tab. As I mentioned, the white background represents the paper on which your drawing will be printed. The dashed line immediately inside the edge of the white area represents the limits of your printer's margins. Finally, the solid rectangle that surrounds your drawing is the outline of the layout viewport. You may also notice the triangular symbol in the lower-left corner of the view.

This is the UCS icon for the Layout tab. It tells you that you're currently in the Layout tab space. The significance of this icon will be clearer in the following exercise.

1. Try selecting part of your drawing. Nothing is selected.

2. Now click the solid rectangle surrounding the drawing, as shown in Figure 14.9. This is the viewport into the Model tab. Notice that you can select it. (When selected, its grips are turned on.)

3. Right-click and select Properties from the pop-up menu. You can see from the Properties dialog box that the viewport is just like any other AutoCAD object with layer, linetype, and color assignments. You can even hide the viewport outline by turning off its layer.

4. Close the Properties dialog box.

5. With the viewport still selected, click the Erase tool in the Modify toolbar. The view of your drawing disappears with the erasure of the viewport. Remember that the viewport is like a window into the drawing you created in the Model tab. Once the viewport is erased, the drawing view goes with it, but not the drawing itself.

FIGURE 14.9:

Selecting the viewport

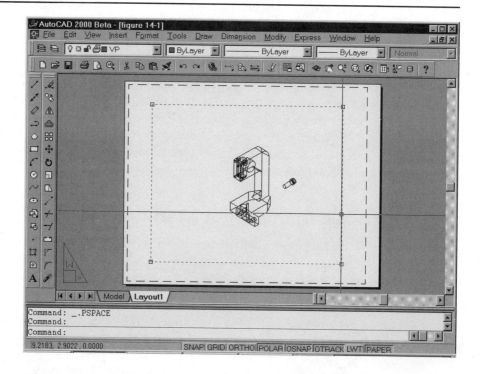

FIGURE 14.9:

Selecting the viewport

6. Type **u↵** or click the Undo button in the Standard toolbar to restore the viewport.

7. Double-click anywhere within the viewport's boundary. Notice that the UCS icon you're used to seeing appears in the lower-left corner of the viewport. The Layout UCS icon disappears.

8. Now try clicking your drawing. You can now select parts of it.

9. Try zooming and panning your view. Changes in your view only take place within the boundary of the viewport.

10. Choose View ➤ Zoom ➤ All or type **z↵a↵** to display the entire drawing in the viewport.

11. Type **ps↵** to return to Layout Space (Paper Space) and **ms↵** to access the space within the viewport.

This exercise shows you the unique characteristics of the Layout tab. The objects within the viewport are inaccessible until you double-click its interior. You can then move about and edit your drawing within the viewport, just as you would while in the Model tab.

The Layout tabs can contain as many viewports as you like, and each viewport can hold a different view of your drawing. Each viewport can be sized and arranged in any way you like or you can even create multiple viewports. Doing so gives you the freedom to lay out your drawing as you would a page in a desktop-publishing program. You can also draw in the Layout tab, or import Xrefs and blocks for title blocks and borders.

Plot Scale in the Layout Tab Viewports

In the first part of this chapter, you plotted your drawing from the Model tab. You learned that to get the plot to fit onto your paper, you either had to use the Scale to Fit option in the Plot Settings tab of the Plot dialog box, or you had to indicate a specific drawing scale, plot area, and drawing orientation.

The Layout tab works in a different way: It is designed to allow you to plot your drawing at a 1-to-1 scale. Instead of specifying the drawing scale in the Plot dialog box as you did when you plotted from the Model tab, your drawing scale is determined by the size of your view in the Layout tab viewport. You can set the viewport view to an exact scale by making changes to the properties of the viewport.

To set the scale of a viewport in a Layout tab, try the following exercise. (If you closed the Figure 14-1 drawing from the previous exercises, reopen it, or use the Figure 14-1 drawing from the CD.)

1. Press the Escape (Esc) key twice to clear any selections. Then click the viewport border to select it.

2. Right-click, and then select Properties from the pop-up menu. The Properties dialog box for the viewport appears.

3. Make sure that the Categorized tab is selected, then locate the Standard Scale option under the Misc category. Go ahead and click the Standard Scale option. The item to its right turns into a list box.

4. Open the list box and select 1:2. The view in the drawing window changes to reflect the new scale for the viewport. Now the drawing fits into the viewport and it is to half-scale.

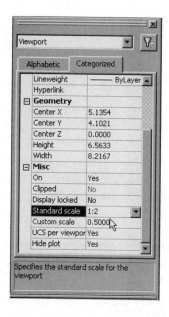

5. Close the Properties dialog box.

6. If the drawing was too large, you could use the viewport grips to enlarge the viewport enough to display all of the drawing.

TIP   You only need to move a single corner grip to resize the layout viewport. As you move a corner grip, notice that the viewport maintains a rectangular shape.

---

7. Choose File ➢ Plot, then at the Plot dialog box, make sure the Scale option in the Plot Settings tab is set to 1:1. Click OK. Your drawing is plotted to scale as it appears in the Layout tab.

8. After reviewing your plot, close the drawing without saving it.

In step 4, you saw that you could choose a scale for a viewport by selecting it from the Properties dialog box. If you look just below the Standard Scale option, you'll see the Custom Scale option. Both of these options work like their counterpart, the Plot Scale group, in the Plot Settings tab of the Plot dialog box.

TIP   Veteran AutoCAD users can still use the View ➤ Zoom ➤ Scale option to control the scale of the viewport view.

Layout tabs and viewports work in conjunction with your plotter settings to give you a better idea of how your plots will look. In fact, there are numerous plotter settings that can dramatically change the appearance of your Layout tab view and your plots. In the next section, you'll learn how some of the plotter settings can enhance the appearance of your drawings. You'll also learn how Layout tabs can display those settings so you can see on your computer screen exactly what will appear on your paper output.

Controlling the Appearance of the Layout Tabs

The Options dialog box offers a set of controls dedicated to the Layout tabs. If you don't like some of the graphics in the Layout tab, you can turn them off. Open the Options dialog box and select the Display tab. You'll see a set of options in the Layout Elements group.

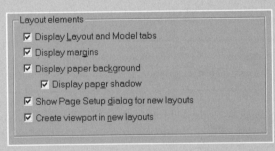

As you can see, you can control the display of the tabs themselves, the margins, the paper background, and the paper shadow. In addition, you can set whether or not AutoCAD automatically creates a viewport or opens the Page Setup dialog box when you open a Layout tab for the first time.

Setting Color, Line Corner Styles, and Shading Patterns with Plot Styles

AutoCAD 2000 introduces a new feature called *plot-style tables*. Plot-style tables are settings that give you control over how objects in your drawing are translated into hard-copy plots. You can control how colors are translated into lineweight and how area fills are converted into shades of gray or screened colors, as well as many other output options. You can also control how the plotter treats each individual object in a drawing.

If you don't use plot-style tables, your plotter will produce output as close as possible to what you see in the drawing editor, including colors. However, you can force your plotter to plot all colors in black, for example. You can also assign a fill pattern or screen to a color. This can be useful for charts and maps that require area fills of different gradations.

The following set of exercises will show you firsthand how plot-style tables can be used to enhance your plotter output. You'll look at how you can adjust the lineweight of selected objects and make color changes to your plotter output.

Choosing between Color and Named Plot-Style Tables

AutoCAD offers two types of plot-style tables: *color* and *named*. Color plot-style tables allow you to assign plotting properties to the AutoCAD colors. For example, you can assign a 0.50 mm pen width to the color red so that anything that is red in your drawing is plotted with a line width of 0.50 mm. You can, in addition, set the pen color to black so that everything that is red in your drawing is plotted in black.

Named plot-style tables let you assign plotting properties directly to objects in your drawing, instead of relying on their color properties. They also allow you to assign plotter properties directly to layers. For example, with named plot styles, you can assign a black pen color and a 0.50 mm pen width to a single circle in a drawing, regardless of its color. Named plot styles are more flexible than color plot styles, but if you already have a library of AutoCAD drawings set up for a specific set of plotter settings, the color plot styles would be a better choice when opening those older files in AutoCAD 2000. This is because color plot styles are more similar to the older method of assigning AutoCAD colors to plotter pens.

You may also want to use color plot-style tables with files that you intend to share with an individual or office that is still using earlier versions of AutoCAD.

WARNING   The type of plot-style table assigned to a drawing depends on the settings in the Output tab of the Options dialog box at the time the file is created. In the case of drawings created in earlier versions of AutoCAD, the type of plot-style table used depends on the settings of the Output tab of the Options dialog box the first time the file is opened in AutoCAD 2000. Once a choice is made between color or named plot styles, your drawing cannot change to the other type of plot style. So choose wisely.

Here's how to set up the plot style type for new and pre-AutoCAD 2000 files.

1.   Open the Options dialog box and select the Plotting tab.

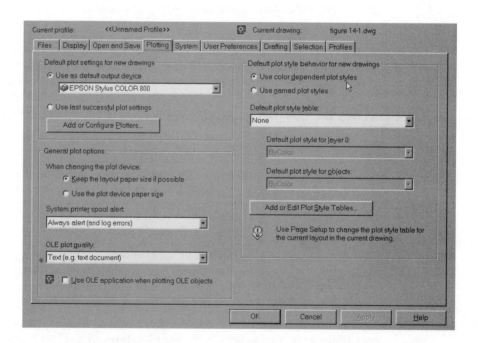

2.   In the Default Plot Style Behavior for New Drawings group, click Use Color Dependent Plot Styles. In an exercise in the section "Using Named Plot-Style Tables" later in this chapter, you'll use the other option.

   3.   Click OK to return to the drawing.

   Next you'll set up a custom color plot-style table.

> **NOTE** Plot-style tables are stored with the `.ctb` or `.stb` filename extension. The files with a `.ctb` extension are color plot-style tables. The files with an `.stb` extension are named plot-style tables.

> **NOTE** Veteran AutoCAD users who use Hewlett-Packard Inkjet plotters may be familiar with many of the settings available in the plot-style table options. These are similar to the settings offered by the Hpconfig command from prior versions of AutoCAD.

Creating a Color Plot-Style Table

   You can have several plot-style table files on hand to quickly apply plot styles to any given plot or Layout tab. Each plot-style table can be set up to create a different look to your drawing. These files are stored in the Plot Styles directory off of the main AutoCAD directory. Take the following steps to create a new plot-style table. You'll use an existing file that was created in Release 14 as an example to demonstrate the plot-style features.

   1.   Open the sample file from the companion CD called Press Base.dwg, then click the Layout1 tab.

> **NOTE** This file's properties have been set to read-only to keep it in R14 format, so you'll receive a warning when you open the drawing. When asked if you would like to Open the file read-only?, select Yes. This means that your changes to this drawing cannot be saved under its current name. If you wish to save this file as an AutoCAD 2000 drawing, use a different name.

   2.   Right-click the Layout1 tab and click Page Setup. In the Page Setup dialog box, click the Plot Device tab. Notice that the Page Setup dialog box is similar

to the Plot dialog box. The main difference is that the Page Setup dialog box has an extra button at the bottom labeled Plot.

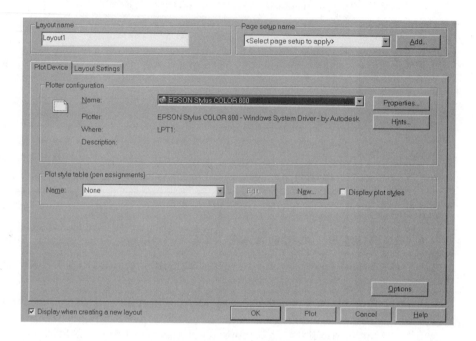

3. In the Plot Style Table (Pen Assignments) group, click the New button. This opens the Add Color-Dependent Plot Syle Table wizard.

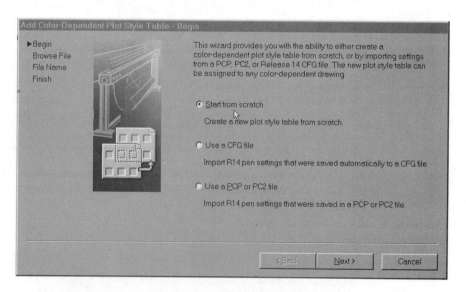

4. Click the Start from Scratch radio button, then click Next. The next page of the wizard asks for a filename. You can also specify whether this new plot-style table you are creating will be the default for all drawings from now on, or whether you want to apply this plot-style table to the current drawing only.

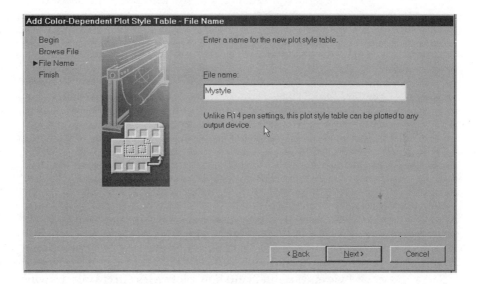

5. Enter **mystyle** for the filename, then click Next. The next page of the wizard lets you edit your plot style and assign the plot style to your current, new, or old drawings. You'll learn about editing plot styles in the next section.

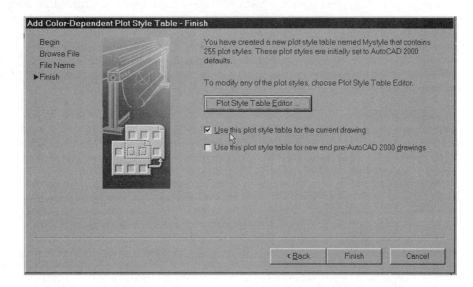

6. Click Finish. You'll return to the Page Setup dialog box. Notice your new plot-style table `Mystyle.ctb` is now in the Name box.

The Add Color-Dependent Plot Style Table wizard allows you to create a new plot-style table from scratch. Or, you can create one based on an AutoCAD R14 `.cfg`, `.pcp`, or `.pc2` file. You can also access the Add Color-Dependent Plot Style Table wizard by choosing File ≻ Plot Style Manager, then double-clicking the Add Color-Dependent Plot Style Table Wizard icon.

The steps shown here are the same whether your drawing is set up for color or named plot styles.

Editing and Using Plot-Style Tables

You now have your own plot-style table. In the next exercise, you'll get to edit the plot style and see firsthand how plot styles affect your drawing.

1. While in the Page Setup dialog box, make sure that the Plot Device tab is selected.

2. The filename `Mystyle.ctb` should appear in the Name drop-down list of the Plot Style Table group. If it isn't, open the Name drop-down list, then locate and select `Mystyle.ctb`.

3. Click the Edit button to the right of the drop-down list. The Plot Style Table Editor appears. Make sure the Form View tab is selected.

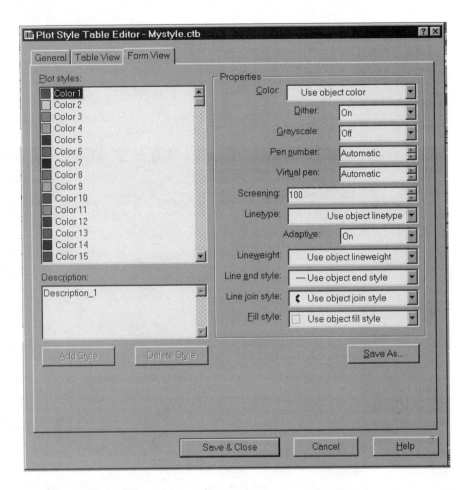

The Plot Style Table Editor dialog box offers three tabs that give you control over how each color in AutoCAD is plotted. The default Form View tab lets you select a color from a list box, then set the properties of that color using the options on the right side of the tab.

NOTE   The Table View tab displays each color as a column of properties. Each column is called a *plot style*. The property names are listed in a column to the far left. While the layout is different, both the table view and the form view offer the same functions.

You'll continue by changing the Lineweight property of the Color 3 (green) plot style. Remember that green is the color assigned to the press base in this drawing.

4. Click the Color 3 listing in the Plot Styles list box

5. Click the Lineweight drop-down list and select 0.50 mm.

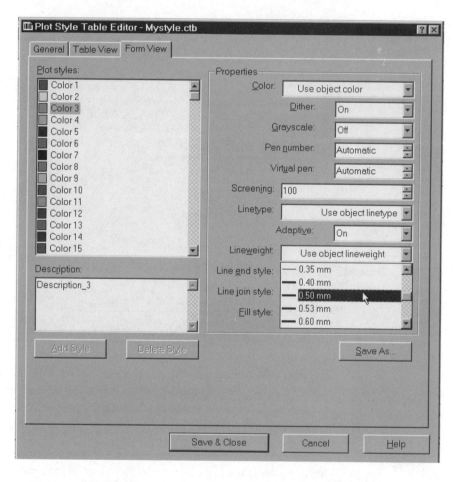

6. Click Save & Close. You'll return to the Page Setup dialog box.

7. Click the Display Plot Styles check box in the Plot Style Table group, then click OK.

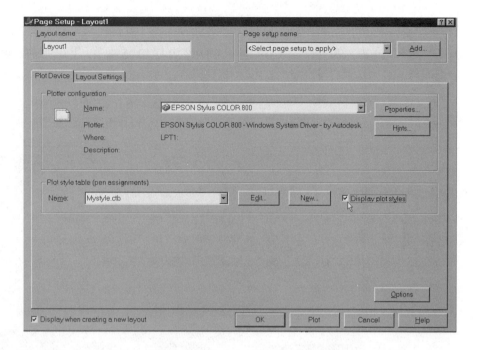

8. You may not notice any change at first. But if you click Full Preview, and then zoom into the preview drawing to enlarge the view, you'll see lines similar to those shown in Figure 14.10.

FIGURE 14.10:

Green object lines with 0.50 mm lineweight

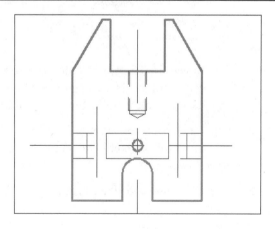

Making Your Plot Styles Visible

You won't see any changes in your drawing yet. You'll need to make one more change to your drawing options.

1. Choose Options ➢ User Preferences. In the User Preferences tab, click the Lineweight Settings button to open the Lineweight Settings dialog box.

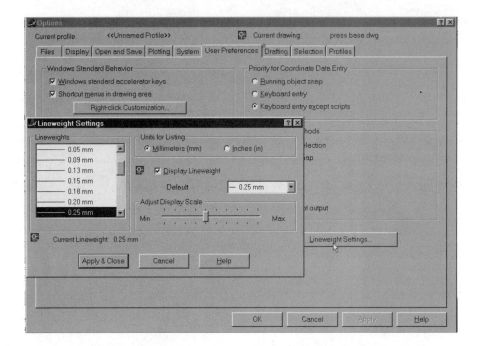

The Lineweight Settings dialog box lets you control the appearances of lineweights in the drawing editor. If lineweights are not showing up, this is the place to look to make them viewable.

2. Click the check box labeled Display Lineweight to turn on this option. This will cause the selected lineweights to be displayed in Model Space and Paper Space also.

3. Click Apply & Close in the dialog box.

4. Choose View ➤ Regen All. Notice that now the green lines have a width or lineweight.

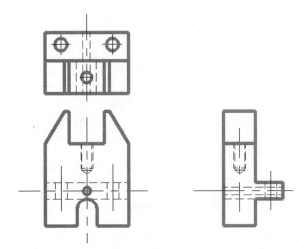

Lineweights Past and Present

In the section "Editing and Using Plot-Style Tables," earlier in this chapter you learned how you can assign a lineweight to an AutoCAD color. In fact, this is the method used in earlier versions of AutoCAD for controlling lineweight. But prior to AutoCAD 2000, there was no way to view the effects of lineweight settings until you produced a printout, nor was there a tool like the Plot Style Manager to help you take control over how AutoCAD colors were plotted. Plot-style tables not only give you a higher degree of control over the translation of AutoCAD colors to final plot, but they also add some additional features. Additionally, the Layout tabs let you proof your color settings before you commit your drawing to paper.

AutoCAD 2000 also lets you assign lineweights through the Layer Properties Manager. Layers have a Lineweight property that can be set inside the Layer Properties Manager dialog box. You can also assign lineweights directly to objects through the Properties dialog box. If you assign lineweights through layers or object properties, you can take advantage of the Use Object Lineweight option in the Plot Style Editor to display and plot the lineweights as you intend them.

Remember that if you want to view any lineweight setting, make sure that you turn on the Display Lineweight option as described in the section "Making Your Plot Styles Visible."

Making Changes to Multiple Plot Styles

Chances are, you'll want to plot your drawing in black and white for most of your work. You can edit your color plot-style table to plot one or all of your AutoCAD colors as black instead of the AutoCAD colors.

You saw in the section "Editing and Using Plot-Style Tables" (earlier in this chapter) how you could open the Plot Style Table Editor from the Page Setup dialog box to edit your color plot-style table. In this exercise, you'll try a different route.

1. Choose File ➢ Plot Style Manager. The Plot Styles window appears. This is a view to the Plot Style directory under the AutoCAD 2000 directory.

2. Locate the file Mystyle.ctb and double-click it. The Plot Style Table Editor appears.

3. In the Plot Style Table Editor dialog box, make sure Form View tab is selected.

4. Click Color 3 in the Plot Style list box.

5. Click the Color drop-down list and select Black.

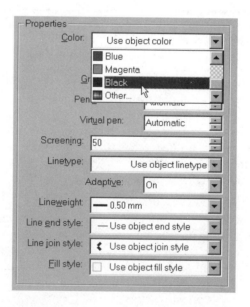

6. Click Save & Close, then close the Plot Styles window.

7. Choose View ➢ Regen All to view your drawing. In the Plot dialog box, click Full Preview. Now the green objects appear black in the Layout tab. Close the preview by pressing [Esc] twice. Now click the Cancel Plot button.

8. Click the Model tab to view your drawing in Model Space. Notice that the objects are still in their original colors. This shows that you haven't actually changed the colors of your objects or layers. You've only changed the color of the plotted output.

Next try changing all of the output colors to black.

1. Follow steps 1 and 2 of the previous exercise to open the `Mystyle.ctb` file again.

2. Make sure that the Form View tab is selected, then click Color 1 in the Plot Style list box.

3. Shift-click Color 9 in the Plot Styles list box. All of the plot styles from Color 1 to Color 9 are selected.

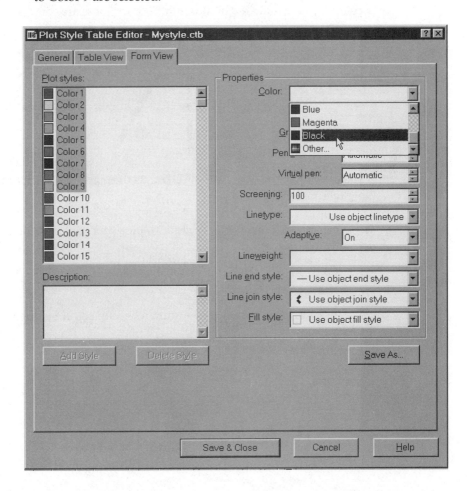

4. Click the Color drop-down list and select Black.

5. Click Save & Close, then click OK to close the Page Setup dialog box.

6. Choose View ➤ Regen All. In the Plot dialog box, click Full Preview. Now the green objects appear black in the Layout tab. Close the preview and click Cancel Plot.

Now when you plot your drawing, you'll get a plot that is composed of black lines entirely.

Setting Up Line Corner Styles

You may notice that the corners of the object lines appear to be notched instead of having a crisp, sharp corner. This will only be apparent in situations where you've used a very wide lineweight. But if this occurs, there are some options.

You can adjust the way AutoCAD draws these corners at plot time through the Plot Style Table Editor.

1. Once again, open the Mystyle.ctb plot-style table, as you did in the previous exercise.

2. Make sure that the Form View tab is selected, then click Color 3 in the Plot Style list box.

3. Click the Line End Style drop-down list and select Square.

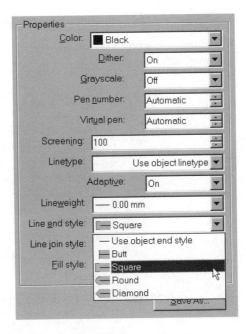

4. Click Save & Close, then click OK to close the Page Setup dialog box.

5. Choose View ➤ Regen All to view your changes. Notice that now the corners meet in a sharp angle.

The Square option in the Line End Style drop-down list extends the endpoints of contiguous lines so that their corners meet in a clean corner instead of a notch. The Line Join Style drop-down list offers a similar group of settings for polylines. For example, you can round the corner of polyline corners using the Round option in the Line Join Style drop-down list.

Setting Up Screen Values for Solid Areas

The last option you'll look at is how to change a color into a screened area. At times, you'll want to add a gray or colored background to an area of your drawing to emphasize that area graphically, as in a focus area in a map or to designate functions in a plan. The setting you're about to use will allow you to create shaded backgrounds.

1. Go to the Page Setup dialog box again and open the Plot Style Table Editor.

2. Select Color 3 from the Plot Styles list box.

3. Go to the Screening list box and double-click the number 100. The number becomes highlighted.

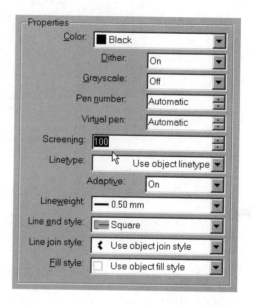

4. Type **50**.

5. Click Save & Close, then click OK in the Page Setup dialog box.

6. Choose View ➤ Regen All. Notice that now the object lines are a shade of gray instead of solid black.

As you can see from these exercises, you turned a wide black line into a gray one. In this example, the Screening option lets you tone down the chosen color from a solid color to one that has 50 percent of its full intensity.

You can use the Screening option in combination with color to obtain a variety of tones. If you need to cover large areas with color, you can use the solid hatch pattern to fill those areas, then use the Screening option in the Plot Style Table Editor to make fine adjustments to the area's color.

TIP   You'll want to know about the Draworder command in conjunction with solid-filled areas. This command lets you control how overlapping areas are displayed. See the section "Controlling Object Visibility and Overlap with Raster Images" in Chapter 16 for more information.

Other Options in the Plot Style Table Editor

You've seen a lot of the plot-style options so far, but there are many others that you may want to use in the future. This section describes those options that were not covered in the previous exercises.

NOTE The options in the Plot Style Table Editor are the same regardless of whether you're editing a color plot-style table or a named plot-style table.

The General Tab

You didn't really look at the General tab of the Plot Style Table Editor in the preceding exercise. Here is a view of the General tab and a description of the options offered there.

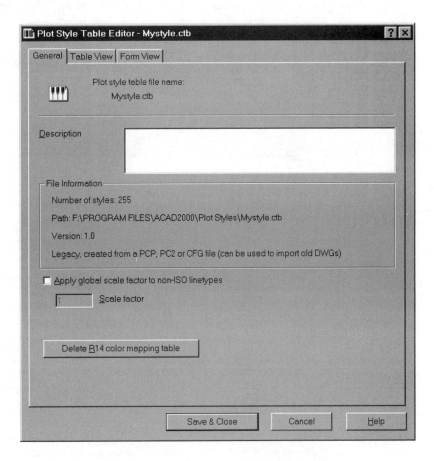

The General tab offers information regarding the plot style you're currently editing. You can enter a description of the style in the Description input box. This can be useful if you plan to include the plot style with a drawing you're sending to someone else for plotting.

The File Information group gives you the basic information on the file location and name, as well as the number of color styles included in the plot-style table.

The Apply Global Scale Factor to Non-ISO Linetypes check box lets you determine whether ISO linetype scale factors are applied to all linetypes. When this item is checked, the Scale Factor input box becomes active, allowing you to enter a scale factor.

The Form View Tab

The Form View tab contains the same settings as the Table View tab but in a different format. Instead of displaying each color as a column of properties, the properties are listed as options along the right side, while the colors are listed in a list box.

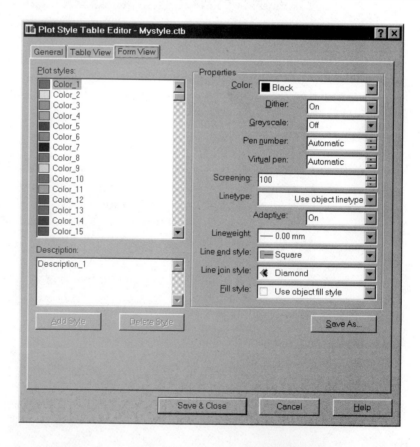

ISO Pen Widths

You may have noticed a setting called ISO Pen Width in the Linetype Manager dialog box discussed in Chapter 4, "Organizing Your Work." Choose Format ➢ Linetype to access this pull-down list. When you select a pen width from that list, the linetype scale is updated to conform to the ISO standard for that width. However, this setting has no effect on the actual plotter output. If you're using ISO standard widths, it's up to you to match the color of the lines to their corresponding widths in the Plot Style Table Editor.

To modify the properties of a color, select the color from the list, then edit the values in the Properties group in the right side of the dialog box. So to change the screen value of the Color 3 style, highlight Color 3 in the Plot Styles list, then double-click the Screening Property input box and enter a new value.

In the preceding sections, you've already seen what the Screening, Color, Lineweight, and Line Join Style options do. Here's a description of the other style properties.

NOTE The names of the properties in the Table View tab are slightly different from those in the Form View tab. The Table View property names are shown enclosed in brackets ([]) in this listing.

Description   Allows you to enter a description for each individual color.

Dither [Enable Dithering]   Enables your plotter to simulate colors beyond the basic 256 colors available in AutoCAD. While this option is desirable when you want to create a wider range of colors in your plots, it can also create some distortions in your plots, including broken, fine lines, and false colors. For this reason, dithering is usually turned off. This option is not available in all plotters.

[Convert to] Grayscale   Converts colors to grayscale.

[Use Assigned] Pen Number   Lets you determine what pen number is assigned to each color in your drawing. This option applies only to pen plotters.

Virtual Pen Number   Allows you to assign AutoCAD colors to a specific virtual pen. Many inkjet and laser plotters offer "virtual pens" to simulate the processes of the old-style pen plotters. Frequently, such plotters offer as many as 255 virtual pens. A virtual pen number is significant if the virtual pens of your plotter can be assigned screening width, end style, and joint styles. You can then use the virtual pen settings instead of the settings in the Plot Style Table Editor. This option is most beneficial for users who already have a library of drawings that are set up for plotters with virtual pen settings.

You can set up your inkjet printer for virtual pens under the Vector Graphics listing of the Device and Documents Setting tab of the Plotter Configuration Editor. See Appendix A, "Hardware and Software Tips," for more on setting up your printer or plotter configuration.

Linetype   Controls linetypes in AutoCAD based on the color of the object. By default, this option is set to Use Object Linetype. This book recommends that you leave this option at its default.

Adaptive Adjustment   Controls how noncontinuous linetypes begin and end. This option is on by default, which forces linetypes to begin and end in a line segment. With the option turned off, the same linetype is drawn without regard for its ending. In some cases, this may produce a line that appears incomplete.

Line End Style   Lets you determine the shape of the ends of simple lines that have lineweights greater than zero.

Line Join Style   Lets you determine the shape of the corners of polylines.

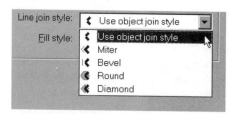

Fill Style   Lets you set up a color to be drawn as a pattern when used in a solid filled area. The patterns appear as follows:

Add Style   Allows you to add more plot styles or colors.

Delete Style   Deletes the selected style.

Save As   Lets you save the current plot-style table.

The Table View Tab

The Table View tab offers the same settings as the Form View tab, only in a different format. Each plot style is shown as a column with the properties of each plot style listed along the left side of the tab. To change a property, click the property in the column.

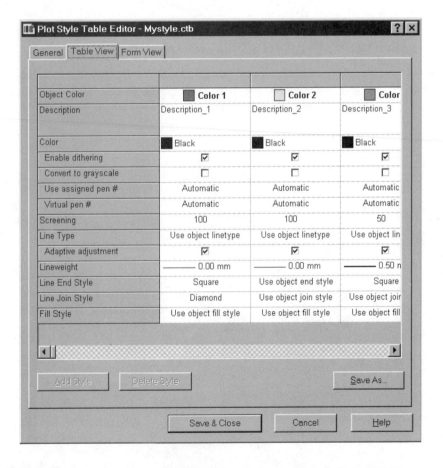

To apply the same setting to all plot styles at once, right-click a setting you want to use from a single plot style, then select Copy from the pop-up menu. Right-click again on the setting again, then select Paste to All Styles.

Assigning Plot Styles Directly to Layers and Objects

So far, you've learned that you can control how AutoCAD translates drawing colors into plotter output. You also have the option to assign plot styles directly to objects or layers. In order to do this, you need to make a few changes to the settings in AutoCAD and you also need to have a named plot-style table. So far, you've only been working with a color plot-style table. In this section you'll learn how to set up AutoCAD with a named plot-style table to assign plots styles to objects, then you'll create a new plot-style table.

Using Named Plot-Style Tables

Before you can start to assign plot styles to objects and layers, you need to make some changes in the AutoCAD Options dialog box. You'll also want to create your own named plot-style table.

1. Choose Tools ➤ Options, then select the Output tab.

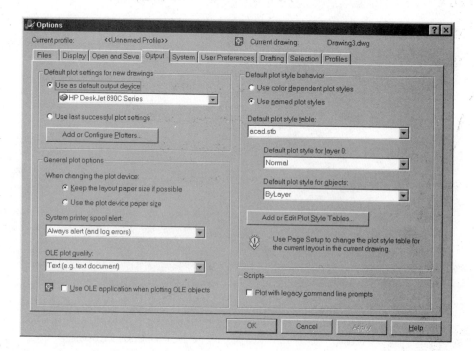